CHILTON'S REPAIR

DATSUN PICK-UPS 1970-81

All Models including:
L16 • L18 • L20B • Z22 Gasoline Engines
SD22 Diesel Engine • 4-Wheel drive

Managing Editor KERRY A. FREEMAN, S.A.E.
Senior Editor RICHARD J. RIVELE, S.A.E.
Editor RICHARD J. RIVELE, S.A.E.

President GARY R. INGERSOLL
Executive Vice President JAMES A. MIADES
Vice President and General Manager JOHN P. KUSHNERICK

CHILTON BOOK COMPANY
Radnor, Pennsylvania
19089

SAFETY NOTICE

Proper service and repair procedures are vital to the safe, reliable operation of all motor vehicles, as well as the personal safety of those performing repairs. This book outlines procedures for servicing and repairing vehicles using safe, effective methods. The procedures contain many NOTES, CAUTIONS and WARNINGS which should be followed along with standard safety procedures to eliminate the possibility of personal injury or improper service which could damage the vehicle or compromise its safety.

It is important to note that repair procedures and techniques, tools and parts for servicing motor vehicles, as well as the skill and experience of the individual performing the work vary widely. It is not possible to anticipate all of the conceivable ways or conditions under which vehicles may be serviced, or to provide cautions as to all of the possible hazards that may result. Standard and accepted safety precautions and equipment should be used when handling toxic or flammable fluids, and safety goggles or other protection should be used during cutting, grinding, chiseling, prying, or any other process that can cause material removal or projectiles.

Some procedures require the use of tools specially designed for a specific purpose. Before substituting another tool or procedure, you must be completely satisfied that neither your personal safety, nor the performance of the vehicle will be endangered.

Although the information in this guide is based on industry sources and is as complete as possible at the time of publication, the possibility exists that the manufacturer made later changes which could not be included here. While striving for total accuracy, Chilton Book Company cannot assume responsibility for any errors, changes, or omissions that may occur in the compilation of this data.

PART NUMBERS

Part numbers listed in this reference are not recommendations by Chilton for any product by brand name. They are references that can be used with interchange manuals and aftermarket supplier catalogs to locate each brand supplier's discrete part number.

ACKNOWLEDGMENTS

The Chilton Book Company expresses its appreciation to the Nissan Motor Corporation in the U.S.A., Carson, California 90248 for their generous assistance.

CONTENTS

Quick Reference Specifications For Your Vehicle

Fill in this chart with the most commonly used specifications for your vehicle. Specifications can be found in Chapters 1 through 3 or on the tune-up decal under the hood of the vehicle.

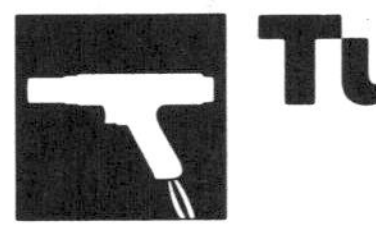 ## Tune-Up

Firing Order____

Spark Plugs:

    Type____

    Gap (in.)____

Point Gap (in.)____

Dwell Angle (°)____

Ignition Timing (°)____

    Vacuum (Connected/Disconnected)_________________________

Valve Clearance (in.)

    Intake___________________________   **Exhaust**___________________________

Capacities

Engine Oil (qts)

    With Filter Change_______________________________________

    Without Filter Change____________________________________

Cooling System (qts)____

Manual Transmission (pts)______________________________________

    Type____

Automatic Transmission (pts)___________________________________

    Type____

Front Differential (pts)____

    Type____

Rear Differential (pts)____

    Type____

Transfer Case (pts)____

    Type____

FREQUENTLY REPLACED PARTS

Use these spaces to record the part numbers of frequently replaced parts.

PCV VALVE	**OIL FILTER**	**AIR FILTER**
Manufacturer__________	Manufacturer__________	Manufacturer__________
Part No.__________	Part No.__________	Part No.__________

General Information and Maintenance

HOW TO USE THIS BOOK

All Datsun Pick-Up models, 1970–81, are covered in this book with procedures specifically labeled as to model year and engine when it makes a difference. The purpose of this book is to cover maintenance and repair procedures that the owner can perform without special tools or equipment. A lot of attention is given to the type of jobs on which the owner can save labor charges and time by doing it him or her self. Jobs which absolutely require special tools, such as transmission overhaul, or which the beginner is unlikely to get right the first time, such as differential adjustment, are purposely not covered.

To use the book properly, each operation must be approached logically, with a clear understanding of the theory behind the work involved. The procedures should be read completely and understood thoroughly before any work is begun. The required tools and supplies should be on hand, and a clean, uncluttered place to work should be available. There is nothing more frustrating than finding yourself one metric bolt short in the middle of your Sunday afternoon repair. So read and plan ahead. To avoid confusion, it is best to complete one job at a time, so that results can be independently evaluated. When reference is made in this book to the "right side" or "left side" of the truck, it should be understood that these positions are to be viewed from the front seat; thus, the left side of the truck is always the driver's side, even when one is facing the truck, as when working on the engine.

We have attempted to eliminate the use of special tools wherever possible, substituting more readily available hand tools. However, in some cases, the special tools are necessary. These can be purchased from your dealer, or from an automotive parts store.

Always be conscious of the need for safety in your work. Never crawl under the truck unless it is firmly supported by jackstands or ramps. Never smoke near or allow flame to get near the battery or fuel system. Keep your clothing, hands and hair clear of the fan and pulleys when working near the engine if it is running. Most importantly, try to be patient, even in the midst of a problem such as a particularly stubborn bolt; reaching for the largest hammer in the garage is usually a cause for later regret and more extensive repair. As you gain experience and confidence, working on your truck will become a source of pride and satisfaction.

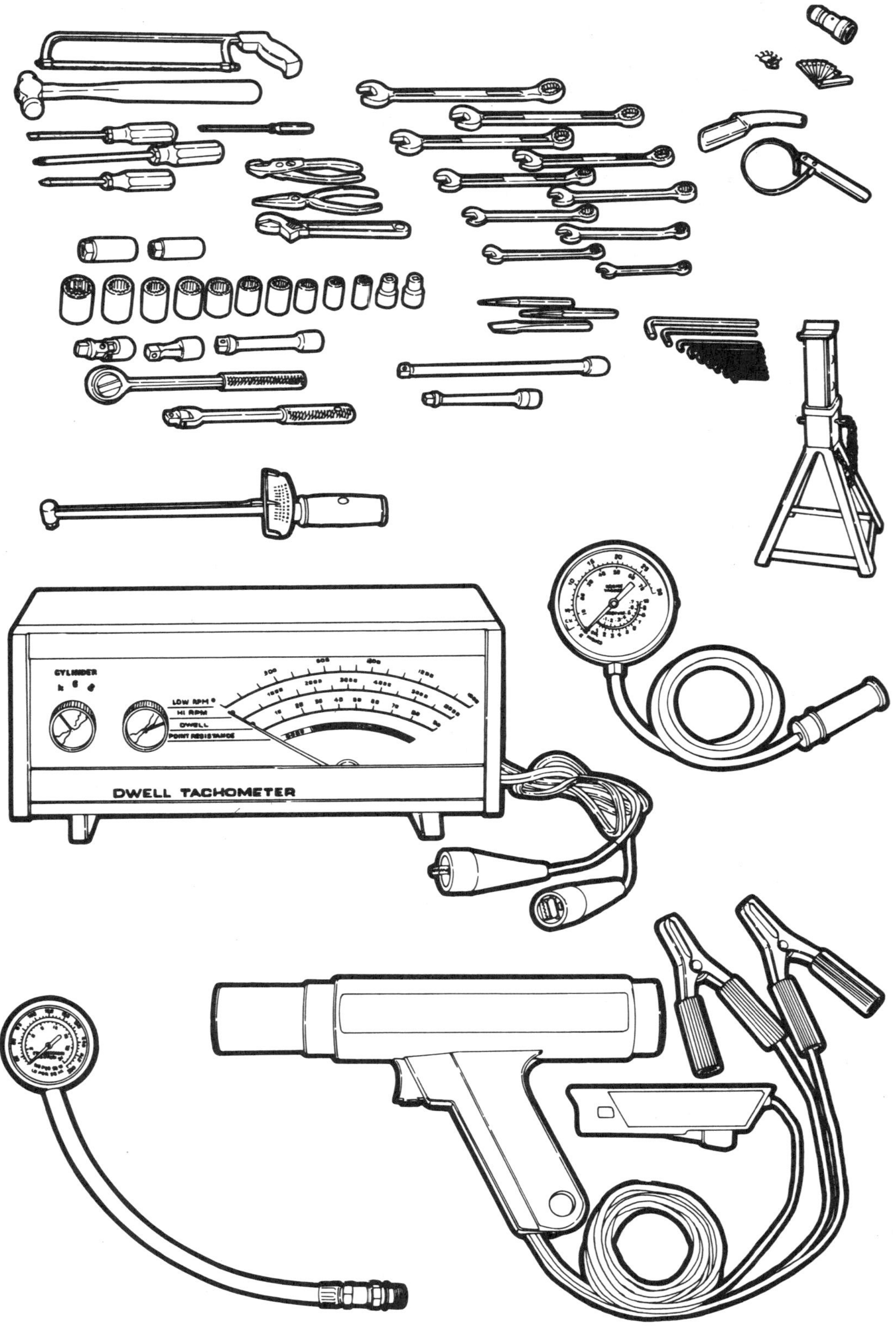

The tools and equipment shown here will handle the majority of the maintenance on a truck

TOOLS AND EQUIPMENT

You will find that virtually every nut and bolt on your Datsun is metric. Therefore, despite various close size similarities, standard inch-size tools will not fit and must not be used. You will need a set of metric wrenches as your most basic tool kit, ranging from about 6 mm to 17 mm in size. In addition to this, and the normal assortment of slot head and phillips screwdrivers, you will need a number of other tools to perform various tune-up and maintenance jobs. It would not be wise to buy a large assortment of tools on the premise that someday they will be needed. Instead, the tools should be acquired one at a time, each for a specific job, both to avoid unnecessary expense and to be certain you have the right tool.

Never buy low cost, "bargain" tools. The higher initial cost of forged steel wrenches, tempered screwdriver blades, and 12 point sockets is easily outweighed by the broken parts and mangled knuckles inflicted by their poorly manufactured counterparts.

Most jobs can be accomplished using the tools on the following list. There will be an occasional need for a special tool, such as snap ring pliers, for some jobs; that need will be mentioned in the text.

1. Set of metric combination (open and box end) wrenches.
2. Slot and phillips head screwdrivers.
3. Metric sockets, and a 13/16 inch spark plug socket; also a ratchet handle if you do not already own one. Metric sockets available in the U.S. are sized to fit the inch-drive (¼, ⅜, ½) ratchet handles already in use. Socket drive extensions in various lengths are essential.
4. Spark plug gap gauge and bending tool.
5. Flat feeler gauges for valve lash and breaker points (if used).
6. Timing light.
7. Dwell/tachometer.
8. Oil filter strap wrench.
9. Two sturdy jackstands.

In addition, a torque wrench will be necessary for some engine and chassis work.

HISTORY

Datsun, then known as D.A.T., began producing cars in 1913. The name D.A.T. was derived from the beginning letters of the founder's last names. The name Datson, for son of D.A.T., was used later and finally evolved into Datsun. The first Datsun was a two seater with motorcycle fenders and a four-speed transmission. In 1933, a reorganization led to the formation of Nissan Motor Company, Ltd., which is now the parent company of Datsun. Datsun cars made their first United States appearance at the 1958 Los Angeles Imported Car Show. Cars and pick-ups were first imported into the U.S. in 1960.

The first Datsun pick-up imported (until 1966) to the U.S. was the model L320 with a 1200 cc engine. Next came the L520 and L521 with 1300 cc engines. In 1970, the model PL521 pick-up with a 1600 cc engine was introduced. In mid-year 1973, the 1800 cc engine was installed in the PL620 pick-up. The PL620 model was introduced in 1972 using the 1600 cc engine. The L20B engine was introduced in 1975. An optional automatic transmission was made available in 1972; a five speed overdrive manual transmission was introduced in 1977. Transistorized ignition was installed on trucks sold in California starting in 1976, and nationwide in 1978. Front disc brakes were also first installed in 1978. In 1980 a 4-wheel drive version was introduced. The SD22 diesel engine was introduced in 1981.

SERIAL NUMBER IDENTIFICATION

Vehicle

The vehicle identification plate is located on the firewall in the engine compartment. The plate gives the vehicle type, engine capacity, maximum engine horsepower, wheelbase, engine number, and truck serial numbers.

The chassis serial number is stamped on the upper face of the right-side frame member.

Vehicle identification plate location on the firewall

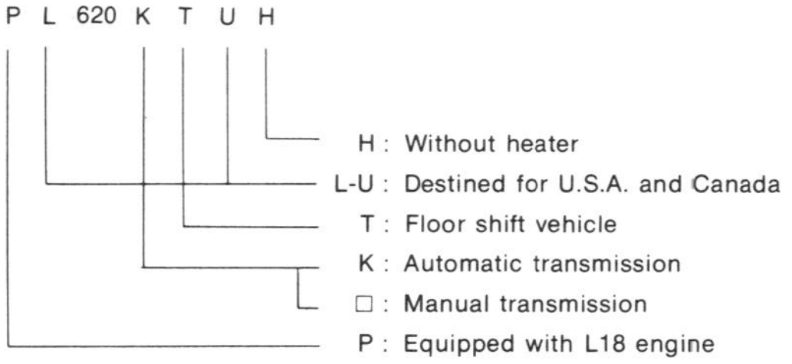

Note: □ means no indication.

Typical vehicle identification model code— 1973–74 shown

The vehicle identification model codes may be interpreted as follows:

• **1970–71:** All models are marked PL521TU. The first letter, P, indicates the L16 engine. The second letter, L, means left hand drive, 521 is the model number. The first suffix letter, T, indicates a floor shift. The second suffix letter, U, indicates a U.S. and Canada specifications model.

• **1972:** All models are marked PL620TU. The only change from 1970–71 is the model number, 620.

• **1973–74:** The prefix letters remain the same, PL, with P indicating the 1600cc engine through mid-1973, or the 1800cc engine thereafter. The next three numbers, 620, are the model number. Four suffix letters are used. The first is either a K or a blank; K indicates an automatic transmission. The second and third suffix letters are T and U, with the same meaning as in earlier years. The fourth suffix letter is an H or blank; H indicates that no heater is installed.

• **1975–76:** Three prefix letters and four suffix letters are used. The first letter is an H, for the L20B engine. The second letter is L, for left hand drive. The third letter is either a G or a blank; G is for long-wheelbase (long bed) trucks. The next three numbers, 620, are the model number. The four suffix letters remain the same as in 1973–74, with

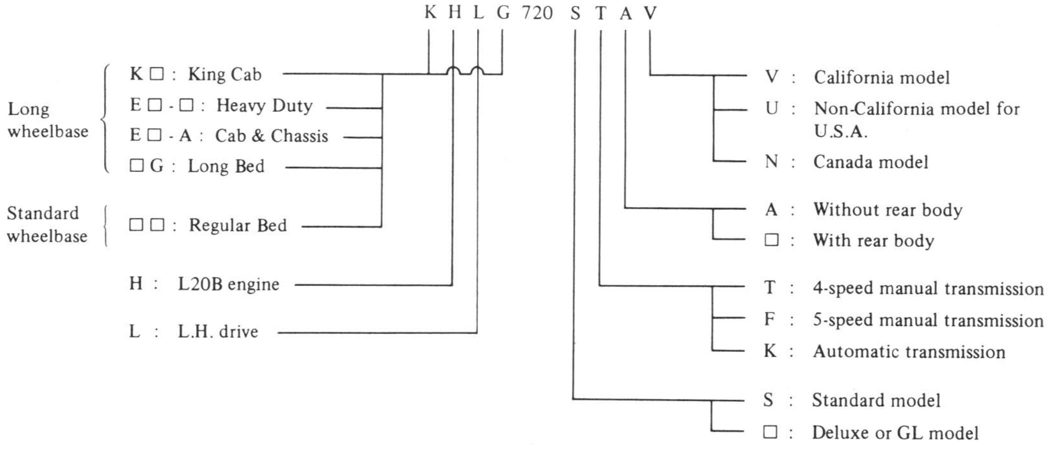

□ : means no indication.

1980–81 2-wd vehicle identification number

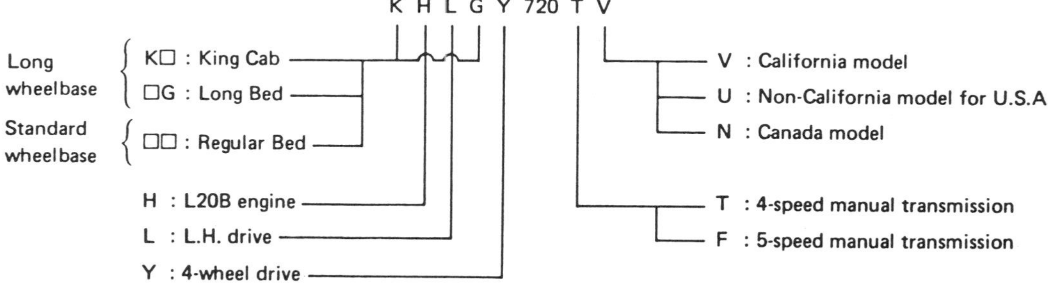

□ : means no indication.

1981 4-wd vehicle identification number

the exception of the last letter in 1976, which may be a V or a blank; V indicates a California model.

• 1977–79: Four prefix and four suffix letters are used. The first letter is a K or a blank; K indicates a king cab model. The second, third, and fourth letters are the same as those used in 1975–76. The next three numbers are the model number, 620. The first suffix letter is either a K, an F, or a blank; K indicates automatic transmission, F indicates a 5-speed, a blank indicates a 4-speed. The second and third suffix letters, T and U, have the same meaning as in earlier years. The last letter is either a V, an N, or a blank; V indicates a California model, N indicates a Canada model, a blank indicates a 49 States model.

• 1980–81: A four letter prefix for 2-wheel drive models and a five letter prefix for 4-wheel drive models followed by a three digit code denoting the truck series is used. The suffix for these years is either four letters for 2-wheel drive or two letters for 4-wheel drive. See the illustrations for an explanation of these serial numbers.

Engine

The gasoline engine serial number is stamped on the right-side of the cylinder block, just below the number four spark plug. The diesel engine serial number is stamped on the left side of the block, just in front of the starter.

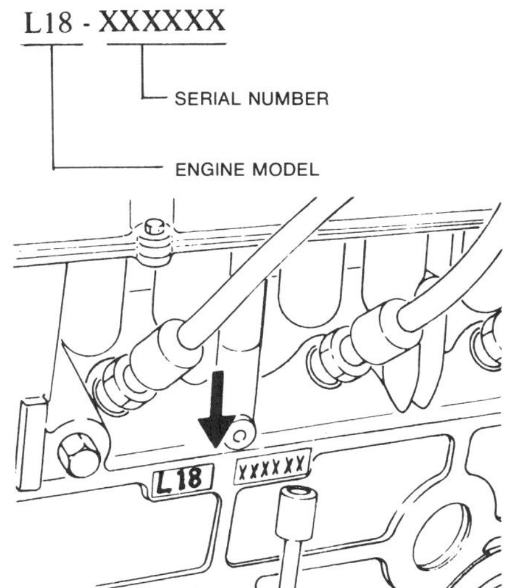

Engine serial number location

Transmission

The transmission serial number is stamped on the front upper face of the transmission case on manual transmissions, or on the right-side of the transmission case on automatic transmissions.

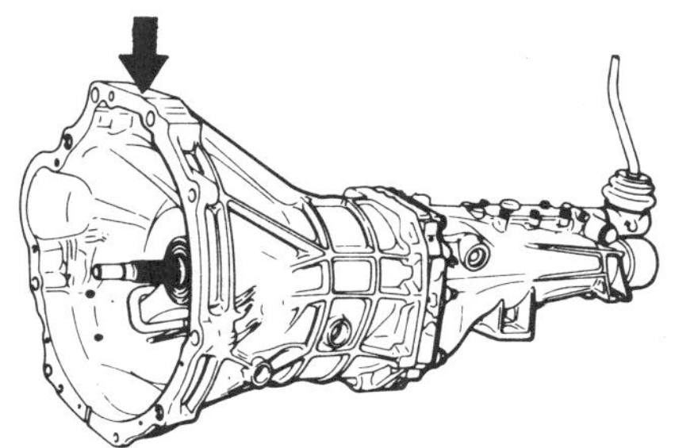

Manual transmission serial number location

Automatic transmission serial number location

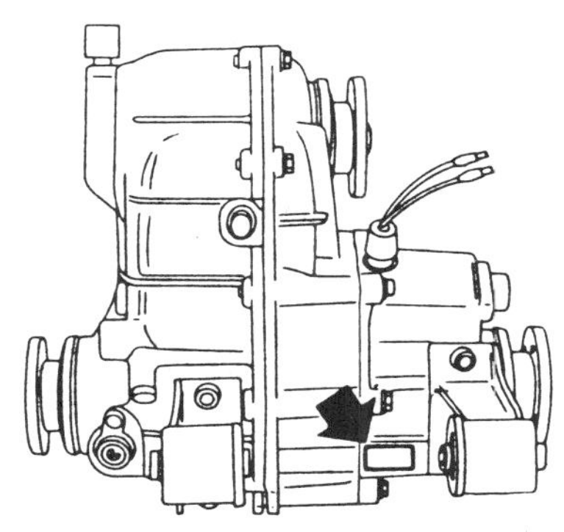

Transer case identification number location

ROUTINE MAINTENANCE

Air Cleaner

The air filter for both gasoline and diesel engines is the paper element type. There is no need to clean the filter element, but it must be replaced every 24,000 miles, or 12,000 miles if the vehicle is operated in dusty areas. To replace the air filter element, simply un-

Replace the air filter every 24,000 miles

screw the wing nut securing the air cleaner lid, lift off the lid, and remove the old filter element. Install the new element in the housing, making sure that it is seated properly and reinstall the lid, tightening the wing nut securely.

PCV Valve

To check the positive crankcase ventilation (PCV) valve, with the engine running, remove the ventilator hose from the PCV valve. If the valve is working, a hissing noise will be heard as air passes through the valve and a strong vacuum should be felt immediately when the valve inlet is blocked with a finger. If the valve is suspected of being plugged, it should be replaced.

Check the hoses and the hose connections for leaks. Disconnect all the hoses and blow them out with compressed air. Replace any hoses which are cracked, damaged, or completely blocked. Make sure that the flame arrester is fully inserted in the hose between the air cleaner and the rocker cover.

Fuel Evaporative Emissions System

Check the evaporation control system every 12,000 miles. Check the fuel and vapor lines and hoses for proper connections and correct routing as well as condition. Replace damaged or deteriorated parts as necessary. Remove and check the operation of the check valve on pre-1975 models in the following manner:

1. With all the hoses disconnected from the valve, apply air pressure to the fuel tank side of the valve. The air should flow through the valve and exit the crankcase side of the valve. If the valve does not behave in the above manner, replace it.

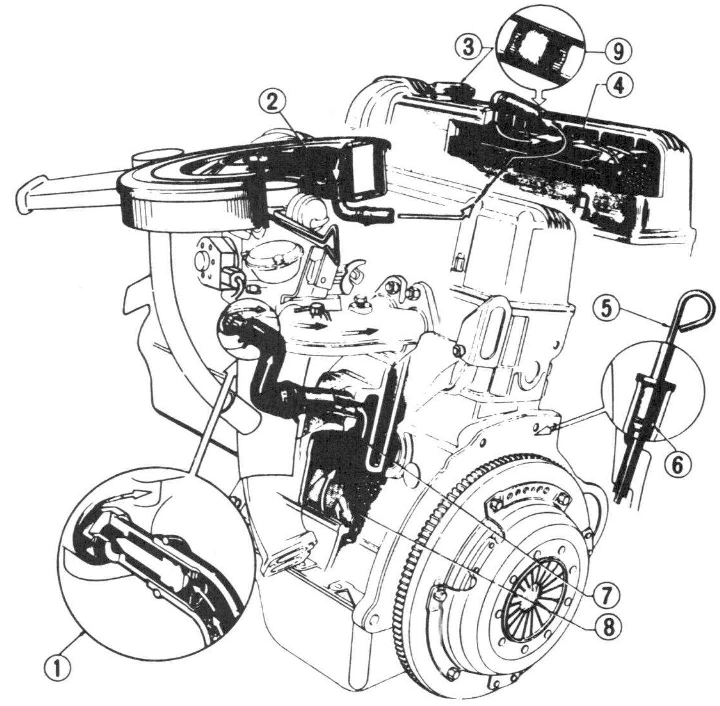

Diagram of the PCV system

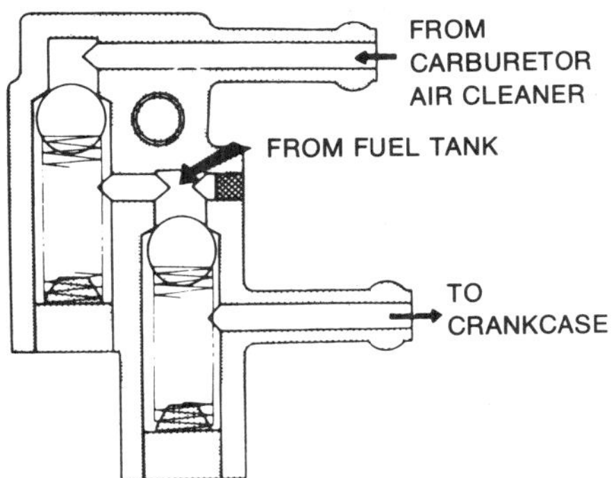

Diagram of the evaporative emission check valve

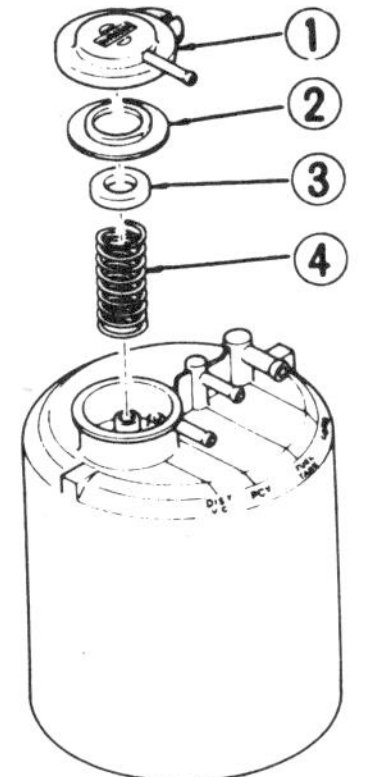

Components of the carbon canister purge control valve—1975 and later models

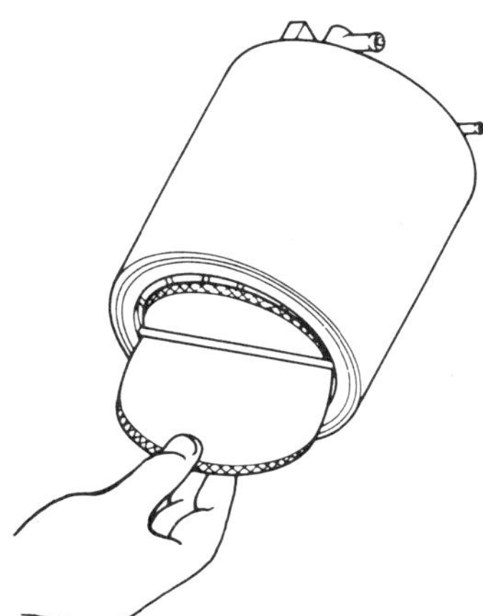

Replacing the carbon canister filter

2. Apply air pressure to the crankcase side of the valve. Air should not pass to either of the other two outlets.

3. When air pressure is applied to the carburetor side of the valve, the air should pass through to exit out the fuel tank and/or the crankcase side of the valve.

On 1975 and later models, the flow guide valve is replaced with a carbon filled canister which stores fuel vapors until the engine is started and the vapors drawn into the combustion chambers and burned.

To check the operation of the carbon canister purge control valve, disconnect the rubber hose between the canister control valve and the T-fitting, at the T-fitting. Apply vacuum to the hose leading to the control valve. The vacuum condition should be maintained indefinitely. If the control valve leaks, remove the top cover of the valve and check for a dislocated or cracked diaphragm. If the diaphragm is damaged, a repair kit containing a new diaphragm, retainer, and spring is available and should be installed.

The carbon canister has an air filter in the bottom of the canister. The filter element should be checked once a year or every 12,000 miles; more frequently if the truck is operated in dusty areas. Replace the filter by pulling it out of the bottom of the canister and installing a new one.

Heat Control Valve

The heat control valve is a thermostatically operated valve used on gasoline engines in the exhaust manifold. It is used only on 1975–77 trucks. It opens when the engine is warming up, to direct hot exhaust gases to the intake manifold, in order to preheat the incoming air/fuel mixture. If it sticks shut, the result will be frequent stalling during warmup, especially in cold or damp weather.

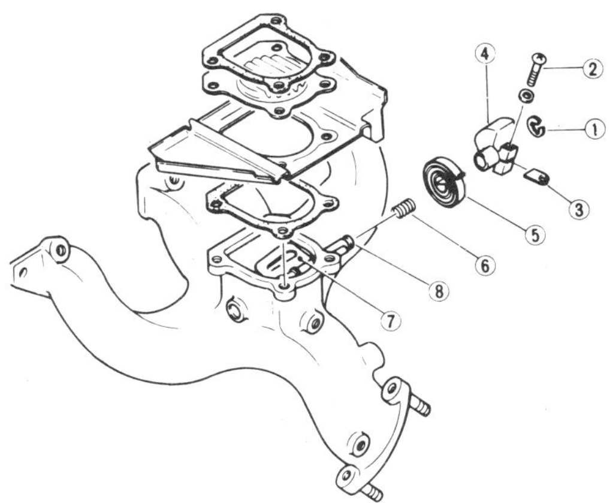

1. Snap ring
2. Lock bolt
3. Key
4. Counterweight
5. Thermostatic spring
6. Coil spring
7. Heat control shaft
8. Valve shaft

Components of the heat control valve

If it sticks open, the result will be a rough idle after the engine is warm.

The heat control valve should be checked for free operation every six months or 6,000 miles. Simply give the counterweight a twirl (engine cold) to ascertain that no binding exists. If the valve sticks, apply a heat control solvent specially formulated for the purpose to the ends of the shaft. This solvent is available in auto parts stores. Sometimes lightly rapping the end of the shaft with a hammer (engine hot) will break it loose. If this fails, the components will have to be removed from the truck for further repairs.

Air Pump Air Filter

1975 and later gasoline engine models have an air pump as one of their emission control components. Regular maintenance for this component includes a check of the drive belt tension at the specified interval, and replacement of the air pump air filter every 24,000 miles. The air filter case is located in the left front of the engine compartment. To replace the air filter, simply unscrew the wing nut(s) securing the cover to the case, withdraw the old filter, install the new one, and reinstall the case. More information on the air pump system can be found in Chapter Four.

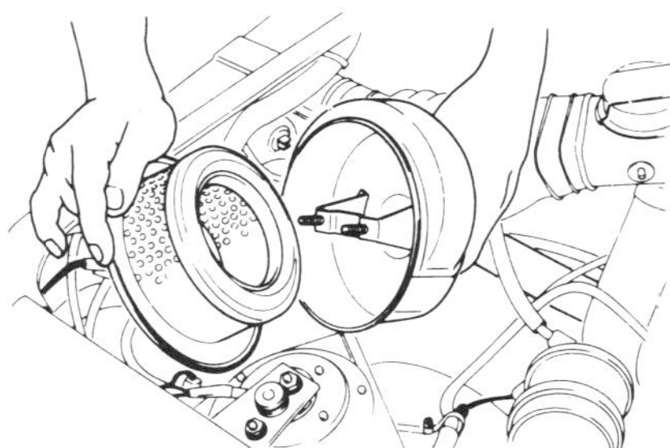

Replace the air pump air filter every 24,000 miles

Belts

Check the belts driving the fan, air pump, air conditioning compressor, and the alternator for cracks, fraying, wear, and tension every 6,000 miles. It is recommended that the belts be replaced every 24 months or 24,000 miles, whichever comes first. Belt deflection at the midpoint of the longest span between pulleys should be not more than 7/16 in. with 22 lbs of pressure applied to the belt.

To adjust the tension on all components except the air conditioning compressor, loosen the pivot and mounting bolts of the component which the belt is driving, then,

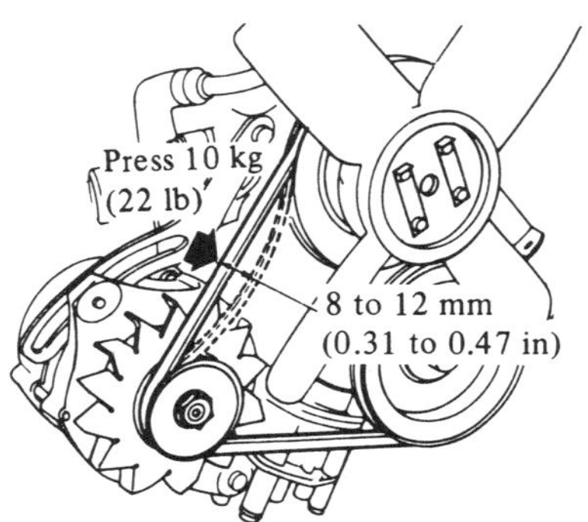

Checking the drive belt tension

using a soft wooden hammer handle, broomstick, or the like, pry the component toward or away from the engine until the proper tension is achieved.

CAUTION: *Do not use a screwdriver or other metal device such as a prybar, or damage to the component may result.*

Tighten the component mounting bolts securely. If a new belt is installed, recheck the tension after about 1,000 miles of driving. Be careful when making tension adjustments not to overdo it. Overtight drive belts will lead to bearing failure.

Belt tension adjustment for the factory–installed air conditioning compressor is made at the idler pulley bracket. Loosen the locknut, then turn the adjusting bolt to move the idler pulley up or down until the belt tension is correct. Retighten the locknut securely.

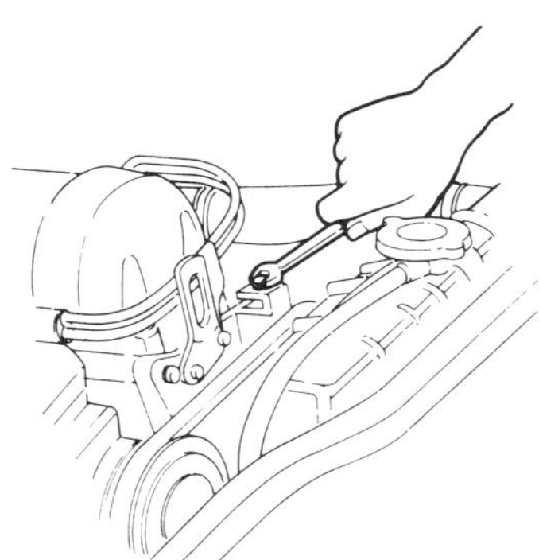

Adjust the air conditioner belt tension at the idler pulley

Air Conditioning System Check

Datsun factory units have a sight glass for checking the refrigerant charge. This is on top of the receiver-dehydrator, located at the right rear of the engine compartment.

CAUTION: *Do not attempt to charge or discharge the refrigerant system unless you*

HOW TO SPOT WORN V-BELTS

V-Belts are vital to efficient engine operation—they drive the fan, water pump and other accessories. They require little maintenance (occasional tightening) but they will not last forever. Slipping or failure of the V-belt will lead to overheating. If your V-belt looks like any of these, it should be replaced.

Cracking or weathering

This belt has deep cracks, which cause it to flex. Too much flexing leads to heat build-up and premature failure. These cracks can be caused by using the belt on a pulley that is too small. Notched belts are available for small diameter pulleys.

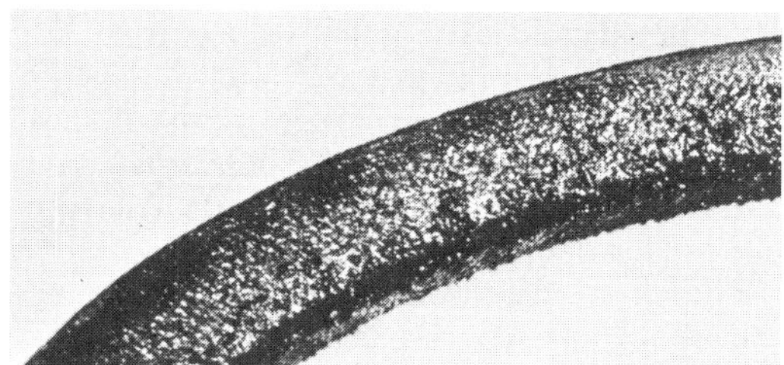

Softening (grease and oil)

Oil and grease on a belt can cause the belt's rubber compounds to soften and separate from the reinforcing cords that hold the belt together. The belt will first slip, then finally fail altogether.

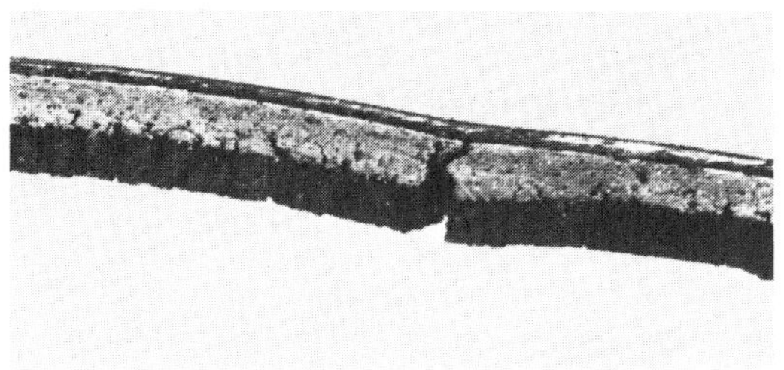

Glazing

Glazing is caused by a belt that is slipping. A slipping belt can cause a run-down battery, erratic power steering, overheating or poor accessory performance. The more the belt slips, the more glazing will be built up on the surface of the belt. The more the belt is glazed, the more it will slip. If the glazing is light, tighten the belt.

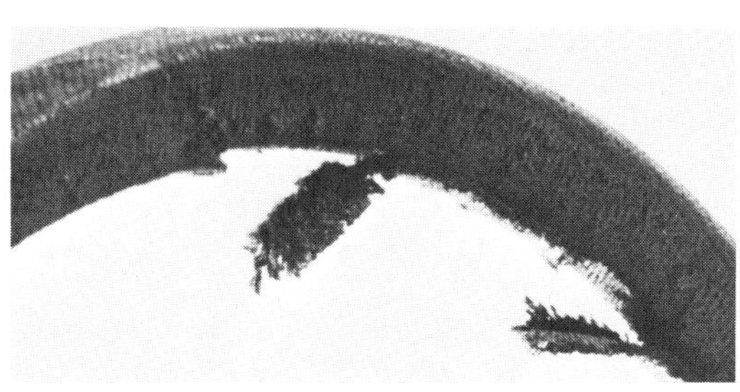

Worn cover

The cover of this belt is worn off and is peeling away. The reinforcing cords will begin to wear and the belt will shortly break. When the belt cover wears in spots or has a rough jagged appearance, check the pulley grooves for roughness.

Separation

This belt is on the verge of breaking and leaving you stranded. The layers of the belt are separating and the reinforcing cords are exposed. It's just a matter of time before it breaks completely.

The arrow points to the receiver dehydrator sight glass

are thoroughly familiar with its operation and the hazards involved. The compressed refrigerant used in the air conditioning system expands and evaporates (boils) into the atmosphere at a temperature of −29.8°C (−21.7°F) or less. This will freeze any surface that it contacts, including your eyes. In addition, the refrigerant decomposes into a poisonous gas in the presence of flame.

NOTE: *If your truck is equipped with an aftermarket air conditioner, the following system check may not apply. You should contact the manufacturer of the unit for instructions on system checks.*

This test works best if the outside air temperature is warm (above 70°F).

1. Place the automatic transmission in Park or the manual transmission in Neutral. Set the parking brake.

2. Run the engine at a fast idle (about 1,500 rpm) either with the help of a friend, or by temporarily readjusting the idle speed screw.

3. Set the controls for maximum cold with the blower on high.

4. Locate the sight glass on top of the receiver-dehydrator. If a steady stream of bubbles are present in the sight glass, the system

is low on charge. Very likely there is a leak in the system.

5. If no bubbles are present, the system is either fully charged or empty. Feel the high and low pressure lines at the compressor. If no appreciable temperature difference is felt, the system is empty, or nearly so.

6. If one hose (high pressure) is warm and the other (low pressure) is cold, the system may be ok. However, you are probably making these tests because there is something wrong with the system, so proceed to the next step.

7. Either disconnect the compressor clutch wire, or have an assistant in the truck turn the fan control on and off to operate the compressor clutch. Watch the sight glass.

8. If bubbles appear when the clutch is disengaged and disappear when it is engaged, the system is properly charged.

9. If the refrigerant takes more than 45 seconds to bubble when the clutch is disengaged, the system is overcharged. This usually causes poor cooling at low speeds.

The air conditioning system should be operated for about five minutes each week, even in winter. This will circulate lubricating oil within the system to prevent the various seals from drying out.

NOTE: *If it is determined that the system has a leak, it should be repaired as soon as possible. Leaks may allow moisture to enter, causing an expensive rust problem.*

Fluid Level Checks

ENGINE OIL

The engine oil should be checked on a regular basis, ideally at each fuel stop. The level can be checked with the engine either hot or cold, although checking the level immediately after stopping will give a false reading, because all of the oil will not yet have drained back into the crankcase. The truck must be resting on a level surface for an accurate reading.

1. Open the hood and locate the dipstick. It is on the right side of the engine. Pull the dipstick from its tube, wipe it clean, and reinsert it.

2. Pull the dipstick again and, holding it horizontally, read the oil level. The oil should be between the "H" and "L" marks. If the oil is below the "L" mark, add oil of the proper viscosity through the capped opening in the front of the valve cover. See the "Oil and Fuel Recommendations" section in this

HOW TO SPOT BAD HOSES

Both the upper and lower radiator hoses are called upon to perform difficult jobs in an inhospitable environment. They are subject to nearly 18 psi at under hood temperatures often over 280°F., and must circulate nearly 7500 gallons of coolant an hour—3 good reasons to have good hoses.

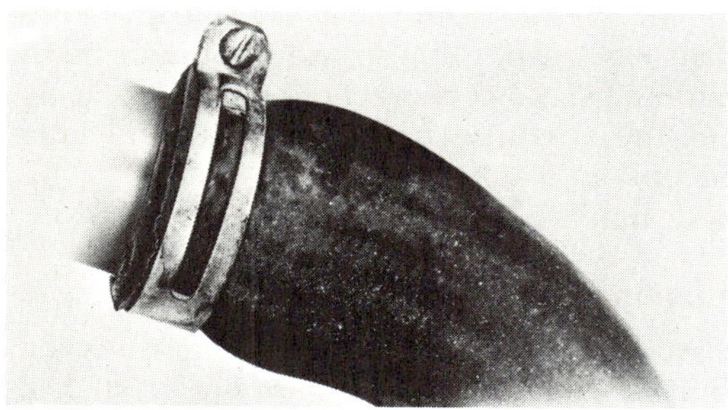

Swollen hose

A good test for any hose is to feel it for soft or spongy spots. Frequently these will appear as swollen areas of the hose. The most likely cause is oil soaking. This hose could burst at any time, when hot or under pressure.

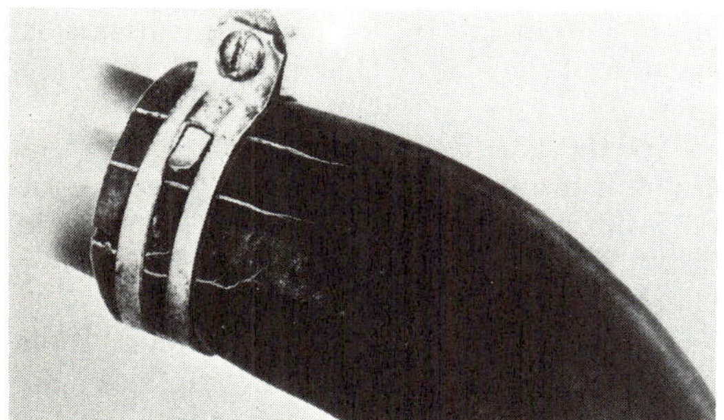

Cracked hose

Cracked hoses can usually be seen but feel the hoses to be sure they have not hardened; a prime cause of cracking. This hose has cracked down to the reinforcing cords and could split at any of the cracks.

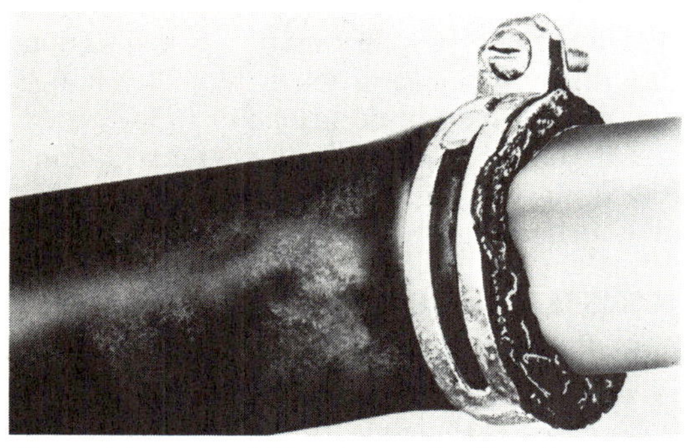

Frayed hose end (due to weak clamp)

Weakened clamps frequently are the cause of hose and cooling system failure. The connection between the pipe and hose has deteriorated enough to allow coolant to escape when the engine is hot.

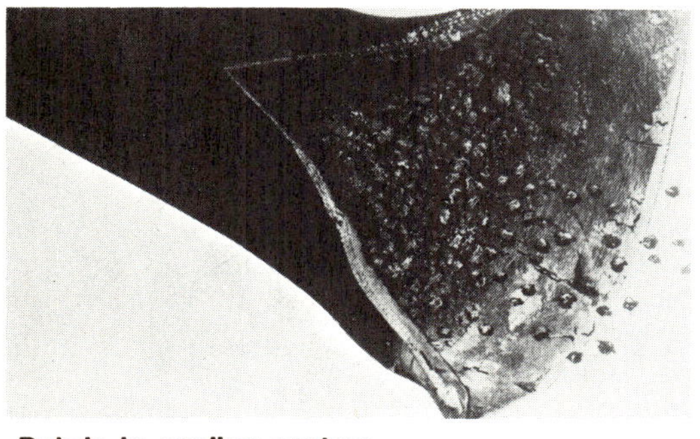

Debris in cooling system

Debris, rust and scale in the cooling system can cause the inside of a hose to weaken. This can usually be felt on the outside of the hose as soft or thinner areas.

The oil level dipstick is on the right side of the engine

Add oil through the valve cover

Chapter for the proper viscosity and oil to use.

3. Replace the dipstick, and check the level after adding oil. Be careful not to overfill the crankcase. Approximately one quart of oil will raise the level from "L" to "H".

TRANSMISSION

Manual

Check the lubricant level at the interval specified in the maintenance chart.

1. With the truck parked on a level surface, remove the filler plug from the left side of the transmission case. The filler plug has a square head.

2. If lubricant begins to trickle out the

hole, there is enough. Otherwise, carefully insert a finger (watch out for sharp threads) and check that the level is up to the edge of the hole.

3. If not, add lubricant through the hole to raise the level to the edge of the filler hole. Most gear lubricants come in a plastic squeeze bottle with a nozzle, making additions easy. You can also use a squeeze bulb. Add API GL-4 gear oil of the proper viscosity (see the viscosity chart under "Oil and Fuel Recommendations").

4. Replace the plug and check for leaks.

Automatic

Check the level of the automatic transmission fluid every 2,000 miles. There is a dipstick at the right rear of the engine under the hood. It has a scale on each side, one for COLD and the other for HOT. The transmission is considered hot after 15 miles of highway driving.

Park the truck on a level surface with the engine running. If the transmission is not hot, shift into drive, low, then Neutral or Park. Set the handbrake and block the wheels.

Remove the dipstick, wipe it clean, then reinsert it firmly. Remove the dipstick and check the fluid level on the appropriate scale. The level should be at the "Full" mark.

If the level is below the "Full" mark, add DEXRON® type automatic transmission fluid through the dipstick tube. Check the level often between additions of fluid, being careful not to overfill the transmission. Overfilling will cause slippage, seal damage, and overheating.

TRANSFER CASE

With the truck parked on level ground, remove the square-headed plug from the side of the case. If the fluid is hot, it should run out of the hole. If the fluid is cold it should

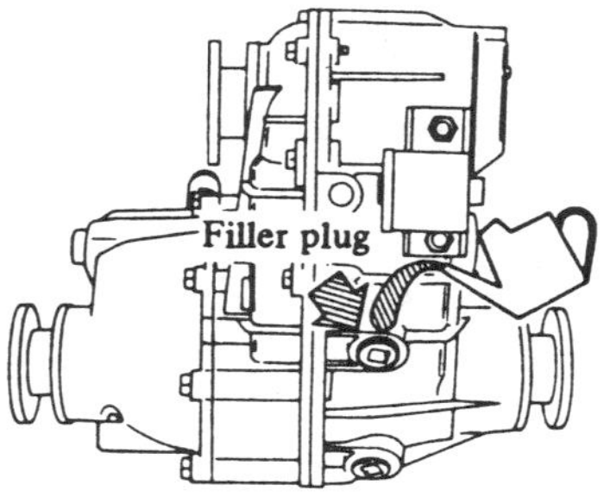

Transfer case fill plug location

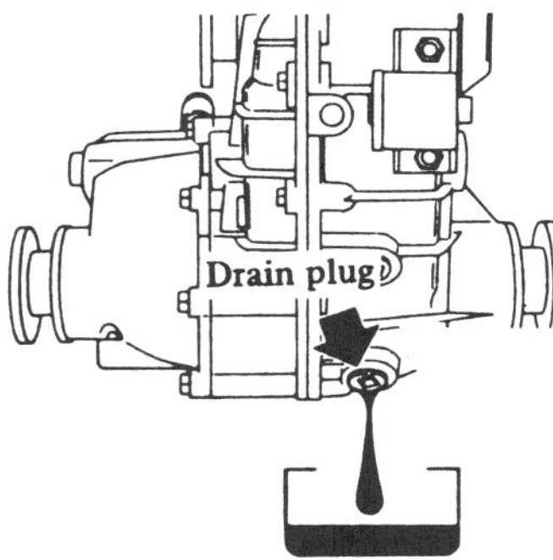

Transfer case drain plug location

be about even with or a little below the hole opening. Check the level with your finger.

If the lubricant needs replenishing, add API GL-4 gear oil through the hole. The level should be checked periodically and changed every 24,000 miles. If operating under severe load conditions, cut the interval for changing by half. If running in deep water, change the lubricant as soon as possible.

BRAKE AND CLUTCH MASTER CYLINDER

Check the levels of brake fluid in the brake and clutch master cylinder reservoirs every 3,000 miles. The fluid level should be maintained to a level not below the bottom line on the reservoirs and not above the top line. Any sudden decrease in the level in any of the reservoirs indicates a probable leak in that particular system and should be checked out.

When making additions of fluid, use only fresh, uncontaminated brake fluid meeting or exceeding DOT 3 standards. Be careful not to spill any brake fluid on painted surfaces, because it eats paint. Do not allow the fluid container or master cylinder reservoirs to remain open any longer than necessary. Brake

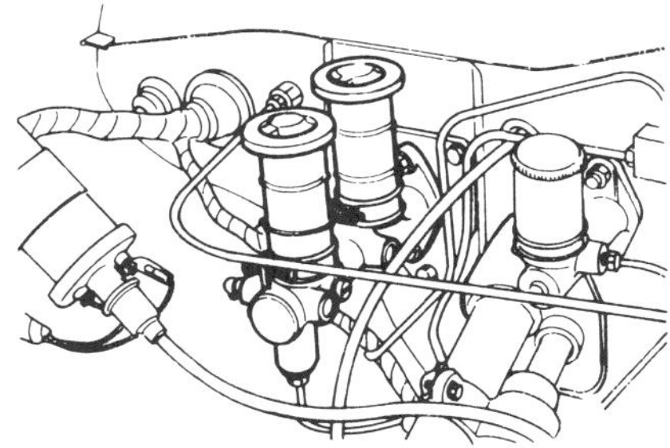

The brake and clutch master cylinders under the hood. The brake master cylinder has two reservoirs starting in 1972.

fluid absorbs moisture from the air, reducing its effectiveness and causing brake line corrosion.

COOLANT

Dealing with the cooling system can be a dangerous matter unless the proper precautions are exercised. It is best to check the coolant level in the radiator when the engine is cold. This is done by removing the radiator cap and seeing that the coolant is within two inches of the bottom of the filler neck. On newer trucks the cooling system has, as one of its components, an expansion tank. If coolant is visible above the "Min" mark on the tank, the level is satisfactory. Always be certain that the filler caps on both the radiator and the reservoir are tightly closed.

Coolant should be visible through the radiator filler neck

In the event that the coolant level must be checked when the engine is warm on engines without the expansion tank, or when no hot coolant is visible within the tank, place a thick rag over the radiator cap and slowly turn the cap counterclockwise until it reaches the first detent. Allow all the hot steam to escape. This will allow the pressure to drop gradually, preventing an explosion of hot coolant. When the hissing noise stops, remove the cap the rest of the way.

If the coolant level is low, add equal amounts of ethylene glycol-based antifreeze and clean water. On models without an expansion tank, add coolant through the radiator filler neck. Fill the expansion tank to the "MAX" level on trucks equipped with that system.

CAUTION: *Never add cold coolant to a hot engine unless the engine is running, to prevent cracking the engine block.*

Capacities

Year	Engine Displacement cc (cu in.)	Crankcase L (qts)		Transmission L (pts)			Transfer Case L (pts)	Rear Drive Axle L (pts)	Front Drive Axle L (pts)	Gas Tank L (gal)	Cooling System L (qts)	
		w/filter	wo/filter	4sp	5sp	Auto					Manual	Auto
1970–71	1595 (97.3)	4.7(5.0)	3.9(4.1)	2.0(4.2)	—	—	—	0.8 (1.7)	—	41(10.8)	6.8 (7.2)	—
1972–73	1595 (97.3)	4.7(5.0)	3.9(4.1)	1.7(3.6)	—	—	—	1.0 (2.1)	—	45(11.8)	6.0 (6.3)	—
1973–74	1770(108.0)	4.8(5.1)	4.1(4.3)	1.7(3.6)	—	5.5(11.7)	—	1.0 (2.1)	—	45(11.8)	6.0 (6.3)	6.0 (6.3)
1975	1952(119.1)	4.8(5.1)	4.1(4.3)	1.7(3.6)	—	5.5(11.7)	—	1.0 (2.1)	—	45(11.8)	6.0 (6.3)	6.0 (6.3)
1976–77	1952(119.1)	4.3(4.5)	3.8(4.0)	1.7(3.6)	2.0(4.2)	5.5(11.7)	—	1.0 (2.1)	—	45(11.8)	8.0 (8.5)	7.8 (8.2)
1978	1952(119.1)	4.3(4.5)	3.8(4.0)	1.7(3.6)	2.0(4.2)	5.5(11.7)	—	1.0 (2.1)	—	45(11.8)	8.9 (9.3)	8.7 (9.2)
1979	1952(119.1)	4.3(4.5)	3.8(4.0)	1.7(3.6)	2.0(4.2)	5.5(11.7)	—	1.0 (2.1)	—	50(13.2)①	8.9 (9.3)	8.7 (9.2)
1980	1952(119.1)	4.3(4.5)	3.8(4.0)	1.7(3.6)	2.0(4.2)	5.5(11.7)	1.4(3.0)	1.25(2.6)	1.0(2.1)	50(13.2)①	8.9 (9.3)	8.7 (9.2)
1981 2-WD	2187(133.5)	4.4(4.6)	3.9(4.12)	1.7(3.6)	2.0(4.2)	5.5(11.7)	—	—	—	50(13.2)①	10.2(10.75)	10.1(10.6)
4WD	2187(133.5)	4.2(4.5)	3.7(3.8)	1.7(3.6)	2.0(4.2)	5.5(11.7)	1.4(3.0)	1.25(2.6)	1.0(2.1)	60(15.8)②	10.2(10.75)	10.1(10.6)
Diesel	2164(132)	5.5(5.8)	—	—	2.0(4.2)	—	—	1.25(2.6)	—	60(15.8)②	N.A.	N.A.

① Long bed: 64(16.8)
② Longbed w/4-WD: 75(19.8)
—Not applicable
N.A.: Information not available at Press time

If the coolant level is chronically low or rusty, refer to the Troubleshooting section at the end of Chapter Two for diagnosis of the problem. Refer to Chapter Three for coolant draining and refilling, which should be done according to the schedule shown in the Maintenance Intervals chart.

DRIVE AXLES

Check the drive axle lubricant every 6,000 miles. Remove the filler plug in the rear center of the axle housing. The lubricant should be up to the bottom of the filler hole with the vehicle resting on a level surface. Add API GL-5 gear oil of the proper viscosity as necessary to bring the lubricant up to the proper level.

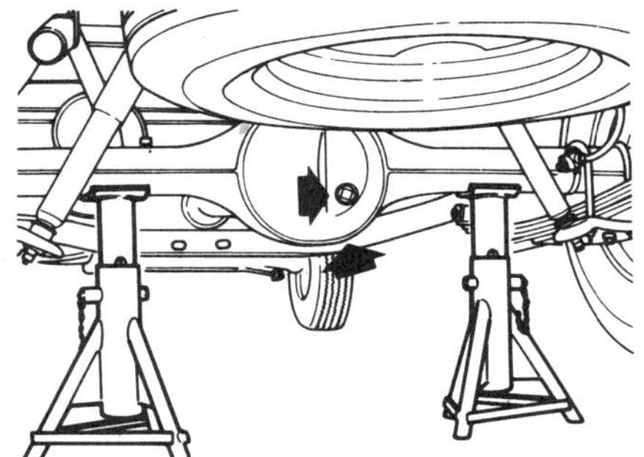

Arrows indicate the locations of the drive axle fill and drain plugs

STEERING GEAR

Check the level of the lubricant in the steering gear every 12,000 miles. If the level is low, check for leakage. An oily film is not considered a leak; solid grease must be present. Change the lubricant every 36,000 miles. Use API GL-4 steering gear lubricant. The capacity of the steering gear is ⅜ of a quart. The lubricant is added and checked through the filler plug hole in the top of the steering gear.

BATTERY

Check the battery electrolyte level at least once a month, more often in hot weather or during periods of extended truck operation. The level should be maintained between the upper and lower levels marked on the battery case, or to the split ring within the well in each cell. If the electrolyte level is low, distilled water should be added until the proper level is reached. Each cell is completely separate from the others, so each must be filled individually. It's a good idea to

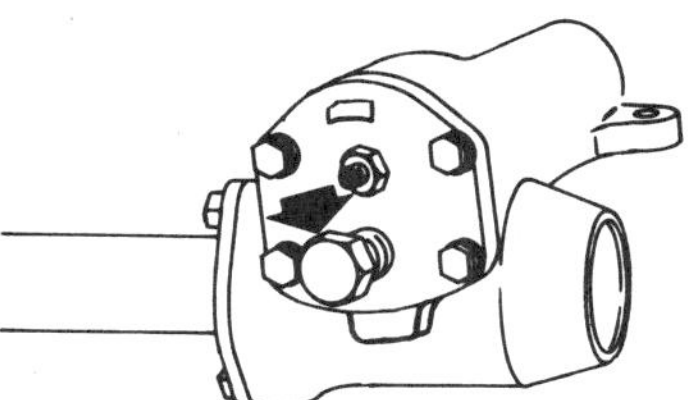

The large arrow points to the steering gear fill plug

Battery electrolyte level

add the distilled water with a squeeze bulb to avoid having electrolyte (sulphuric acid) splash out. If water is frequently needed, and the battery is not leaking, the most likely cause is overcharging, caused by a faulty voltage regulator.

At least once a year check the specific gravity of the battery. It should be between 1.22 and 1.28 at room temperature. Also at that time clean and tighten the terminal clamps and apply a thin external coating of petroleum jelly to the terminals. This will help retard corrosion. The terminals can be cleaned with a stiff wire brush or with a terminal cleaner made for the purpose. These are inexpensive and can be purchased in most auto parts stores.

If water is added in freezing weather, the truck should be driven several miles to allow the water to mix with the electrolyte. Otherwise the battery could freeze.

If the battery becomes corroded, or if electrolyte should splash out during additions of water, a mixture of baking soda and water will neutralize the corrosion. This should be washed off after making sure that the caps are tight. Rinse the solution off with cold water.

If a charger is used to charge the battery while the battery is in the truck, disconnect the battery cables, negative (ground) cable first.

When replacing a battery, it is important that the replacement have an output rating equal to or greater than original equipment. See Chapter Three for details on battery replacement.

CAUTION: *If you get battery acid in your eyes or on your skin, rinse it off immedi-*

ately with lots of water. Go to a doctor if it gets in your eyes.

The gases formed inside the battery cells are highly explosive. Never check the level of the electrolyte in the presence of flame or when smoking.

Tires

The importance of proper tire inflation cannot be overemphasized. A tire employs air under pressure as part of its structure. It is designed around the supporting strength of a gas at a specified pressure. For this reason improper inflation drastically reduces the tire's ability to perform as it was intended. Tire pressures should be checked regularly with a reliable pressure gauge. Too often the gauge on the end of the air hose at your corner garage is not accurate enough because it suffers too much abuse.

Always check tire pressure when the tires are cold, as pressure increases with temperature. If you must move the vehicle to check the tire inflation, do not drive more than a mile before checking. A cold tire is one that has not been driven on for a period of about three hours.

CAUTION: *Never exceed the maximum tire pressure embossed on the tire.*

When buying new tires, you should keep the following points in mind, especially if you are switching to larger tires or a different profile series (50, 60, 70, 78):

1. All four tires should be of the same construction type. Radial, bias, or bias-belted tires should not be mixed.

2. The wheels must be the correct width for the tire. Tire dealers have charts of tire and wheel rim compatibility. A mismatch can cause sloppy handling and rapid tread wear. The tread width should match the rim width (inside bead to inside bead) within an inch. For radial tires, the rim width should be 80% or less of the tire (not tread) width.

3. The height (mounted diameter) of the new tires can change speedometer accuracy, engine speed per given road speed, fuel mileage, acceleration, and ground clearance. Tire manufacturers furnish full measurement specifications.

4. The spare tire should be usable, at least for low speed operation, with the new tires.

5. There shouldn't be any body interference when the truck is loaded, on bumps or in turning.

So that the tires wear more uniformly, it is recommended that the tires be rotated every 6,000 miles. This can be done when all four tires are of the same size and load rating capacity. Any abnormal wear should be investigated and the cause corrected.

Radial tires should not be cross-switched; they'll last longer if their direction of rotation is not changed. Truck-type tires sometimes have directional tread, indicated by arrows molded into the sidewalls; the arrow shows the direction of rotation. They will wear very rapidly if reversed. Studded snow tires will lose their studs if their direction of rotation is reversed.

NOTE: *Mark the wheel position or direction of rotation on radial tires or studded snow tires before removal.*

CAUTION: *Avoid overtightening the lug nuts or the brake disc or drum may become permanently distorted. Alloy wheels can be cracked by overtightening. Always tighten the nuts in a criss-cross pattern.*

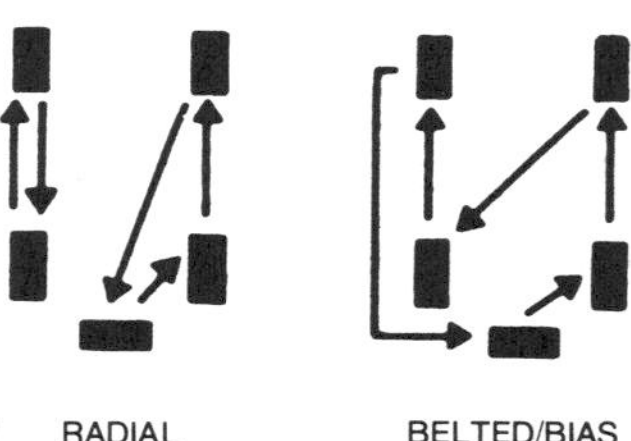

These patterns should be followed when rotating tires; the spare can be bypassed, if desired.

TIRE LIFE AND SAFETY

Common sense and good driving habits will afford maximum tire life. Fast starts and stops, and hard cornering are hard on tires and will shorten their useful life span. If you start at normal speeds, allow yourself sufficient time to stop, and take corners at a reasonable speed, the life of your tires will increase greatly. Also make sure that you don't overload your vehicle or run with incorrect pressure in the tires. Both of these practices increase tread wear.

Inspect your tires frequently. Be especially careful to watch for bubbles in the tread or side wall, deep cuts, or underinflation. Remove any tires with bubbles. If the cuts are so deep that they penetrate to the cords, discard the tire. Also look for uneven tread wear patterns that indicate that the front end is out of alignment or that the tires are out of balance.

Fuel Filter

The gasoline fuel filter is located on the right inner fender in the engine compartment. It is a disposable cartridge type, and should be replaced every 24,000 miles.

Fuel filter location

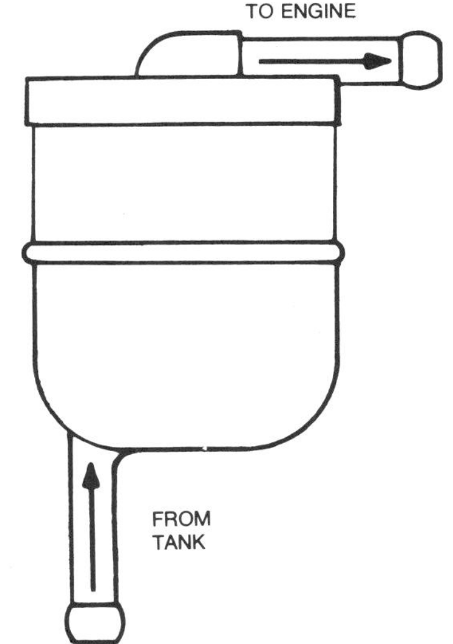

Diagram showing inlet (bottom) and outlet (top) lines of the fuel filter

To replace the filter, loosen the clamps on the fuel lines and slide the clamps down the hoses past the point to which the filter pipes extend. Gently twist and pull on the hoses to remove them from the filter. Be careful, because some gas will spill from the bottom fuel line. Remove the old filter from the clip, install a new filter, and replace the hoses and clamps. Be certain that the line from the fuel tank connects to the fuel inlet, and that the carburetor line connects to the outlet. Start the engine and check for leaks.

The diesel fuel filter is located on the upper right front of the engine. It is the canister type with a paper cartridge inside. To replace the filter, unbolt the top of the canister and lift out the old filter. Insert the new filter using new gaskets and O-rings. Replace the top. The filter should be replaced every 6,000 miles.

LUBRICATION

Oil and Fuel Recommendations
OIL

The SAE (Society of Automotive Engineers) grade number indicates the viscosity of the engine oil, and thus its ability to lubricate at a given temperature. The lower the SAE grade number, the lighter the oil; the lower the viscosity, the easier it is to crank the engine in cold weather.

The API (American Petroleum Institute) designation indicates the classification of engine oil for use under given operating conditions. Only oils designated for use "Service SE or SF" should be used. Oils of the SE or SF type perform a variety of functions inside the engine in addition to the basic function as a lubricant. Through a balanced system of metallic detergents and polymeric dispersants, the oil prevents the formation of high and low temperature deposits, and also keeps sludge and dirt particles in suspension. Acids, particularly sulfuric acid, as well as other byproducts of combustion, are neutralized. Both the SAE grade number and the API designation can be found on the top of the oil can.

NOTE: *Non-detergent or straight mineral oils must never be used.*

Oil viscosities should be chosen from those oils recommended for the lowest anticipated temperatures during the oil change interval.

Multi-viscosity oils offer the important advantage of being adaptable to temperature extremes. They allow easy starting at low temperatures, yet give good protection at high speeds and engine temperatures. This is a decided advantage in changeable climates or in long distance touring.

FUEL

The gasoline engine in the Datsun pick-up is designed to operate on regular leaded fuel,

Maintenance Intervals Chart

Intervals are for number of months or thousands of miles, whichever comes first.

NOTE: *Heavy-duty operation (trailer towing, prolonged idling, severe stop and start driving) should be accompanied by a 50% increase in maintenance. Cut the interval in half for these conditions.*

Maintenance	1970–71	1972	1973	1974	1975–77	1978–81
Air cleaner (Replace)	24	24	24	24	24	24
PCV valve						
Check	12	12	12	12	12	12
Replace	12	12	12	12	24	24
Carbon canister filter (Replace)	—	—	—	—	24	24
Heat riser (Check and lubricate)	—	—	—	—	6	—
Belt tension (Adjust)	6	6	12	12	12	12
Engine oil (Change)	3	3	3	4	6	6
Engine oil filter (Change)	6	6	6	8	6	6
Fuel filter, gasoline (Change)	24	24	24	24	24	24
Fuel filter, diesel (Change)	—	—	—	—	—	6
Manual transmission and transfer case						
Check	3	3	3	4	6	6
Change	30	30	30	36	24	24
Automatic transmission						
Check	—	3	3	4	6	6
Change	—	30	30	36	—	—
Drive axles						
Check	3	3	3	4	6	6
Change	30	30	30	36	24	24
Front wheel bearings (Clean and repack)	12	12	12	12	24	24
Engine coolant (Change)	24	24	24	24	24	24
Steering gear (Check)	6	6	6	6	12	12
Chassis lubrication						
Linkage and suspension	3	3	3	4	6	12
Ball joints	12	12	12	24	24	24
Driveshaft lubrication	30	30	30	36	24	24
Rotate tires	6	6	6	8	12	12
Valve lash (Check and adjust)	6	6	12	12	12	12
Brake and clutch fluid (Check)	3	3	3	4	6	12
Air pump filter (Change)	—	—	—	—	24	24

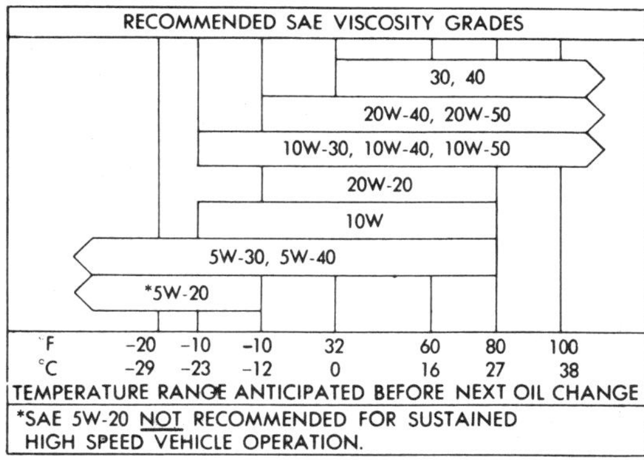

Gasoline engine oil viscosity usage

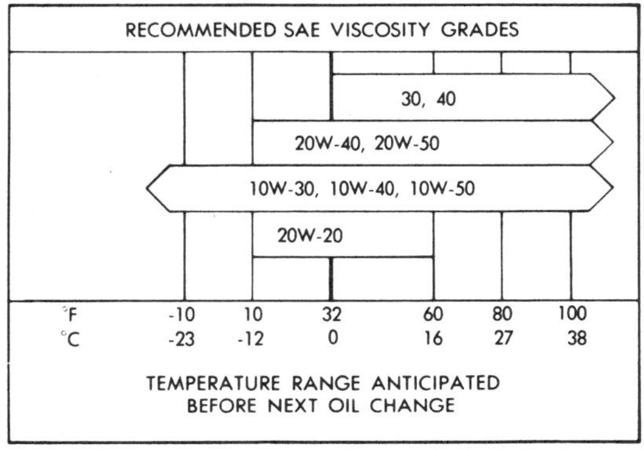

Diesel engine oil viscosity usage

Recommended Lubricants

Lubricant	Classification
Engine Oil	Gasoline: API SE or SF Diesel: API CD or CC
Manual Transmission and Transfer case	API GL-4
Automatic Transmission	DEXRON® or DEXRON II®
Differentials	API GL-5
Wheel Bearings	NLGI #2
Chassis Grease	NLGI #2
Driveshafts	NLGI #2
Brake Fluid	DOT 3
Clutch Fluid	DOT 3
Steering Gear	API GL-4
Antifreeze	Ethylene Glycol

1970–71, or regular, low-lead, or unleaded fuel 1972 and later, with the exception of 1976 and later pick-ups sold in California and all 1980–81 trucks, which have a catalytic converter. Converter-equipped trucks must use unleaded fuel. The use of leaded fuel will plug the catalyst rendering it inoperative, and will increase the exhaust back pressure to the point where engine output will be severely reduced. In all cases, the minimum octane rating of the fuel used must be at least 91 RON.

Use of a fuel too low in octane (a measurement of anti-knock quality) will result in spark knocks. Since many factors affect operating efficiency, such as altitude, terrain, and air temperature and humidity, knocking may result even though the recommended fuel is being used. If persistent knocking occurs, it may be necessary to switch to a slightly higher grade of gasoline. Continuous or heavy knocking may result in engine damage.

NOTE: *Your engine's fuel requirement can change with time, mainly due to carbon buildup, which changes the compression ratio. If your engine pings, knocks, or runs on, switch to a higher grade of fuel, if possible, and check the ignition timing. If your engine requires unleaded fuel, sometimes changing brands will cure the problem. If it is necessary to retard timing from specifications, don't change it more than a few degrees. Retarded timing will reduce power output and fuel mileage, and will increase engine temperature.*

The diesel engine in the Datsun pick-up is designed to operate on SAE #2 fuel oil year-round. In severely cold weather, #1 fuel oil may be used, if available. #1 fuel oil is more expensive and harder to find. In cold weather, any good brand of commercially available diesel fuel additive may be used in conjunction with #2 fuel oil.

Oil Changes
ENGINE

If you purchased your Datsun pick-up new, the engine oil should be changed at the end of the first 1,000 miles. The engine oil should then be changed according to the schedule in the Maintenance Intervals chart. You should also make it a practice to change the filter at

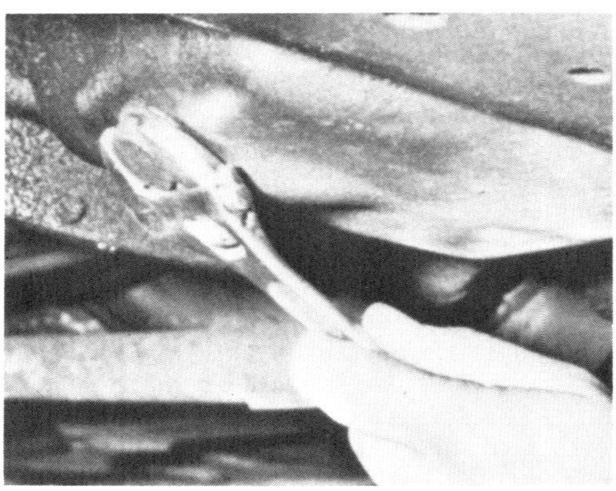

The drain plug is in the right corner of the oil pan

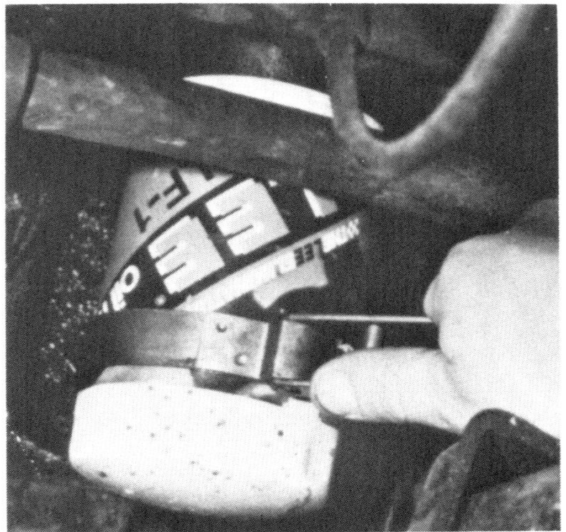

Use an oil filter wrench to remove the old filter. Install the new one by hand.

Coat the new filter gasket with clean oil

every oil change. These intervals ought to be halved when the truck is operated under severe conditions, such as dusty conditions, trailer towing, prolonged high speeds, off-road driving, or repeated short trips in freezing temperatures.

1. Operate the truck until the engine is at normal operating temperature. A run to the parts store for oil and a filter will accomplish this. If the engine is not hot when the oil is changed, most of the acids and contaminants will remain inside the engine.

2. Shut off the engine, and slide a pan of at least six quarts capacity under the oil pan. Throw-away aluminum roasting pans can be used for this.

3. Remove the drain plug from the engine oil pan, after wiping the plug area clean. The drain plug is the bolt inserted at an angle into the right side of the oil pan.

4. The oil from the engine will be HOT. It will probably not be possible to hold onto the drain plug. You may have to let it fall into the pan and fish it out later. Allow all the oil to drain completely. This will take a few minutes.

5. The drain plug is magnetic. Wipe it off thoroughly, removing any traces of metal particles. Pay particular attention to the threads. Replace it, and tighten it snugly. Every fourth oil change, replace the drain plug gasket.

6. The oil filter is on the right side of the engine. On gasoline engines use a filter wrench to loosen it. These are available at auto parts stores, among other sources. On diesel engines, a canister type filter, with replaceable paper cartridge is used. Place the drain pan on the ground under the filter. Remove and discard the old filter. It will be VERY HOT, so be careful.

7. If the spin-on oil filter is on so tightly that it collapses under pressure from the wrench, drive a long punch or a nail through it, across the diameter and as close to the base as possible, and use this as a lever to unscrew it. Make sure you are turning it counterclockwise.

8. On gasoline engines, clean off the oil filter mounting surface with a rag. Apply a thin film of clean engine oil to the filter gasket. On diesel engines, use new gaskets and O-rings with a new filter.

9. With spin-on filters screw the filter on by hand until the gasket makes contact. Then tighten it by hand an additional ½ to ¾ of a turn. Do not overtighten.

10. Remove the filler cap on the rocker (valve) cover, after wiping the area clean.

11. Add the correct number of quarts of oil specified in the Capacities Chart. If you

don't have an oil can spout, you will need a funnel. Be certain you do not overfill the engine, which can cause serious damage. Replace the cap.

12. Check the oil level on the dipstick. It is normal for the level to be a bit above the full mark. Start the engine and allow it to idle for a few minutes.

CAUTION: *Do not run the engine above idle speed until it has built up oil pressure, indicated when the oil light goes out.* Check around the filter and drain plug for any leaks.

13. Shut off the engine, allow the oil to drain for a minute, and check the oil level.

After completing this job, you will have several quarts of filthy oil to dispose of. The best thing to do with it is to funnel it into old plastic milk containers or bleach bottles. Then, you can either pour it into the recycling barrel at the gas station (if you're on good terms with the attendant), or put the containers into the trash.

TRANSMISSION

Manual Transmission and/or Transfer Case

Change the oil according to the schedule in the Maintenance Intervals chart. You may also want to change it if you have bought your truck used, or if it has been driven in water deep enough to reach the case.

1. The oil should be hot before it is drained. If the truck is driven until the engine is at normal operating temperature, the oil should be hot enough.

2. Remove the filler plug from the side of the case to provide a vent.

3. The drain plug is located on the bottom of the case. Place a pan under the drain plug and remove it.

CAUTION: *The oil will be HOT. Push up against the threads as you unscrew the plug to prevent leakage.*

4. Allow the oil to drain completely. Clean off the plug and replace, tightening it until it is just snug.

5. Fill the case with gear oil through the filler plug hole. Use API service GL-4 gear oil of the proper viscosity (see the Viscosity Chart in this Chapter for recommendations). This oil usually comes in a squeeze bottle with a long nozzle. If yours isn't, you can use a rubber squeeze bulb of the type used for kitchen basting to squirt the stuff in. Refer to the Capacities Chart for the amount of oil needed.

6. The oil level should come right up to the edge of the filler hole. You can stick your finger in to verify this. Watch out for sharp threads.

7. Replace the filler plug. Dispose of the old oil in the same manner as old engine oil. Take a drive in the truck, stop, and check for leaks.

Automatic

The fluid should be changed according to the schedule in the Maintenance Intervals chart. If the truck is normally used in severe service, such as start-and-stop driving, trailer towing, or the like, the interval should be halved. The fluid must be hot before it is drained; a 20 minute drive should accomplish this.

1. There is no drain plug; the fluid pan must be removed. Partially remove the pan screws until the pan can be pulled down at one corner. Place a container under the transmission, lower a rear corner of the pan, and allow the fluid to drain.

2. After draining, remove the pan screws completely, and remove the pan and gasket.

3. Clean the pan thoroughly and allow it to air dry. If you wipe it out with a rag you risk leaving bits of lint in the pan which will clog the tiny hydraulic passages in the transmission.

4. Install the pan using a new gasket. If you decide to use sealer on the gasket apply it only in a very thin bead running to the outside of the pan screw holes. Tighten the pan screws evenly in rotation from the center outwards, to 3–5 ft. lbs.

5. It is a good idea to measure the amount of fluid drained to determine how much fresh fluid to add. This is because some parts of the transmission, such as the torque converter, will not drain completely, and using the dry refill amount specified in the Capacities Chart may lead to overfilling. Fluid is added through the dipstick tube. Make sure that the funnel, hose, or whatever you are using is completely clean and dry before pouring transmission fluid through it. Use DEXRON® or DEXRON® II automatic transmission fluid.

6. Replace the dipstick after filling. Start the engine and allow it to idle. Do NOT race the engine.

7. After the engine has idled for a few minutes, shift the transmission slowly through the gears, then return the lever to Park. With the engine idling, check the fluid

level on the dipstick. It should be between the H and L marks. If below L, add sufficient fluid to raise the level to between the marks.

8. Drive the truck until the transmission is at operating temperature. The fluid should be at the H mark. If not, add sufficient fluid until this is the case. Be careful not to overfill; overfilling causes slippage, overheating, and seal damage.

NOTE: *If the drained fluid is discolored (brown or black), thick, or smells burnt, serious transmission problems due to overheating should be inspected by a transmission specialist to determine the cause.*

DRIVE AXLES

The axle lubricant should be changed according to the schedule in the Maintenance Intervals chart; you may also want to change it if you have bought your truck used, or if it has been driven in water deep enough to reach the axle.

1. Park the truck on a level surface. Place a pan of at least 2 quarts capacity underneath the drain plug. The drain plug is located on the bottom of the differential housing, just to the right of center. Remove the plug.

2. Allow the lubricant to drain completely.

3. Install the drain plug. Tighten it so that it will not leak, but do not overtighten.

4. Refill the axle housing with API GL-5 gear oil of the proper viscosity to the level of the oil fill hole.

5. Install the filler plug.

Chassis Greasing

There are two schedules for regular chassis greasing. The steering linkage and front suspension components should be greased at regular intervals of 3, 4, 6, or 12 thousand miles, depending on the year of manufacture. Ball joints must be lubricated at 12 or 24 thousand mile intervals. The Maintenance Interval chart gives the correct schedule for your truck.

Steering and suspension lubrication points have conventional grease fittings already installed. Ball joint lubrication points are covered with threaded plugs. To grease the ball joints, the plugs are removed and grease fittings are installed. The fittings must be removed and replaced by the plugs afterwards.

If you wish to perform this operation yourself, you should purchase a cartridge-type grease gun and a cartridge of NLGI No. 2 multi-purpose lithium base grease. You will also need to purchase grease fittings for the ball joints; these should be obtained from your Datsun dealer, since metric threaded grease fittings are still a rare commodity in most auto supply stores.

To grease the ball joints, remove the plug and install a grease fitting. Push the nozzle of the grease gun down firmly onto the fitting and, while holding it there, pump a few shots of grease into the fitting. Force sufficient grease into the joint to cause the old grease to be expelled, but be careful not to rupture the rubber boots, where installed. Once this has been accomplished, remove the fitting and replace the plug.

Front suspension and steering linkage lubrication points are greased in the same way, except for the installation of grease fittings, since these are already present.

Wheel Bearings

The following is for 2-wheel drive only. For 4-wheel drive vehicles, see the front drive axle section in Chapter 7.

Only the front wheel bearings require periodic service. The lubricant to use is high temperature disc brake wheel bearing grease meeting NLGI No. 2 specifications. (This grease should be used even if your Datsun is equipped with drum brakes; it has superior protection characteristics.) This service is recommended at the specified period in the Maintenance Intervals chart, or whenever the truck has been driven in water up to the hubs.

You will not need any special tools for this job, although the use of a torque wrench is strongly recommended for accurate measurement of bearing preload. Procedures are basically the same for either drum or disc brakes.

The most important thing to remember when working with the wheel bearings is that although they are basically durable, in some ways they are remarkably fragile. Mishandling, grit, misalignment, scratches, improper preload, etc. will quickly destroy any roller bearing, no matter how well hardened during manufacture.

1. Loosen the wheel nuts, raise the truck, and remove the wheel and tire. Remove the brake drum or brake caliper, following the procedure in Chapter 9.

2. It is not necessary to remove the drum or disc from the hub. The outer wheel bear-

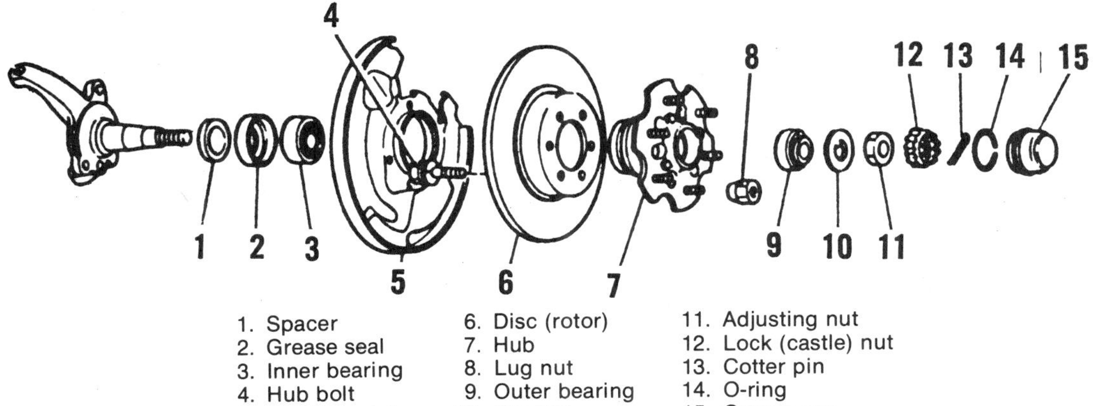

1. Spacer
2. Grease seal
3. Inner bearing
4. Hub bolt
5. Backing plate
6. Disc (rotor)
7. Hub
8. Lug nut
9. Outer bearing
10. Washer
11. Adjusting nut
12. Lock (castle) nut
13. Cotter pin
14. O-ring
15. Grease cap

Exploded view of the 2-wd hub and bearings with disc brakes. Drum brakes are similar

ing will come off with the hub. Simply pull the hub and disc or drum assembly toward you off the spindle. Be sure to catch the bearing before it falls to the ground.

3. From the inner side of the hub, remove the inner grease seal, and lift the inner bearing from the hub. Discard the grease seal.

4. Clean the bearings in solvent and allow them to air dry. You risk leaving bits of lint in the races if you dry them with a rag. Clean the grease cap, nuts, spindle, and the races in the hub thoroughly, and allow the parts to dry.

5. Inspect the bearings carefully. If they are worn, cracked, brinelled, pitted, burned, scored, etc., they should be replaced, along with the bearing cups in which they run in the hub. Do not mix old and new parts.

6. If the cups are worn, remove them from the hub, by using a brass rod as a drift.

To install:

7. If the old cups were removed, install the new inner and outer cups into the hub, using either a tool made for the purpose, or a socket or piece of pipe of a large enough diameter to press on the outside rim of the cup only.

CAUTION: *Use care not to cock the bearing cups in the hub. If they are not fully seated, the bearings will be impossible to adjust properly.*

8. Pack the inside area of the hub and cups with grease, according to the illustration given. Pack the inside of the grease cap while you're at it, but do not install the cap into the hub.

9. Pack the inner bearing with grease. Place a large glob of grease into the palm of one hand and push the inner bearing through it with a sliding motion. The grease must be forced through the side of the bearing and in between each roller. Continue until the grease begins to ooze out the other side through the gaps between the rollers; the

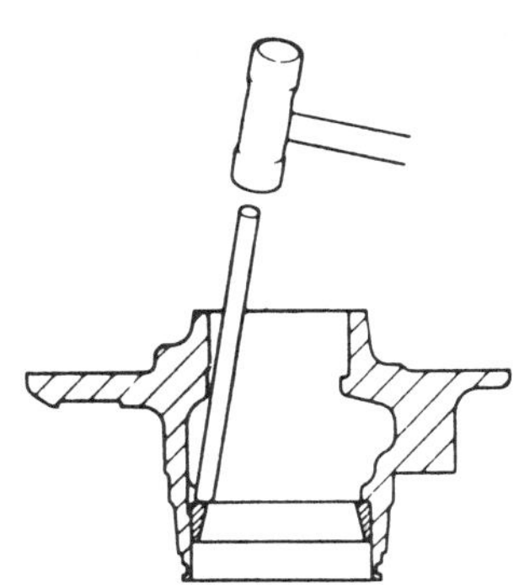

Drive worn bearing cups from the hub with a soft drift and a hammer

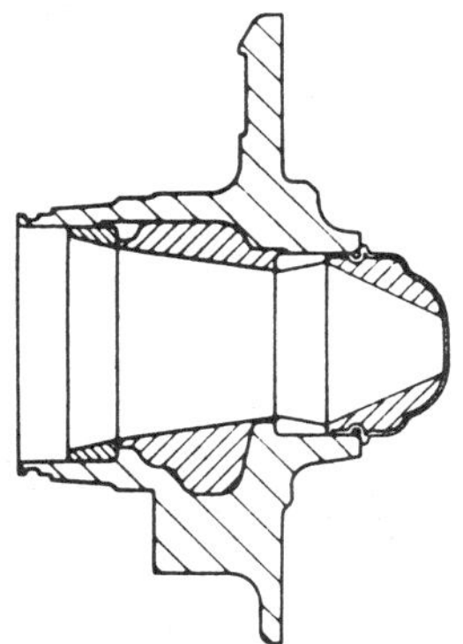

Fill the shaded portion of the hub and grease cap with grease; also coat the cups with grease

bearing must be completely packed with grease. Install the inner bearing into its cup in the hub, then press a new grease seal into place over it.

10. Install the hub and rotor or drum assembly onto the spindle. Pack the outer bearing with grease in the same manner as the inner bearing, then install the outer bearing into place in the hub.

11. Apply a thin coat of grease to the washer and the threaded portion of the spindle, then loosely install the washer and adjusting nut. Go on to the bearing preload adjustment following.

BEARING PRELOAD ADJUSTMENT

1. While turning the hub forward, tighten the adjusting nut to 22–29 ft. lbs.

2. Rotate the hub a few more times to snug down the bearings.

3. Retighten the nut to 22–29 ft. lbs. Unscrew the adjusting nut ⅛ of a turn (45 degrees). Install the lock nut (castellated nut) and snug it down against the adjusting nut until one of its grooves lines up with the hole in the spindle. It is okay to tighten the adjusting nut up to 15 degrees to allow the lock nut holes to align. Install a new cotter pin, bending its ends around the lock nut.

4. Install the wheel and a couple of lug nuts. Check the axial play of the wheel by shaking it back and forth. The bearing free play should feel close to zero, but the wheel should spin freely. Be sure the brake shoes are not dragging against the drum, if your truck has drum brakes.

5. If the bearing play is correct, with drum brakes you can install the grease cap and the rest of the lug nuts. With disc brakes, remove the wheel, replace the caliper, then install the wheel and grease cap.

PUSHING AND TOWING

All trucks sold with manual transmissions through 1975, and 1976 and later 49 States and Canada trucks with manual transmissions can be push started. 1976 and later trucks sold in California with a catalytic converter may not be push started, since doing so might cause the converter to explode. Trucks with an automatic transmission may not be push started. Check to make sure that the bumpers of both vehicles are aligned so neither will be damaged. Be sure that all electrical system components are turned off (headlights, heater, blower, etc.). Turn on the ignition switch. Place the shift lever in Third or Fourth and push in the clutch pedal. At about 15 mph, signal the driver of the pushing vehicle to fall back, depress the accelerator pedal, and release the clutch pedal slowly. The engine should start.

When your truck is doing the pushing or pulling, make sure that the two bumpers match so you won't damage the vehicle you are to push. Another good idea is to put an old tire in between the two vehicles. If the bumpers don't match, perhaps you should tow the other vehicle. If the other vehicle is just stuck, use First gear to slowly push it out. Tell the driver of the other vehicle to go slowly too. Try to keep your vehicle right up against the other vehicle while you are pushing. If the two vehicles do separate, stop and start over again instead of trying to catch up and ramming the other vehicle. Also try, as much as possible, to avoid riding or slipping the clutch. When the other vehicle gains enough traction, it should pull away from your vehicle.

If you have to tow the other vehicle, make sure that the tow chain or rope is sufficiently long and strong, and that it is attached securely to both vehicles at a strong place. Attach the chain at a point on the frame or as close to it as possible. Once again, go slowly and tell the other driver to do the same. Warn the other driver not to allow too much slack in the line when he gains traction and can move under his own power. Otherwise he may run over the tow line and damage both vehicles.

Your Datsun may be towed with all four wheels on the ground as long as the transmission and rear axle are in good working order. Place the transmission in Neutral, and be certain that the parking brake is released. Turn the ignition key to "Off," not "Lock." If your truck has an automatic transmission, it may not be towed in this manner for more than six miles, nor at speeds exceeding 20 mph.

If the transmission or rear axle is damaged, or if the towing distance exceeds six miles (automatic transmission only), the truck should be towed from the rear, with the rear wheels raised off the ground.

JUMP STARTING

Jump starting is the only way to start an automatic transmission model with a weak bat-

JUMP STARTING A DEAD BATTERY

The chemical reaction in a battery produces explosive hydrogen gas. This is the safe way to jump start a dead battery, reducing the chances of an accidental spark that could cause an explosion.

Jump Starting Precautions

1. Be sure both batteries are of the same voltage.
2. Be sure both batteries are of the same polarity (have the same grounded terminal).
3. Be sure the vehicles are not touching.
4. Be sure the vent cap holes are not obstructed.
5. Do not smoke or allow sparks around the battery.
6. In cold weather, check for frozen electrolyte in the battery.
7. Do not allow electrolyte on your skin or clothing.
8. Be sure the electrolyte is not frozen.

Jump Starting Procedure

1. Determine voltages of the two batteries; they must be the same.
2. Bring the starting vehicle close (they must not touch) so that the batteries can be reached easily.
3. Turn off all accessories and both engines. Put both cars in Neutral or Park and set the handbrake.
4. Cover the cell caps with a rag—do not cover terminals.
5. If the terminals on the run-down battery are heavily corroded, clean them.
6. Identify the positive and negative posts on both batteries and connect the cables in the order shown.
7. Start the engine of the starting vehicle and run it at fast idle. Try to start the car with the dead battery. Crank it for no more than 10 seconds at a time and let it cool off for 20 seconds in between tries.
8. If it doesn't start in 3 tries, there is something else wrong.
9. Disconnect the cables in the reverse order.
10. Replace the cell covers and dispose of the rags.

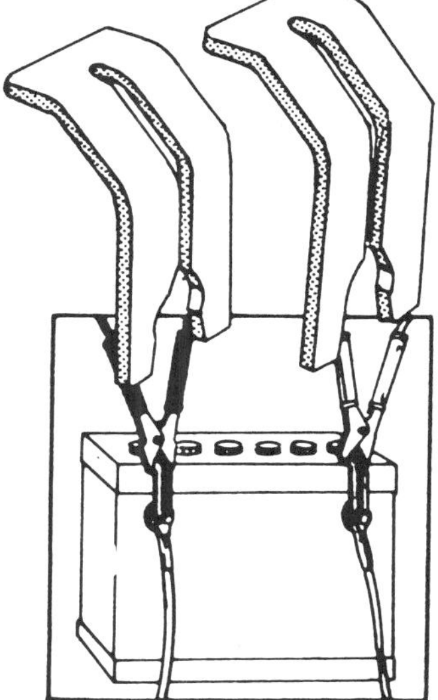

Side terminal batteries occasionally pose a problem when connecting jumper cables. There frequently isn't enough room to clamp the cables without touching sheet metal. Side terminal adaptors are available to alleviate this problem and should be removed after use

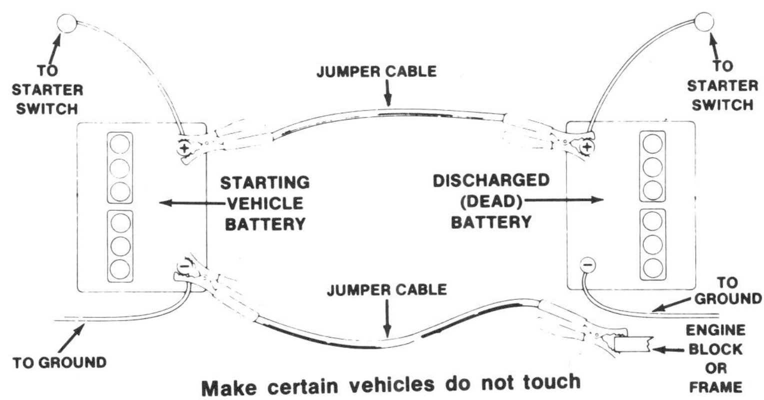

Make certain vehicles do not touch

This hook-up for negative ground cars only

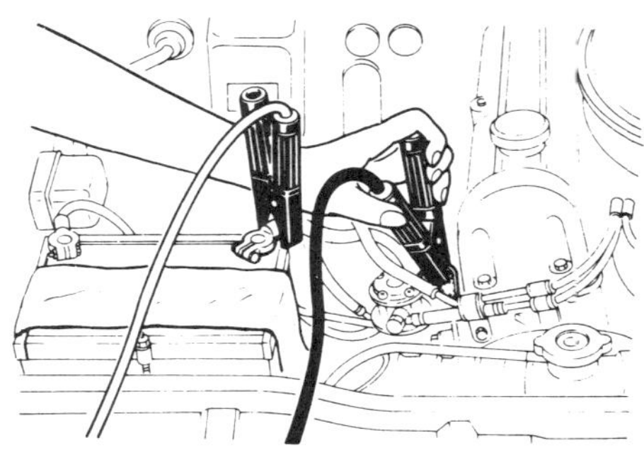

Connect the negative cable from the booster battery to the engine, not the negative battery terminal

tery, and the best method for a manual transmission model.

CAUTION: *Do not attempt this procedure on a frozen battery; it will probably explode.*

The battery in the other vehicle must be a 12 volt, negatively grounded one. Do not attempt to jump start your Datsun with a 24 volt power source; serious electrical damage will result.

1. Turn off all electrical equipment. Place the automatic transmission in Park or the manual in Neutral and set the parking brake.

2. Make sure that the two vehicles are not touching. It is a good idea to keep the engine running in the booster vehicle.

3. Remove the caps from both batteries and cover the openings with cloths.

4. Attach one end of a jumper cable to the positive (+) terminal of the booster battery. The red cable is usually positive. Attach the other end to the positive terminal of the discharged battery.

CAUTION: *Be very careful about these connections. An alternator and regulator can be destroyed in a remarkably short time if battery polarity is reversed.*

5. Attach one end of the other cable (the black one) to the negative (−) terminal of the booster battery. Attach the other end to a ground point such as the alternator bracket on the engine of the truck being started. Do not connect it to the battery.

CAUTION: *Be careful not to lean over the battery while making this last connection.*

6. If the engine will not start, disconnect the batteries as soon as possible. If this is not done, the two batteries will soon reach a state of equilibrium, with both too weak to start an engine. This is no problem if the engine of the booster vehicle is running fast enough to keep up the charge. Lengthy cranking can also damage the starter.

7. Reverse the procedure exactly to remove the humper cables. Discard the rags, because they may have acid on them.

NOTE: *It is recognized that some or all of the precautions outlined in this procedure are often ignored with no harmful results. However, the procedure outlined is the only fully safe, foolproof one.*

JACKING AND HOISTING

The Datsun pick-up may be jacked at points under the frame or under the rear axle housing; never under front suspension components.

It is imperative that strict safety precautions be observed when raising your truck with a jack and in the subsequent supporting of the vehicle. Severe damage to the vehicle may result, not to mention the possible physical harm that may be imparted to yourself or some other person if you fail to exercise religious caution.

The jack should be positioned squarely on the frame of the vehicle behind the front wheel when raising the front end. The vehicle should be parked, in gear, on level ground and the emergency brake set. A block

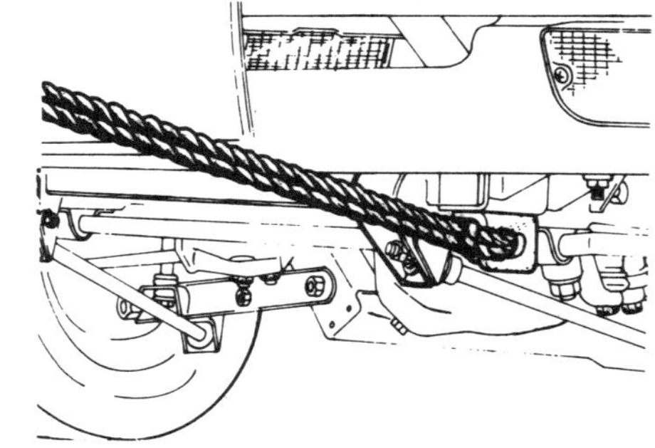

Front towing point

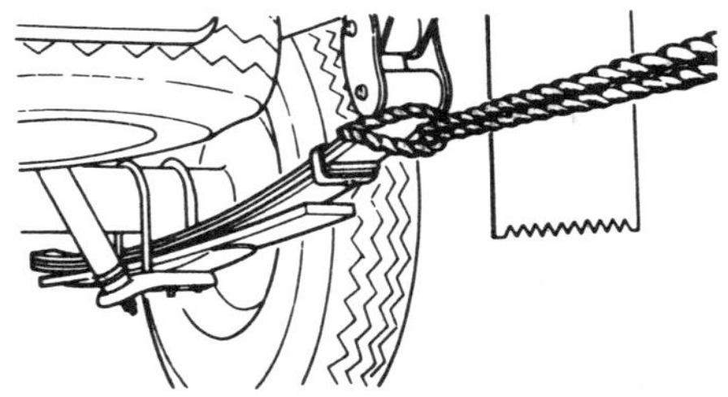

Rear towing point

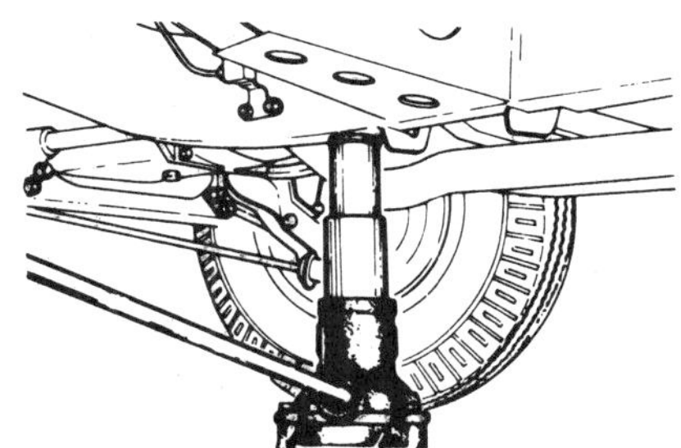

Front jacking point

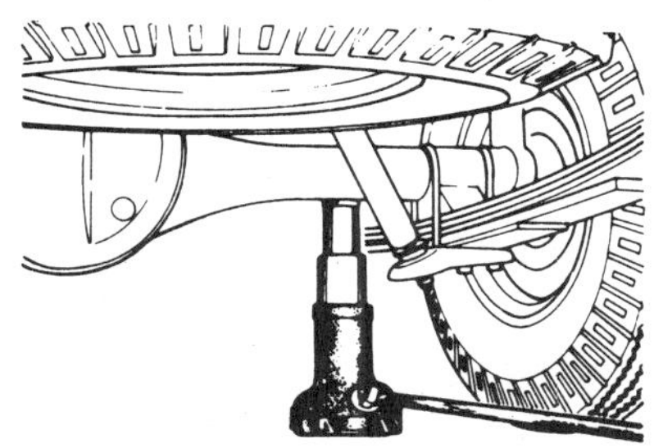

Rear jacking point

should be positioned behind the opposite rear wheel. When raising the rear of the truck, position the jack close to the inside edge of the rear leaf spring on the rear axle housing. Again, a block should be placed at the opposite front wheel.

Whenever you plan to work under the truck you must support it on jackstands or ramps. Never use cinder blocks or stacks of wood to support a vehicle which you will be underneath for even a few minutes. Small hydraulic, screw, or scissors jacks are satisfactory for raising the truck. Drive-on trestles, or ramps, are also a handy and safe way to both raise and support the truck. These can be bought or constructed from suitably heavy steel or wood.

If the Datsun is to be raised with a hoist such as those used at service stations, the pads of the hoist should be positioned on the frame rails of the truck, and never on any suspension member or underbody panel.

Tune-Up

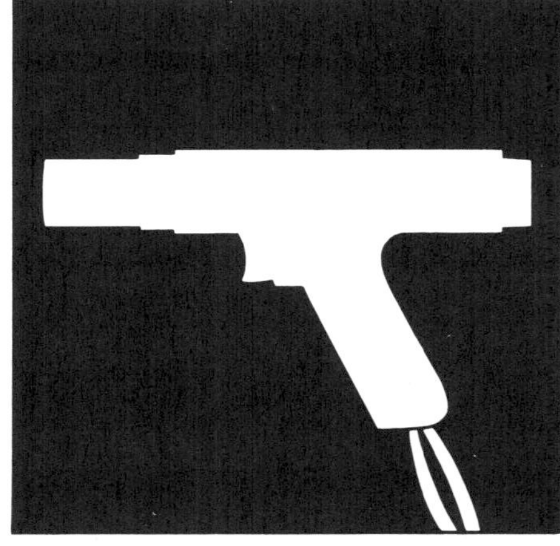

TUNE-UP PROCEDURES—

GASOLINE ENGINES

In order to extract the full measure of performance and economy from your engine it is essential that it be properly tuned at regular intervals. A regular tune-up will keep your Datsun's engine running smoothly and will prevent the annoying minor breakdowns and poor performance associated with an untuned engine.

> NOTE: *All Datsun pick-ups use a conventional breaker points ignition system throuh 1975. 1970–73 models utilize a dual points system for emission control purposes. 1976 and later trucks sold in California, and all 1978 and later models sold in the U.S. use a fully transistorized ignition system. 1976–77 49 States and Canada models use the single breaker points system used in all models in 1974 and 1975.*

A complete tune-up should be performed every 12,000 miles or twelve months, whichever comes first. This interval should be halved if the truck is operated under severe conditions, such as trailer towing, prolonged idling, continual stop and start driving, or if starting or running problems are noticed. It is assumed that the routine maintenance described in Chapter 1 has been kept up, as this will have a decided effect on the results of a tune-up. All of the applicable steps of a tune-up should be followed in order, as the result is a cumulative one.

If the specifications on the tune-up sticker in the engine compartment of your Datsun disagree with the "Tune-Up Specifications" chart in this chapter, the figures on the sticker must be used. The sticker often reflects changes made during the production run.

Spark Plugs

Spark plugs ignite the air and fuel mixture in the cylinder as the piston reaches the top of the compression stroke. The controlled explosion that results forces the piston down, turning the crankshaft and the rest of the drive train.

The average life of a spark plug is 12,000 miles. This is, however, dependent on a number of factors: the mechanical conditions of the engine; the type of fuel; driving conditions; and the driver.

When you remove the spark plugs, check their condition. They are a good indicator of the condition of the engine. It is a good idea to remove the spark plugs every 6,000 miles to keep an eye on the mechanical state of the engine.

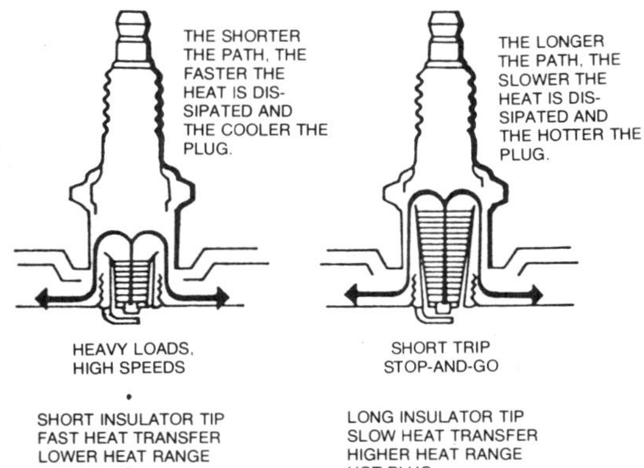

Spark plug heat range

A small deposit of light tan or gray material (or rust red with unleaded fuel) on a spark plug that has been used for any period of time is to be considered normal. Any other color, or abnormal amounts of deposit, indicates that there is something amiss in the engine.

The gap between the center electrode and the side or ground electrode can be expected to increase not more than 0.001 in. every 1,000 miles under normal conditions.

When a spark plug is functioning normally or, more accurately, when the plug is installed in an engine that is functioning properly, the plugs can be taken out, cleaned, regapped, and reinstalled in the engine without doing the engine any harm.

When, and if, a plug fouls and begins to misfire, you will have to investigate, correct the cause of the fouling, and either clean or replace the plug.

There are several reasons why a spark plug will foul and you can learn which is at fault by just looking at the plug. A few of the most common reasons for plug fouling, and a description of the fouled plug's appearance, are listed in the "Color" section which also offers solutions to the problems.

Spark plugs suitable for use in your Datsun's engine are offered in a number of different heat ranges. The amount of heat which the plug absorbs is determined by the length of the lower insulator. The longer the insulator, the hotter the plug will operate; the shorter the insulator, the cooler it will operate. A spark plug that absorbs (or retains) little heat and remains too cool will accumulate deposits of lead, oil, and carbon, because it is not hot enough to burn them off. This leads to fouling and consequent misfiring. A spark plug that absorbs too much heat will have no

deposits, but the electrodes will burn away quickly and, in some cases, preignition may result. Preignition occurs when the spark plug tips get so hot that they ignite the fuel/air mixture before the actual spark fires. This premature ignition will usually cause a pinging sound under conditions of low speed and heavy load. In several cases, the heat may become high enough to start the fuel/air mixture burning throughout the combustion chamber rather than just to the front of the plug. In this case, the resultant explosion will be strong enough to damage pistons, rings, and valves.

In most cases the factory recommended heat range is correct; it is chosen to perform well under a wide range of operating conditions. However, if most of your driving is long distance, high speed travel, you may want to install a spark plug one step colder than standard. If most of your driving is of the short trip variety, when the engine may not always reach operating temperature, a hotter plug may help burn off the deposits normally accumulated under those conditions.

REMOVAL

1. Number the wires so that you won't cross them when you replace them.

2. Remove the wire from the end of the spark plug by grasping the wire by the rubber boot. If the boot sticks to the plug, remove it by twisting and pulling at the same time. Do not pull the wire itself or you will damage the core.

3. Use a $^{13}/_{16}$ in. spark plug socket to loosen all of the plugs about two turns.

NOTE: *The cylinder head is cast from aluminum. Remove the spark plugs when the*

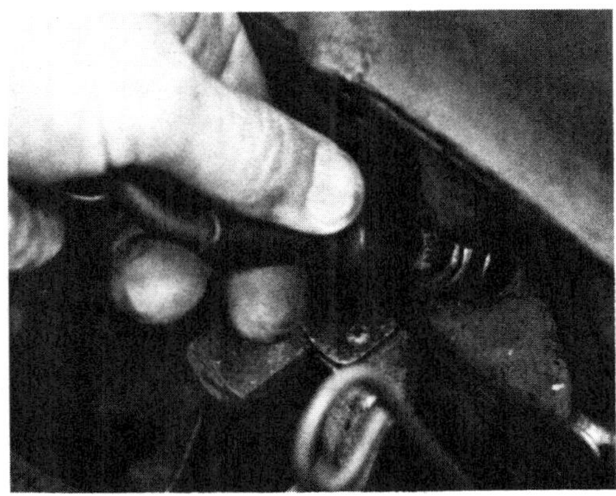

Twist and pull on the boot, not the wire

Gasoline Engine Tune-Up Specifications

Year	Engine Displacement cu in. (cc)	Spark Plug		Distributor		Ignition Timing (deg)		Intake Valve Opens (deg)	Fuel Pump Pressure (psi)	Compression Pressure (psi) ▲	Idle Speed (rpm)		Valve Clearance (in.) ●	
		Type	Gap (in.)	Point Dwell (deg)	Point Gap (in.)	MT	AT				MT	AT	In	Ex
1970–71	4-97.4 (1595)	B6ES	0.033	52	0.020	10B	—	12B	2.6–3.4	171②	700	—	0.010	0.012
1972	4-97.4 (1595)	B6ES	0.032	52	0.020	7B	7B	12B	2.6–3.4	171②	700	600①	0.010	0.012
1973	4-97.4 (1595)	B6ES	0.030	52	0.020	8B	8B	12B	2.6–3.4	171③	800	650①	0.010	0.012
	4-108 (1770)	B6ES	0.030	52	0.020	5B	5B	12B	2.6–3.4	171③	800	650①	0.010	0.012
1974	4-108 (1770)	B6ES	0.032	52	0.020	12B	12B	16B	2.6–3.4	171③	800	650①	0.010	0.012
1975	4-119 (1952)	BP6ES	0.034	52	0.020	12B④	12B	16B	3.0–3.8	171③	750	650①	0.010	0.012
1976–77	4-119 (1952)	BP6ES⑤	0.033⑥	52	0.020⑦	12B④	12B	16B	3.0–3.8	171③	750	650①	0.010	0.012

1978	4-119 (1952)	BP6ES-11	0.041	—	[7]	12B	12B	16B	3.0–3.9[8]	171[3]	600	600[1]	0.010	0.012
1979	4-119 (1952)	BP6ES-11	0.041	—	[9]	12B	12B	16B	3.0–3.9[8]	171[3]	650	630[1]	0.010	0.012
1980	4-119 (1952)	BP6ES-11	0.041	—	[9]	12B[10]	12B[10]	16B	3.0–3.9[8]	171[3]	600	600	0.010	0.012
1981	4-2187 (133.5)	BP6ES[11]	0.033	—	[9]	5B	5B	16B	3.0–3.9	171[3]	650[12]	650	0.012	0.012

● Measured with engine hot
▲ Lowest reading must be at least 80% of the highest
B Before top dead center
[1] Transmission in Drive
[2] 159 psi minimum
[3] 128 psi minimum
[4] 10° B in California
[5] BP6ES-11, Transistor ignition; BPR6ES, Canada, 1977 and later with conventional ignition
[6] 0.041, Transistor ignition
[7] 0.008–0.016, Transistor ignition air gap
[8] 4.6 or less with electric fuel pump (air conditioned models)
[9] 0.012–0.020, Transistor ignition air gap
[10] Calif. Heavy Duty models: 10B
[11] Canada: BPR6ES
[12] 4-WD: 800
NOTE: *Part numbers in this chart are not recommendations by Chilton for any product by brand name.*

Diesel Engine Tune-Up Specifications

Injector Opening Pressure (psi)	Low Idle	Valve Clearance (in.)		Intake Valve Opens (deg.)	Injection Timing rpm	Firing Order
		Intake	Exhaust			
1422.5	700	.014	.014	28	20 BTDC	1-4-3-2

engine is cold, if possible, to prevent damage to the threads.

If removal of the plugs is difficult, apply a few drops of penetrating oil or silicone spray to the area around the base of the plug, and allow it a few minutes to work.

4. If compressed air is available, apply it to the area around the spark plug holes. Otherwise use a rag or a brush to clean the area. Be careful not to allow any foreign material to drop into the spark plug holes.

5. Remove the plugs by unscrewing them the rest of the way from the engine.

INSPECTION

Check the plugs for deposits and wear. If they are not going to be replaced, clean the plugs thoroughly. Remember that any kind of deposit will decrease the efficiency of the plug. Plugs can be cleaned on a spark plug cleaning machine, which can sometimes be found in service stations, or you can do an acceptable job of cleaning with a stiff brush. If the plugs are cleaned, the electrodes must be filed flat. Use an ignition points file, not an emery board or the like, which will leave deposits. The electrodes must be filed perfectly flat with sharp edges; rounded edges reduce the spark plug voltage by as much as 50%.

Check spark plug gap before installation.

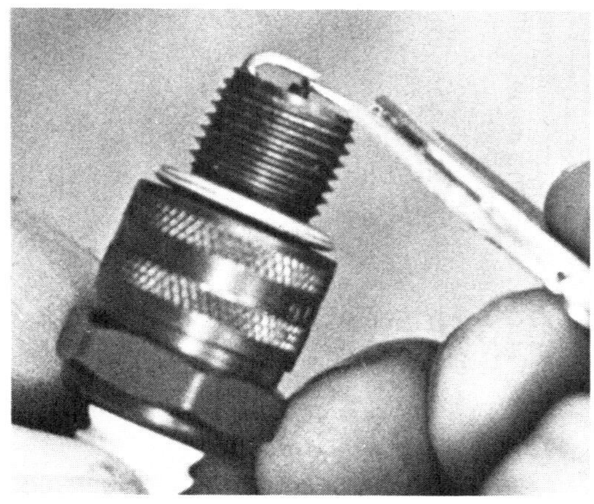

Check the gap with a wire gauge

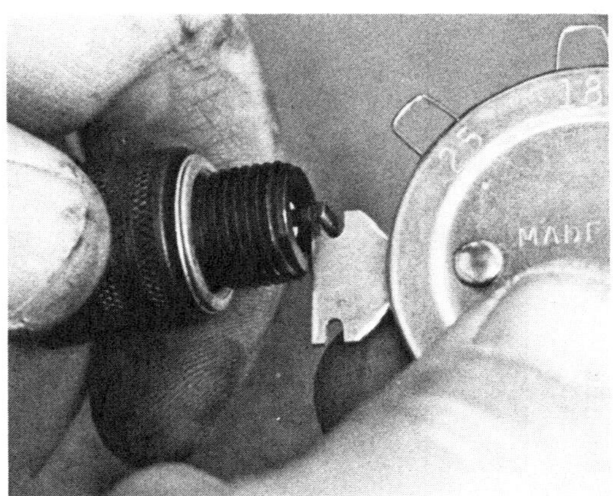

Adjust by carefully bending the side electrode

The ground electrode must be parallel to the center electrode and the specified size wire gauge should pass through the gap with a slight drag. Always check the gap on new plugs, too; they are not always correctly set at the factory. Do not use a flat feeler gauge when measuring the gap. Because the reading will be inaccurate. Wire gapping tools usually have a bending tool attached. Use that to adjust the side electrode until the proper distance is obtained. Absolutely never bend the center electrode. Also, be careful not to bend the side electrode too far or too often; it may weaken and break off within the engine, requiring removal of the cylinder head to retrieve it.

INSTALLATION

1. Lubricate the threads of the spark plugs with a drop of oil. Install the plugs and tighten them hand-tight. Take care not to cross-thread them.

2. Tighten the spark plugs with the socket. Do not apply the same amount of force you would use for a bolt; just snug them in. If a torque wrench is available, tighten to 11–15 ft. lbs.

3. Install the wires on their respective plugs. Make sure the wires are firmly connected. You will be able to feel them click into place.

CHECKING AND REPLACING SPARK PLUG CABLES

At every tune-up, visually inspect the spark plug cables for burns, cuts, or breaks in the insulation. Check the boots and the nipples on the distributor cap and coil. Replace any damaged wiring.

Every 36,000 miles or so, the resistance of the wires should be checked with an ohmmeter. Wires with excessive resistance will cause misfiring, and may make the engine difficult to start in damp weather. Generally, the useful life of the cables is 36,000–50,000 miles.

To check resistance, remove the distributor cap, leaving the wires attached. Connect one lead of an ohmmeter to an electrode within the cap; connect the other lead to the corresponding spark plug terminal (remove it from the plug for this test). Replace any wire which shows a resistance over 50,000 ohms. Generally speaking, however, resistance should not be over 30,000 ohms, and 50,000 ohms must be considered the outer limit of acceptability. Test the high tension lead from the coil by connecting the ohmmeter between the center contact in the distributor cap and either of the primary terminals of the coil. If resistance is more than 25,000 ohms, remove the cable from the coil and check the resistance of the cable alone. Anything over 15,000 ohms is cause for replacement. It should be remembered that resistance is also a function of length; the longer the cable, the greater the resistance. Thus, if the cables on your truck are longer than the factory originals, resistance will be higher, quite possibly outside these limits.

When installing new cables, replace them one at a time to avoid mixups. Start by replacing the longest one first. Install the boot firmly over the spark plug. Route the wire over the same path as the original. Insert the nipple firmly into the tower on the cap or the coil.

Breaker Points and Condenser

The points function as a circuit breaker for the primary circuit of the ignition system. The ignition coil must boost the 12 volts of electrical pressure supplied by the battery to as much as 25,000 volts in order to fire the plugs. To do this, the coil depends on the points and the condenser to make a clean break in the primary circuit.

The coil has both primary and secondary circuits. When the ignition is turned on, the battery supplies voltage through the coil and onto the points. The points are connected to ground, completing the primary circuit. As the current passes through the coil, a magnetic field is created in the iron center core of the coil. When the cam in the distributor turns, the points open, breaking the primary circuit. The magnetic field in the primary circuit of the coil then collapses and cuts through the secondary circuit windings around the iron core. Because of the physical principle called "electromagnetic induction," the battery voltage is increased to a level sufficient to fire the spark plugs.

When the points open, the electrical charge in the primary circuit tries to jump the gap created between the two open contacts of the points. If this electrical charge were not transferred elsewhere, the metal contacts of the points would melt and the gap between the points would start to change rapidly.

The function of the condenser is to absorb excessive voltage from the points when they open and thus prevent the points from become pitted or burned.

INSPECTION OF THE POINTS

1. Disconnect the high-tension wire from the top of the distributor and the coil.

2. The distributor cap is retained by two spring clips. Insert a screwdriver under their ends and release them. Lift off the cap with the spark plug wires attached. Inspect the inside of the cap. Wipe it clean with a rag and check for burned contacts, cracks and carbon tracks. A carbon track shows as a dark line running from one terminal to another. It cannot be successfully removed, so replace the cap if it has one of these. Generally, a cap and rotor will last 36,000 miles.

3. Remove the rotor from the distributor shaft by pulling it straight up. Examine the condition of the rotor. If it is cracked or the metal tip is excessively worn or burned, it should be replaced. Clean the metal tip with a clean cloth, but don't file it.

4. Pry open the contacts of the points with a screwdriver and check the condition of the contacts. If they are excessively worn, burned or pitted, they should be replaced.

5. If the points are in good condition, adjust them and replace the rotor and the distributor cap. If the points need to be replaced, follow the replacement procedure given next.

REPLACEMENT OF THE BREAKER POINTS AND CONDENSER

1. Remove the cap and rotor as outlined in steps 1–3 of the preceding section.

2. On single points distributors (1974–77), loosen the two screws securing the points. Use a magnetic screwdriver to avoid losing a screw down the distributor. Loosen the screw in the side of the distributor and slip the points wire out. Remove the point set.

On dual points distributors (1970–73) you will first have to decide if the tools available to you will enable you to remove the two condensers on the outside of the distributor body. If you have a flexible screwdriver, or one bent at a 90° angle, you should be able to get at them. Otherwise, it may be best to remove the distributor from the engine for points and condenser replacement.

To remove the distributor, first note and mark the position of the distributor on the small timing scale on the front of the distributor. Then mark the position of the rotor in relation to the distributor body. Do this by simply replacing the rotor on the distributor shaft and marking the spot on the distributor body where the rotor is pointing. Remove the small bolt at the rear of the distributor, and lift the distributor out of the block. DO NOT CRANK THE ENGINE WITH THE DISTRIBUTOR REMOVED.

3. On dual points distributors, loosen the two mounting screws which secure the points. Do not loosen the factor pre-set phase adjusting screw (#3 in the photograph). Loosen the screws retaining the points wires and remove each points set. The bottom of

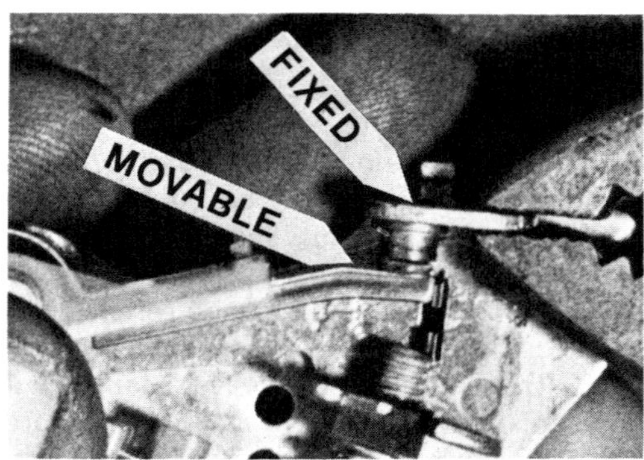

Bend the fixed point only until the faces are square

the points set is slotted; thus, it is not necessary to completely remove the points mounting screws.

4. On single points distributors, the condenser is held in place by the same screw as the points mounting screw, and its lead is attached at the same place as the points wire. Simply remove the condenser.

On dual points distributors, the two condensers are mounted on the outside of the distributor body. Note that the two are different. When new condensers are installed, they must be replaced in the same relationship as the originals. This is important, because they have different electrical capacities. Remove the condenser mounting screws, loosen the condenser lead screws, and remove the condensers.

5. Before installing the new points and condenser(s), place a matchhead-sized dab of grease on the distributor shaft cam and smear it evenly around the cam. Do not use oil, because it will lead to rapid point burning.

6. Install the new points set(s) and condenser(s). Be sure you are putting the dual points condensers in their proper locations. The different mounting ears on each make them pretty much idiot-proof. Tighten the condenser mounting screws, but leave the points screws slightly loose.

7. Check that the faces of the points meet squarely. If not, the *fixed* mount can be bent slightly with gentle force and a set of needlenose pliers. Do not bend the movable contact.

8. The point gap must be adjusted next. The gap is adjusted with the rubbing block of the points resting on one of the four high spots of the distributor cam. To get it there, the engine can be rotated by bumping the starter with the ignition key, or the crank-

On dual point distributors, #1 and #2 are the mounting screws. Do not loosen #3, the phase adjusting screw

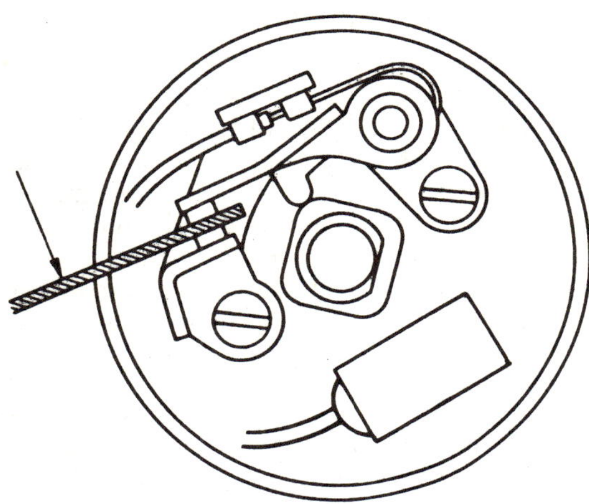

The arrow indicates the flat feeler gauge used to measure the point gap. Be sure that the rubbing block rests on the high spot of the cam, as shown.

shaft can be turned with a wrench on the crankshaft pulley bolt; this is easier to do with the spark plugs removed.

If the distributor is removed (dual points) the distributor shaft can be rotated until the points blocks are resting on the high spots of the cam. It won't matter if you move the distributor shaft; it can only go back into the engine one way. Just note the position from which it is moved, and move it back there prior to replacing the distributor in the engine.

9. Insert a 0.020 in. thick flat feeler gauge between the points. A slight drag should be felt. If no drag can be felt, or if the gauge cannot be inserted at all, insert a screwdriver into the eccentric adjusting screw, or into the notch provided for adjustment, and use it to open or close the gap between the points until it is correct.

10. When the gap is set, tighten the points screws, and then recheck the gap. Sometimes it takes three or four tries to get it correct, so don't feel frustrated if they seem to move around on you a little. It is not easy to feel the correct gap, either. Use gauges 0.002 in. larger and smaller than 0.020 as a test. If the points are spread slightly by a 0.022 in. gauge, and not touched at all by a 0.018 in. gauge, the setting should be right.

11. After all the adjustments are complete, pull a clean piece of tissue or a white business card between the points to clear any bits of grit.

12. On dual points distributors, if the distributor was removed, reset the shaft and rotor to their original positions, and install the

distributor. Note that the slot for the oil pump drive is tapered and will only fit one way. Be sure the marks you made earlier line up, and tighten the distributor hold down bolt.

13. Replace the rotor and distributor cap, and snap on the clips. If you have a dwell meter (recommended) you should next set the dwell. Otherwise, go on to the ignition timing.

ADJUSTMENT OF THE BREAKER POINTS WITH A DWELL METER
Single Point Distributor

The dwell angle is the number of degrees of distributor cam rotation through which the points remain closed (conducting electricity). Increasing the point gap decreases dwell, while decreasing the gap increases dwell.

The dwell angle may be checked with the distributor cap and rotor installed and the engine running, or with the cap and rotor removed and the engine cranking at starter speed. The meter gives a constant reading with the engine running. With the engine cranking, the meter will fluctuate between zero degrees dwell and the maximum figure for that setting. Never attempt to adjust the points when the ignition is on, or you may receive a shock.

1. Connect a meter as per the manufacturer's instructions (usually one lead to the distributor terminal of the coil and the other lead to a ground). Zero the meter, if necessary.

NOTE: *If the dwell meter does not have a four cylinder scale, multiply the eight cylinder reading by two.*

2. Check the dwell by either the cranking method, or with the engine running. If the setting is incorrect, the points must be adjusted.

CAUTION: *Keep your hands, hair and clothing clear of the engine fan and pulleys. Be sure the wires from the dwell meter are routed out of the way. If the engine is running, block the front wheels, put the transmission in Neutral, and set the parking brake.*

3. To change the dwell angle, turn the ignition off, loosen the points hold down screw and adjust the point gap; increase the gap to decrease dwell, and vice versa. Tighten the hold down screw and check the dwell angle with the engine cranking. If it seems to be correct, replace the cap and rotor and check

dwell with the engine running. Readjust as necessary.

4. Run the engine speed up to about 2,500 rpm, and then let the speed drop abruptly; the dwell reading should not change. If it does, a worn distributor shaft, bushing or cam, or a worn breaker plate is indicated. The parts must be inspected and replaced, if necessary.

5. After adjusting dwell angle, go on to the Ignition Timing section following. Ignition timing must be checked after adjusting the point gap, as a 1° increase in dwell results in an ignition timing retard of 2°, and vice versa.

Dual Point Distributor

Adjust the point gap of a dual point distributor with a dwell meter as follows:

1. Disconnect the wiring harness of the distributor from the engine wiring harness.

2. Using a jumper wire (a length of wire with an alligator clip at each end), connect the black wire of the engine side of the harness to the black wire of the distributor side of the harness (advance points).

3. Start the engine and observe the reading on the dwell meter. Shut the engine off and adjust the points accordingly as previously outlined for single point distributors.

4. Disconnect the jumper wire from the black wire of the distributor side of the wiring harness and connect it to the yellow wire (retard points).

5. Adjust the point gap as necessary in the manner previously discussed.

6. After the dwell of both sets of points is correct, remove the jumper wire and connect the engine-to-distributor wiring harness securely.

Use the terminals provided (arrows) for jumper wire connection

Electronic Ignition

Datsun pick-ups sold in California beginning in 1976 are equipped with electronic ignition, and all 1978 and later trucks are equipped with the system. The 1978 system differs somewhat from the earlier system; the 1979 and later system is markedly different.

The electronic ignition differs from its conventional counterpart only in the distributor component area. The secondary side of the ignition system is the same as a conventional breaker points system.

Located in the distributor, in addition to the normal ignition rotor, is a four spoke rotor (reluctor) which rests on the distributor shaft where the breaker points cam is found on earlier systems. A pick-up coil, consisting of a magnet, coil, and wiring, rests on the "breaker plate" next to the reluctor. The system also uses a transistor ignition unit, located on the right side of the firewall in the passenger compartment, through 1978, 1979 and later models have an integrated circuit (IC) ignition unit, which is mounted on the side of the distributor. In addition, 1979 and later models use a ring-type pick-up coil, which surrounds the reluctor, rather than the arm-type coil used through 1978.

When a reluctor spoke is not aligned with the pick-up coil, it generates large lines of flux between itself, the magnet, and the pick-up coil. This large flux variation results in a high generated voltage in the pick-up coil, preventing current from flowing to the pick-up coil. When a reluctor spoke lines up with the pick-up coil, the flux variation is low— thus, zero voltage is generated, allowing current to flow to the pick-up coil. Ignition primary current is then cut off by the electronic unit, allowing the field in the ignition coil to collapse, inducing high secondary voltage in the conventional manner. The high voltage then flows through the distributor to the spark plug, as usual.

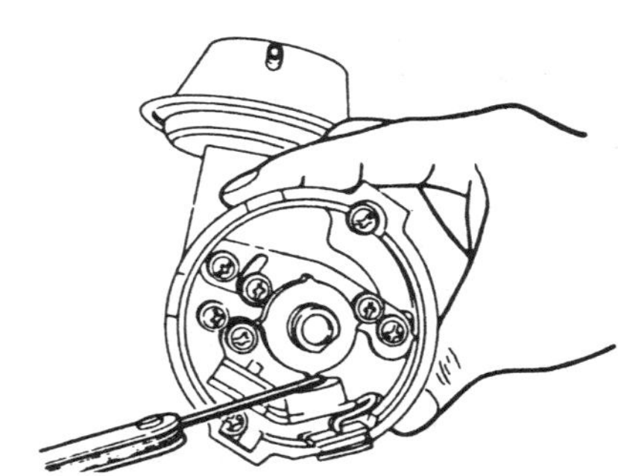
1976–78 air gap adjustment

Because no points or condenser are used, and because dwell is determined by the electronic unit, no adjustments are necessary. Ignition timing is checked in the usual way, but unless the distributor is disturbed it is not likely to ever change very much.

Service consists of inspection of the distributor cap, rotor, and ignition wires, replacing when necessary. These parts can be expected to last for at least 40,000 miles. In addition, the reluctor air gap should be checked periodically.

1. The distributor cap is held on by two clips. Release them with a screwdriver and lift the cap straight up and off, with the wires attached. Inspect the cap for cracks, carbon tracks, or a worn center contact. Replace it if necessary, transferring the wires one at a time from the old cap to the new.

2. Pull the ignition rotor (not the spoked reluctor) straight up to remove. Replace it if its contacts are worn, burned, or pitted. Do not file the contacts. To replace, press it firmly onto the shaft. It only goes on one way, so be sure it is fully seated.

3. Before replacing the ignition rotor, check the reluctor air gap. Use a non-magnetic feeler gauge. Rotate the engine until a reluctor spoke is aligned with the pick-up coil (either bump the engine around with the starter, or turn it with a wrench on the crankshaft pulley bolt). The gap should measure 0.008–0.016 in. through 1978, or 0.012–0.020 in. for 1979. Adjustment, if necessary, is made by loosening the pick-up coil mounting screws and shifting its position on the "breaker plate" either closer to or farther from the reluctor. On 1979 models, center the pick-up coil (ring) around the reluctor. Tighten the screws and recheck the gap.

4. Inspect the wires for cracks or brittleness. Replace them one at a time to prevent crosswiring, carefully pressing the replacement wires into place. The cores of electronic wires are more susceptible to breakage than those of standard wires, so treat them gently.

TROUBLESHOOTING

1976–78

The main differences between the 1976–77 and 1978 systems are: (1) the 1976–77 system uses an external ballast resistor located next to the ignition coil, and (2) the earlier system uses a wiring harness with individual eyelet connectors to the electronic unit, while the later system uses a multiple plug connector.

You will need an accurate voltmeter and ohmmeter for these tests, which must be performed in the order given.

1. Check all connections for corrosion, looseness, breaks, etc., and correct if necessary. Clean and gap the spark plugs.

2a. Disconnect the harness (connector or plug) from the electronic unit. Turn the ignition switch On. Set the voltmeter to the DC 50v range. Connect the positive (+) voltmeter lead to the black/white wire terminal, and the negative (−) lead to the black wire terminal. Battery voltage should be obtained. If not, check the black/white and black wires for continuity; check the battery terminals for corrosion; check the battery state of charge.

2b. Next, connect the voltmeter + lead to the blue wire and the − lead to the black wire. Battery voltage should be obtained. If not, check the blue wire for continuity; check the ignition coil terminals for corrosion or looseness; check the coil for continuity. On 1976–77 models, also check the external ballast resistor.

3. Disconnect the distributor harness wires from the ignition coil ballast resistor on 1976–77 models, leaving the ballast resistor-to-coil wires attached. On 1978 models, disconnect the ignition coil wires. Connect the

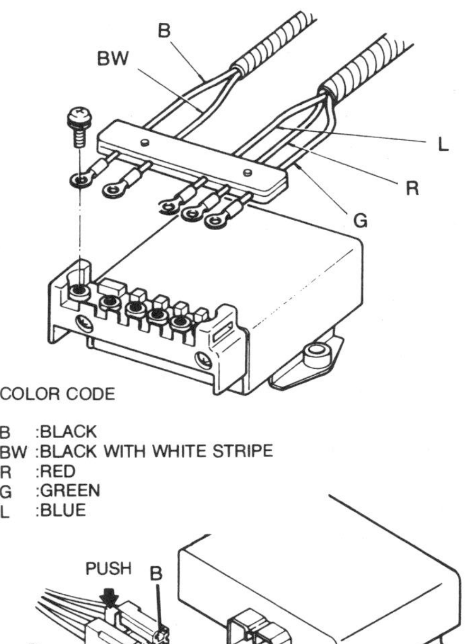

Electronic control unit connections: Upper, 1976–77; Lower, 1978.

leads of an ohmmeter to the ballast resistor outside terminals (at each end) through 1977; resistance should be 1.6–2.0 ohms. In 1978, connect the ohmmeter to the coil terminals: resistance should be 0 ohms. If more than 2.0 ohms, 1976–77, or 1.8 ohms, 1978, replace the coil.

4. Disconnect the harness from the electronic control unit. Connect an ohmmeter to the red and the green wire terminals. Resistance should be 720 ohms. If far more or far less, replace the distributor pick-up coil.

5. Disconnect the anti-dieseling solenoid connector (see Chapter 4). Connect a voltmeter to the red and green terminals of the electronic control harness. When the starter is cranked, the needle should deflect slightly. If not, replace the distributor pick-up coil.

6. Reconnect the ignition coil and the electronic control unit. Leave the anti-dieseling solenoid wire disconnected. Unplug the high tension lead (coil to distributor) from the distributor and hold it ⅛–¼ in. from the cylinder head with a pair of insulated pliers and a heavy glove. When the engine is cranked, a spark should be observed. If not, check the lead, and replace the electronic control unit.

7. Reconnect all wires.

1976–77: connect the voltmeter + lead to the blue electronic control harness connector and the − lead to the black wire. The harness should be attached to the control unit.

1978: connect the voltmeter + lead to the − terminal of the ignition coil and the − lead to ground.

As soon as the ignition switch is turned ON, the meter should indicate battery voltage. If not, replace the electronic control unit.

1979 and Later

1. Make a check of the power supply circuit. Turn the ignition OFF. Disconnect the connector from the top of the IC unit. Turn the ignition ON. Measure the voltage at each terminal of the connector in turn by touching the probe of the positive lead of the voltmeter to one of the terminals, and touching the probe of the negative lead of the voltmeter to a ground, such as the engine. In each case, battery voltage should be indicated. If not, check all wiring, the ignition switch, and all connectors for breaks, corrosion, discontinuity, etc., and repair as necessary.

2. Check the primary windings off the ignition coil. Turn the ignition OFF. Disconnect the harness connector from the negative coil terminals. Use an ohmmeter to measure the resistance between the positive and negative coil terminals. If resistance is 0.84–1.02 ohms, the coil is OK. Replace if far from this range.

If the power supply, circuits, wiring, and coil are in good shape, check the IC unit and pick-up coil, as follows:

3. Turn the ignition OFF. Remove the distributor cap and ignition rotor. Use an ohmmeter to measure the resistance between the two terminals of the pick-up coil, where they attach to the IC unit. Measure the resistance by reversing the polarity of the probes. If approximately 400 ohms are indicated, the pick-up coil is OK, but the IC unit is bad and must be replaced. If other than 400 ohms are measured, go to the next Step.

4. Be certain the two pin connector to the IC unit is secure. Turn the ignition ON. Measure the voltage at the ignition coil negative terminal. Turn the ignition OFF.

CAUTION: *Remove the tester probe from the coil negative terminal before switching the ignition OFF, to prevent burning out the tester.*

If zero voltage is indicated, the IC unit is bad and must be replaced. If battery voltage is indicated, proceed.

5. Remove the IC unit from the distributor:

a. Disconnect the battery ground (negative) cable.

b. Remove the distributor cap and ignition rotor.

c. Disconnect the harness connector at the top of the IC unit.

d. Remove the two screws securing the IC unit to the distributor.

e. Disconnect the two pick-up coil wires from the IC unit.

CAUTION: *Pull the connectors free with a pair of needlenosed pliers. Do not pull on the wires to detach the connectors.*

f. Remove the IC unit.

6. Measure the resitance between the terminals of the pick-up coil. It should be approximately 400 ohms. If so, the pick-up coil is OK, and the IC unit is bad. If not approximately 400 ohms, the pick-up coil is bad and must be replaced.

7. With a new pick-up coil installed, install the IC unit. Check for a spark at one of the spark plugs (see Step 4.1 in the Troubleshooting Section at the end of this Chapter). If a good spark is obtained, the IC unit is OK. If not, replace the IC unit.

Ignition Timing

NOTE: *For diesel engine injection pump timing, see the fuel system section in Chapter 4.*

Ignition timing is the measurement, in degrees of crankshaft rotation, of the point at which the spark plugs fire in each of the cylinders. It is measured in degrees before or after Top Dead Center (TDC) of the compression stroke.

Because it takes a fraction of a second for the spark plug to ignite the mixture in the cylinder, the spark plug must fire a little before the piston reaches TDC. Otherwise, the mixture will not be completely ignited as the piston passes TDC and the full power of the explosion will not be used by the engine.

The timing measurement is given in degrees of crankshaft rotation before the piston reaches TDC (BTDC). If the setting for the ignition timing is 5° BTDC, the spark plug must fire 5° before each piston reaches TDC. This only holds true, however, when the engine is at idle speed.

As the engine speed increases, the pistons go faster. The spark plugs have to ignite the fuel even sooner if it is to be completely ignited when the piston reaches TDC. To do this, the distributor has a means to advance the timing of the spark as the engine speed increases. This is accomplished by centrifugal weights within the distributor and a vacuum diaphragm, mounted on the side of the distributor. On Datsun pick-ups, it is not necessary to disconnect the vacuum line from the diaphragm when the ignition timing is being set.

If the ignition is set too far advanced (BTDC), the ignition and expansion of the fuel in the cylinder, will occur too soon and tend to force the piston down while it is still traveling up. This causes engine ping. If the ignition spark is set too far retarded, after TDC (ATDC), the piston will have already passed TDC and started on its way down when the fuel is ignited. This will cause the piston to be forced down for only a portion of its travel. This will result in poor engine performance and lack of power.

Timing marks consist of a notch on the rim of the crankshaft pulley and a scale of degrees attached to the front of the engine. The notch corresponds to the position of the piston in the number 1 cylinder. A stroboscopic (dynamic) timing light is used, which is hooked into the circuit of the No. 1 cylinder spark plug. Every time the spark plug fires, the timing light flashes. By aiming the timing light at the timing marks, the exact position of the piston within the cylinder can be read, since the stroboscopic flash makes the mark on the pulley appear to be standing still. Proper timing is indicated when the notch is aligned with the correct number on the scale. There are three basic types of timing light available. The first is a simple neon bulb with two wire connections (one for the spark plug and one for the plug wire, connecting the light in a series). This type of light is quite dim, and must be held closely to the marks to be seen, but it is quite inexpensive. The second type of light operates from the truck battery. Two alligator clips connect to the battery terminals, while a third wire connects to the spark plug with an adapter. This type of light is more expensive, but the xenon bulb provides a nice bright flash which can even be seen in sunlight. The third type replaces the battery source with 110 volt house current. Some timing lights have other functions built into them, such as dwell meters, tachometers, or remote starting switches. These are convenient, in that they reduce the tangle of wires under the hood, but may duplicate the functions of tools you already have.

If your Datsun has electronic ignition, you should use a timing light with an inductive pickup. This pickup simply clamps onto the No. 1 plug wire, eliminating the adapter. It is not susceptible to crossfiring or false triggering, which may occur with a conventional light, due to the greater voltages produced by electronic ignition.

IGNITION TIMING ADJUSTMENT

NOTE: *Refer to Chapter 4 "Emission Controls and Fuel System" for the procedure to time diesel engines and to check and adjust the phase timing of the two sets of points in 1970–73 models.*

1. Set the dwell of the breaker points to the proper specification.

2. Locate the timing marks on the crankshaft pulley and the front of the engine.

3. Clean off the timing marks, so that you can see them.

4. Use chalk or white paint to color the mark on the crankshaft pulley and the mark on the scale which will indicate the correct timing when aligned with the notch on the crankshaft pulley.

5. Attach a tachometer to the engine.

6. Attach a timing light to the engine, according to the manufacturer's instructions. If the timing light has three wires, one, usually green or blue, is attached to the No. 1 spark plug with an adapter. The other wires are connected to the battery. The red wire goes to the positive side of the battery and the black wire is connected to the negative terminal of the battery.

7. Leave the vacuum line connected to the distributor vacuum diaphragm.

8. Check to make sure that all of the wires clear the fan and then start the engine. Allow the engine to reach normal operating temperature.

CAUTION: *Block the front wheels and set the parking brake. Shift the manual transmission to Neutral or the automatic to Drive. Do not stand in front of the truck when making adjustments!*

9. Adjust the idle to the correct setting.

10. Aim the timing light at the timing marks. If the marks which you put on the pulley and the engine are aligned when the light flashes, the timing is correct. Turn off the engine and remove the tachometer and the timing light. If the marks are not in alignment, proceed with the following steps.

11. Turn off the engine.

12. Loosen the distributor lockbolt just enough so that the distributor can be turned with a little effort.

13. Start the engine. Keep the wires of the timing light clear of the fan.

14. With the timing light aimed at the pulley and the marks on the engine, turn the distributor in the direction of rotor rotation to retard the spark, and in the opposite direction of rotor rotation to advance the spark.

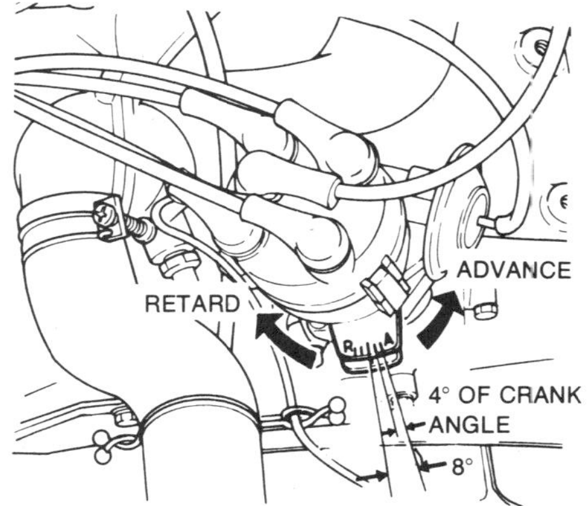

Ignition timing marks on the distributor. Each graduation represents a change of four degrees of crankshaft rotation

Timing scale on the front of the engine

Align the marks on the pulley and the engine with the flashes of the timing light.

15. Tighten the distributor lockbolt and recheck the timing.

Valve Lash

Valve adjustment determines how far the valves enter the cylinder and how long they stay open and closed.

If the valve clearance is too large, part of the lift of the camshaft will be used in removing the excessive clearance. Consequently, the valve will not be opening for as long as it should. This condition has two effects; the valve train components will emit a tapping sound as they take up the excessive clearance and the engine will perform poorly because the valves don't open fully and allow the proper amount of gases to flow into and out of the engine.

If the valve clearance is too small, the intake valves and the exhaust valves will open too far and they will not fully seat on the cylinder head when they close. When a valve seats itself on the cylinder head, it does two things: it seals the combustion chamber so that none of the gases in the cylinder escape and it cools itself by transferring some of the heat it absorbs from the combustion in the cylinder to the cylinder head and to the engine's cooling system. If the valve clearance is too small, the engine will run poorly because of the gases escaping from the combustion chamber. The valves will also become overheated and will warp, since they cannot transfer heat unless they are touching the valve seat in the cylinder head.

NOTE: *While all valve adjustments must be made as accurately as possible, it is better to have the valve adjustment slightly loose than slightly tight, as a burned valve may result from overly tight adjustments.*

ADJUSTMENT

All Except 1981 Gasoline Engine

1. The valves are adjusted with the engine at normal operating temperature. Oil temperature, and the resultant parts expansion, is much more important than water temperature. Run the engine for at least fifteen minutes to ensure that all the parts have reached their full expansion. After the engine is warmed up, shut it off.

2. Purchase either a new gasket or some silicone gasket seal before removing the camshaft cover. Note the location of any wires and hoses which may interfere with cam cover removal, disconnect them and move them aside. Then remove the bolts which hold the cam cover in place and remove the cam cover.

3. Place a wrench on the crankshaft pulley bolt and turn the engine over until the valves for No. 1 cylinder are closed. When both cam lobes are pointing up, the valves are closed. If you have not done this before, it is a good idea to turn the engine over slowly several times and watch the valve action until you have a clear idea of just when the valve is closed.

4. Check the clearance of the intake and exhaust valves. You can differentiate between them by lining them up with the tubes of the intake and exhaust manifolds. The correct size feeler gauge should pass between the base circle of the cam and the rocker arm

with just a slight drag. Be sure the feeler gauge is inserted *straight* and not on an angle.

5. If the valves need adjustment, loosen the locking nut and then adjust the clearance with the adjusting screw. You will probably find it necessary to hold the locking nut while you turn the adjuster. After you have the correct clearance, tighten the locking nut and recheck the clearance. Remember, it's better to have them too loose than too tight, especially exhaust valves.

6. Repeat this procedure until you have checked and/or adjusted all the valves. Keep in mind that all that is necessary is to have the valves closed and the camshaft lobes pointing up. It is not particularly important what stroke the engine is on.

7. Install the cam cover gasket, the cam cover, and any wires and hoses which were removed.

1981 Z22 Engine

1. The valves must be adjusted with the engine warm, so start the car and run the engine until the needle on the temperature gauge reaches the middle of the gauge. After the engine is warm, shut it off.

2. Purchase either a new gasket or some silicone gasket sealer before removing the

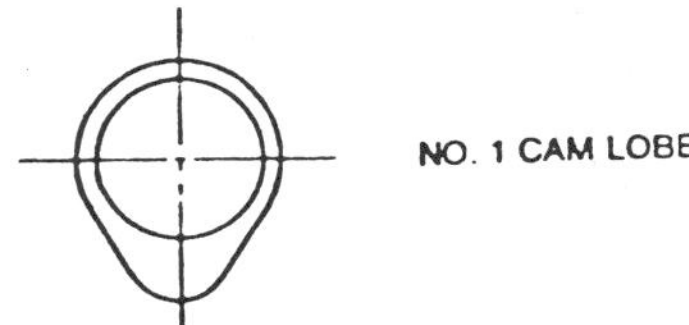

Z-series engines: cam lobe pointing straight down

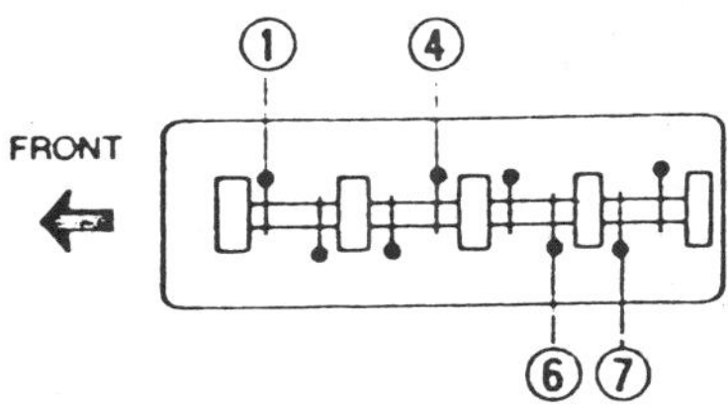

Primary adjustment, Z-series engines

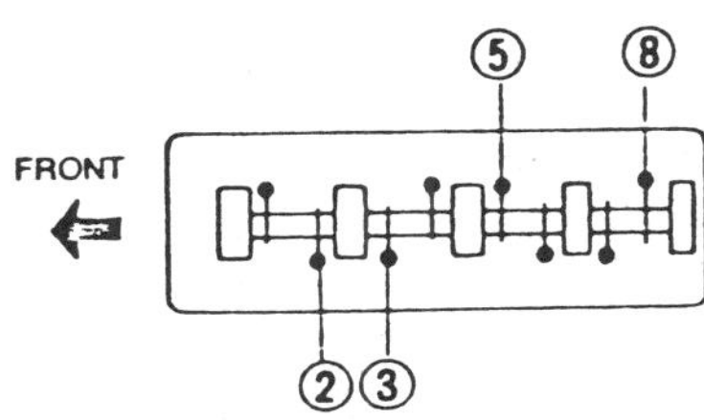

Secondary adjustment, Z-series engines

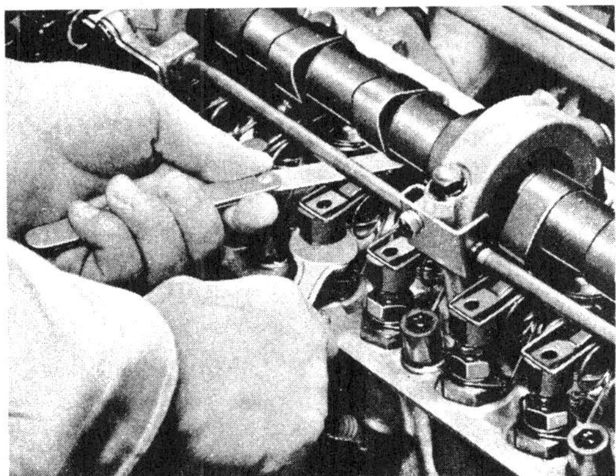

Valve lash adjustment, 1970–80 engines

camshaft cover. Counting on the old gasket to be in good shape is a losing proposition; always use new gaskets. Note the location of any wires and hoses which may interfere with cam cover removal, disconnect them and move them to one side. Remove the bolts holding the cover in place and remove the cover. Remember, the engine will be hot, so be careful.

3. Place a wrench on the crankshaft pulley bolt and turn the engine over until the first cam lobe behind the camshaft timing chain sprocket is pointing straight down.

NOTE: *If you decide to turn the engine by "bumping" it with the starter, be sure to disconnect the high tension wire from the coil(s) to prevent the engine from accidentally starting and spewing oil all over the engine compartment.*

CAUTION: *Never attempt to turn the engine by using a wrench on the camshaft sprocket bolt; there is a one to two turning ratio between the camshaft and the crankshaft which will put a tremendous strain on the timing chain.*

4. See the illustration marked "Primary adjustment" and adjust valves (1), (4), (6), and (7) to 0.012 in. using a flat-bladed feeler gauge. The feeler guage should pass between the valve stem end and the rocker arm screw with a very slight drag. Insert the feeler gauge *straight,* not at an angle.

5. If the clearance is not within specified value, loosen the rocker arm lock nut and turn the rocker arm screw to obtain the proper clearance. After correct clearance is obtained, tighten the lock nut.

6. Turn the engine over so that the first cam lobe behind the camshaft timing chain sprocket is pointing straight up and adjust the valves marked (2), (3), (5), and (8) in the "Secondary adjustment" illustration. They, too, should have a clearance of 0.012 in.

7. Install the cam cover gasket, the cam cover and any wires and hoses which were removed.

Carburetor

This section contains only tune-up adjustment procedures for carburetors. Descriptions, adjustments, and overhaul procedures for carburetors can be found in the "Fuel System" section.

When the engine in your Datsun is running, the air-fuel mixture from the carburetor is being drawn into the engine by a partial vacuum which is created by the movement of the pistons downward on the intake stroke. The amount of air-fuel mixture that enters into the engine is controlled by the throttle plates in the bottom of the carburetor. When the engine is not running the throttle plates are closed, completely blocking off the bottom of the carburetor from the inside of the engine. The throttle plates are connected by the throttle linkage to the accelerator pedal in the passenger compartment of the Datsun. When you depress the pedal, you open the throttle plates in the carburetor to admit more air-fuel mixture to the engine.

When the engine is not running, the throttle plates are closed. When the engine is idling, it is necessary to have the throttle plates open slightly. To prevent having to hold your foot on the pedal when the engine is idling, an idle speed adjusting screw is added to the carburetor linkage.

The idle adjusting screw contacts a lever (throttle lever) on the outside of the carburetor. When the screw is turned, it either opens or closes the throttle plates of the carburetor, raising or lowering the idle speed of the engine. This screw is called the curb idle adjusting screw.

A special mixture circuit is incorporated into the carburetor to enable the engine to run smoothly at idle. This circuit is controlled by the mixture screw, which determines the amount of fuel admitted at idle.

IDLE SPEED AND MIXTURE ADJUSTMENT

1. Start the engine and run it until it reaches operating temperature.

2. Allow the engine idle seed to stabilize by running the engine at idle for at least one minute.

3. If it hasn't already been done, check and adjust the ignition timing to the proper setting.

4. Turn off the engine and connect a tachometer to the engine.

5. Disconnect and plug the air hose between the three way connector and the check valve, if equipped. Start the engine. With the transmission in Neutral, check the idle speed on the tachometer. If the reading on the tachometer is correct, turn the idle adjusting screw clockwise with a screwdriver to increase the idle speed and counterclockwise to decrease it.

6. With an automatic transmission in

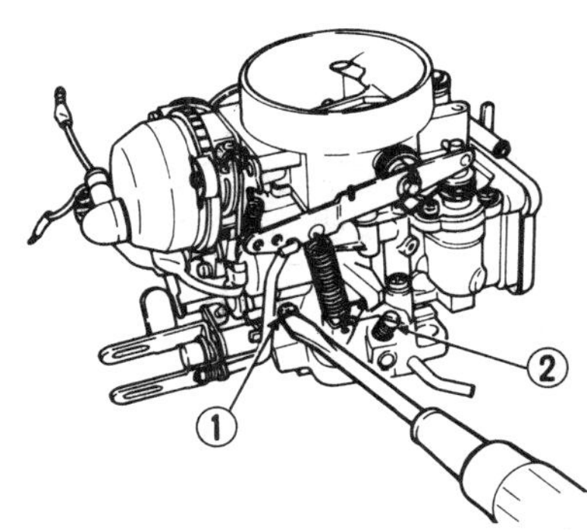

1. Idle speed adjusting screw
2. Air/fuel mixture adjusting screw

Idle speed and mixture adjustment

Drive (wheels chocked and parking brake applied) or a manual transmission in Neutral, turn the mixture screw out until the engine rpm starts to drop due to an overly rich mixture.

7. Turn the screw in past the starting point until the engine rpm start to drop because of a too lean mixture. On 1975–77 models, turn the mixture screw in until the idle speed drops 60–70 rpm with manual transmission, or 15–25 rpm with automatic transmission (in Drive). On 1978 models, the rpm drop should be 45–55 rpm for all trucks. For 1979 and later the rpm drop should be 45–55 rpm with manual transmission, or 25–35 rpm with automatic transmission (in Drive). If the mixture limiter cap will not allow this adjustment, remove it, make the adjustment, and reinstall it. Go on to Step 10 for 1975–81 trucks.

8. On 1970–74 models, turn the mixture screw back out to the point midway between the two extreme positions where the engine began losing rpm to achieve the fastest and smoothest idle.

9. Adjust the curb idle speed to the proper specification, on 1970–74 models.

10. Install the air hose. If the engine speed increases, reduce it with the idle speed screw.

NOTE: *To be sure that the vehicle is complying with emission laws, have the exhaust checked with a "CO" meter. The percentages of CO should be 3% for 1970–71, 2% 1972, 1.5% 1973–74, 2% 1975–77, and 1% 1978–81 at idle speed.*

NOTE: *Idle limiter caps are installed on the mixture adjusting screws so that an incorrect adjustment cannot be made. If a satisfactory idle cannot be obtained within the range of the limiter caps, remove them and make the adjustment as outlined above. Reinstall the limiter caps so that the cap can be turned only ⅛ of a turn counterclockwise before it reaches the stop. Have the engine checked with a CO meter after making the adjustment.*

Engine and Engine Rebuilding

ENGINE ELECTRICAL

Distributor

REMOVAL

1. Remove the high-tension wires from the distributor cap terminal towers, noting their positions to assure correct reassembly. Number the wires with pieces of adhesive tape if they are not already numbered.

2. Disconnect the distributor wiring harness.

3. Disconnect the vacuum line(s).

4. Unlatch the two distributor cap retaining clips and remove the distributor cap.

5. Note the position of the rotor in relation to the base. Scribe a mark on the base of the distributor and on the engine block to facilitate reinstallation. Align the marks with the direction the metal tip of the rotor is pointing.

6. Remove the bolt which holds the distributor to the engine.

7. Lift the distributor assembly from the engine.

INSTALLATION

1. Insert the distributor shaft and assembly into the engine. Line up the mark on the distributor and the one on the engine with the metal tip of the rotor. Make sure that the vacuum advance diaphragm is pointed in the same direction as it was pointed originally. This will be done automatically if the marks on the engine and the distributor are lined up with the rotor.

2. Install the distributor hold-down bolt and clamp. Leave the screw loose enough so that you can move the distributor with heavy hand pressure.

3. Connect the distributor wiring harness. Install the distributor cap on the distributor housing. Secure the distributor cap with the spring clips.

4. Install the spark plug wires. Make sure that the wires are pressed all the way into the top of the distributor cap and firmly onto the spark plug.

5. Adjust the point dwell and set the ignition timing.

NOTE: *If the crankshaft has been turned or the engine disturbed in any manner (i.e., disassembled and rebuilt) while the distributor was removed, or if the marks were not drawn, it will be necessary to initially time the engine. Follow the procedure given below.*

1. It is necessary to place the No. 1 cylinder in the firing position to correctly install the distributor. To locate this position, the ignition timing marks on the crankshaft front pulley are used.

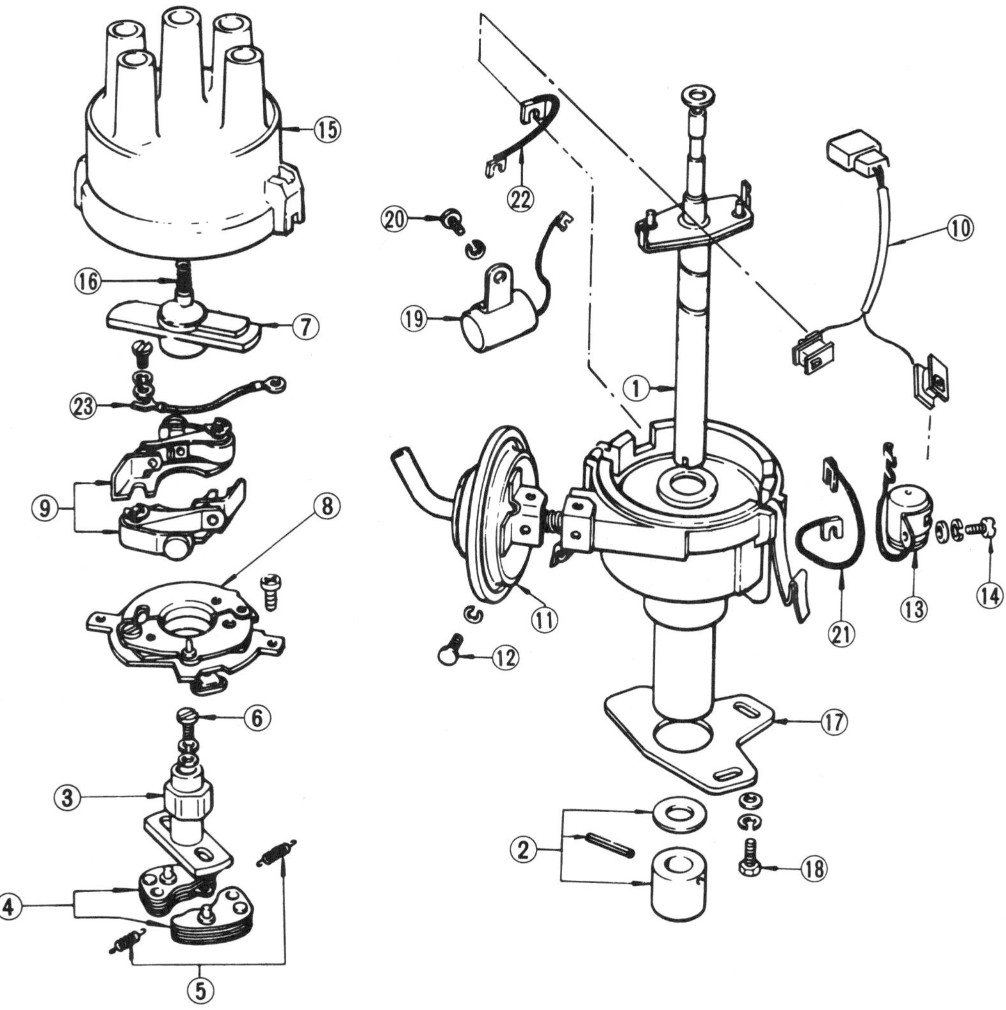

1. Shaft assembly
2. Collar set assembly
3. Cam assembly
4. Governor weight assembly
5. Governor spring set
6. Screw
7. Rotor head assembly
8. Breaker assembly
9. Contact set
10. Connector assembly
11. Vacuum control assembly
12. Screw
13. Condenser assembly
   (for advanced points)
14. Screw
15. Distributor cap assembly
16. Carbon point assembly
17. Fixing plate
18. Bolt
19. Condenser assembly
   (for retarded points)
20. Screw
21. Lead wire assembly
   (for advanced points)
22. Lead wire assembly
   (for retarded points)
23. Ground wire assembly

Exploded view of the 1970–73 distributor

2. Remove the No. 1 cylinder spark plug. Turn the crankshaft until the piston in the No. 1 cylinder is moving up on the compression stroke. This can be determined by placing your thumb over the spark plug hole and feeling the air being forced out of the cylinder. Stop turning the crankshaft when the timing marks that are used to time the engine are aligned.

3. Oil the distributor housing lightly where the distributor bears on the cylinder block.

4. Install the distributor so that the rotor, which is mounted on the shaft, points toward the No. 1 spark plug terminal tower position when the cap is installed. Of course you won't be able to see the direction in which the rotor is pointing if the cap is on the distributor. Lay the cap on the top of the distributor and make a mark on the side of the

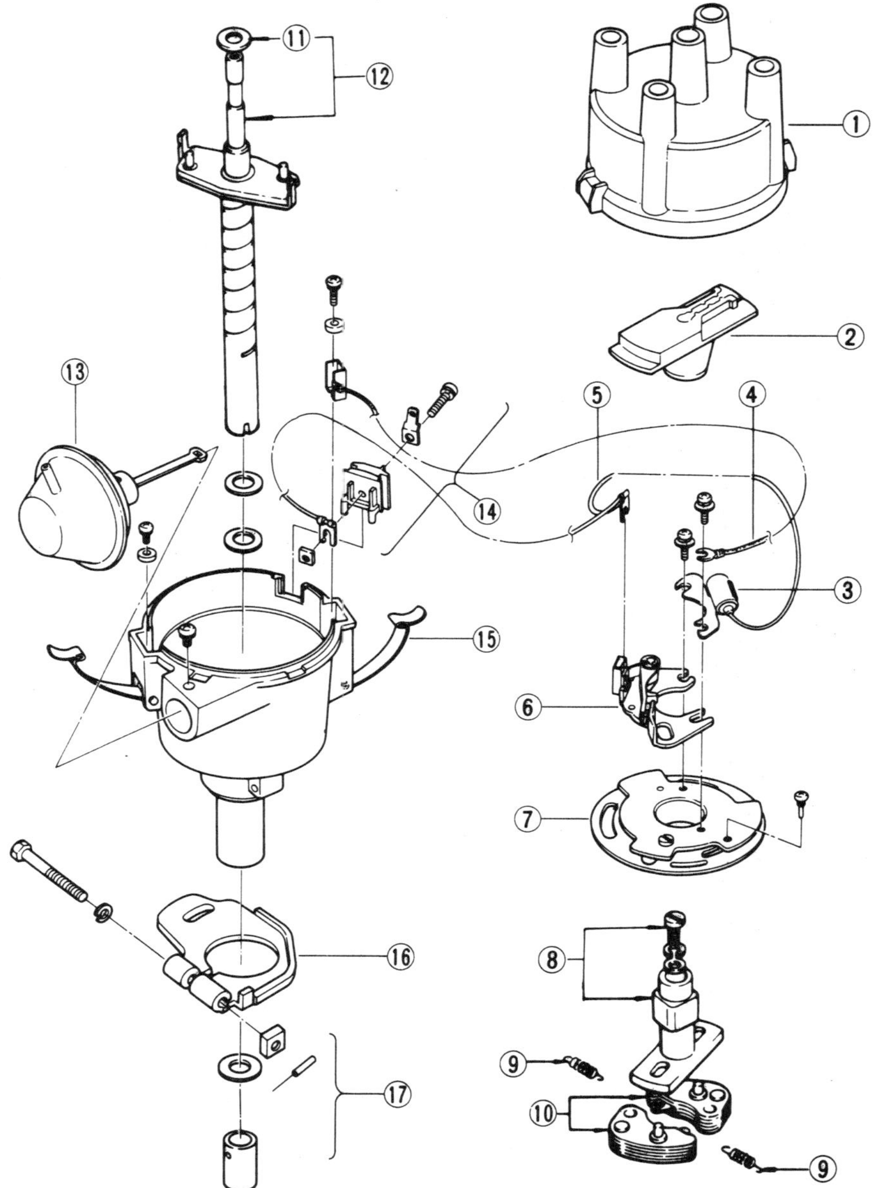

1. Cap assembly
2. Rotor head assembly
3. Condenser assembly
4. Ground wire assembly
5. Lead wire assembly
6. Contact set
7. Breaker plate assembly
8. Cam assembly
9. Governor spring
10. Governor weight
11. Thrust washer
12. Shaft assembly
13. Vacuum control assembly
14. Terminal assembly
15. Clamp
16. Fixing plate
17. Collar set

Exploded view of the 1974 distributor; later point-type models are the same except for the fixing plate.

distributor housing just below the No. 1 spark plug terminal. Make sure that the rotor points toward that mark when you install the distributor.

5. When the distributor shaft has reached the bottom of the hole, move the rotor back and forth slightly until the driving lug on the end of the shaft enters the slots cut in the end of the oil pump shaft and the distributor assembly slides down into place.

6. When the distributor is correctly installed, the breaker points should be in such

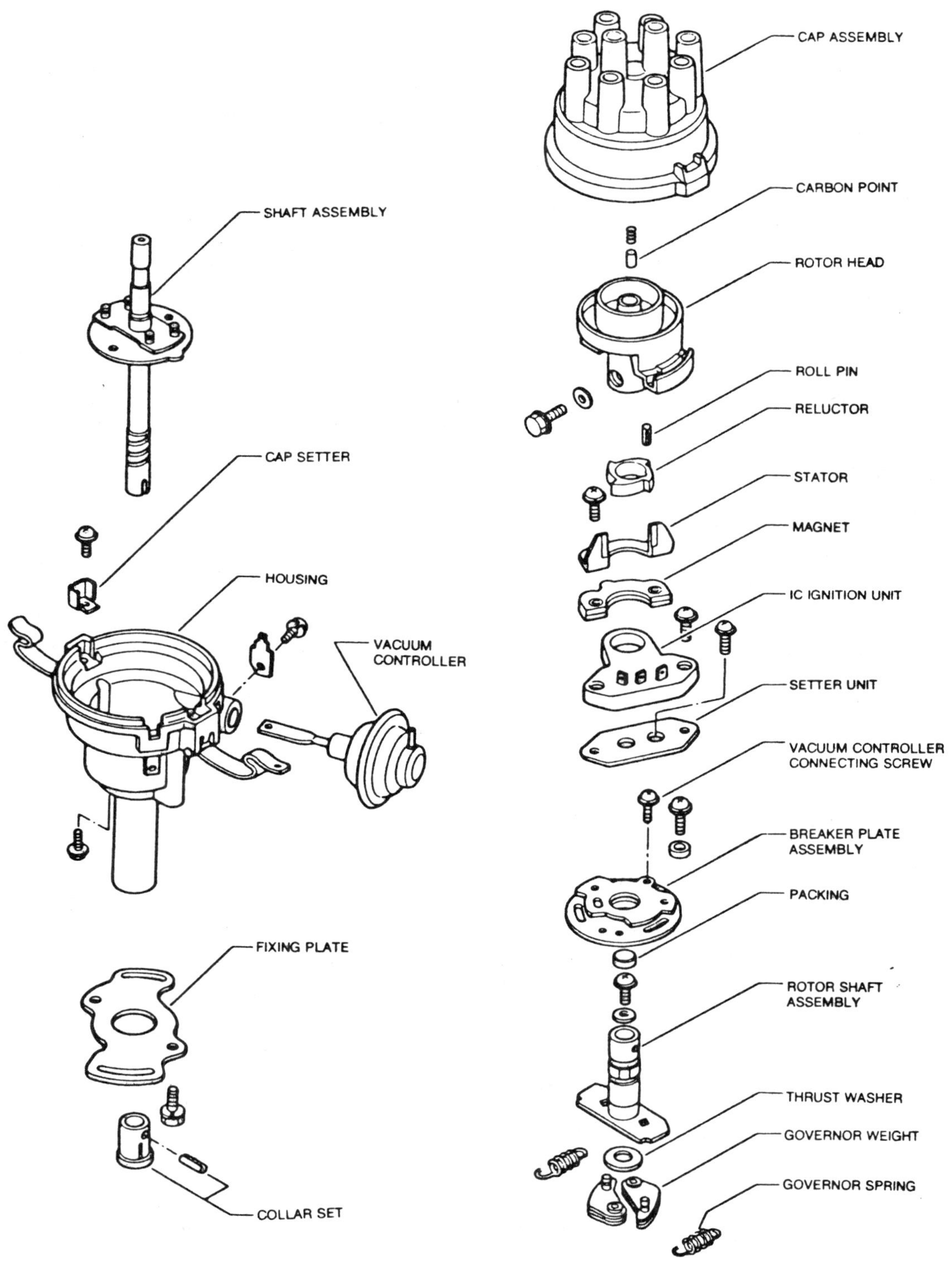

Z22 engine distributor

a position that they are just ready to break contact with each other. This is accomplished by rotating the distributor body after it has been installed in the engine. Once again, line up the marks that you made before the distributor was removed from the engine.

7. Install the distributor hold-down bolt.
8. Install the spark plug into the No. 1

spark plug hole and continue from Step 3 of the distributor installation procedure.

Firing Order

Always number the spark plug wires before removal, if they are not already labeled, to prevent cross-wiring.

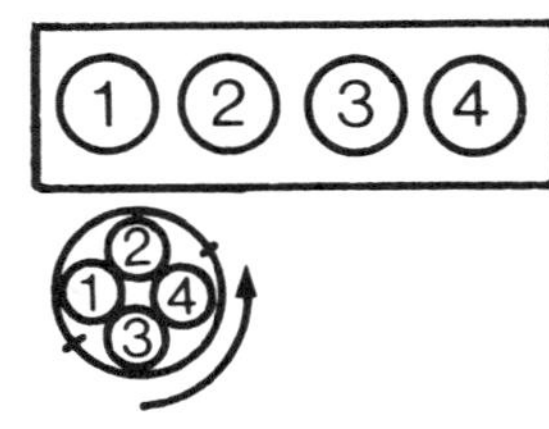

1970–73 L16, L18

1974 L18

L20B

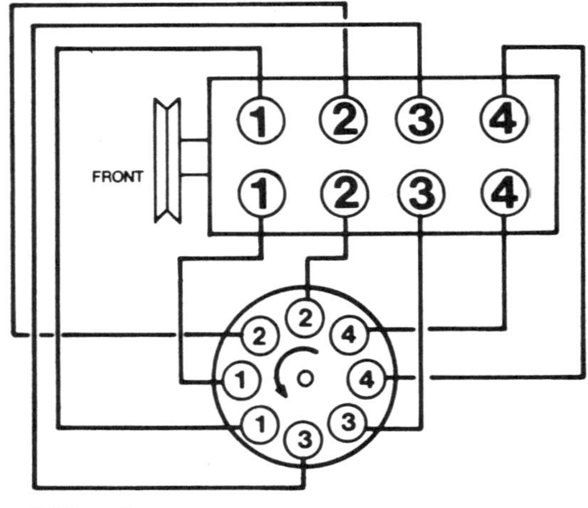

Z22 engine

Alternator

Datsun pick-ups are equipped with either 30 amp (1970–72) or 35 amp (1973–76) alternators with electromechanical, adjustable voltage regulators through 1976. 1977 models are the same, except that trucks with factory air conditioning have 38 amp alternators. 1978–80 models use 35 amp alternators (38 amp with A/C), but have a transistorized, non-adjustable regulator integral with the alternator. 1981 models have a 40, 50, or 60 amp unit, with the diesel featuring a 70 amp unit.

ALTERNATOR PRECAUTIONS

To prevent damage to the alternator and regulator, the following precautionary measures must be taken when working with the electrical system.

1. Never reverse battery connections. Always check the battery polarity visually. This is to be done before any connections are made to be sure that all of the connections correspond to the battery ground polarity of the truck.

2. Booster batteries for starting must be connected properly. Make sure that the positive cable of the booster battery is connected to the positive terminal of the battery which is getting the boost.

3. Disconnect the battery cables before using a fast charger; the charger has a tendency to force current through the diodes in the opposite direction for which they were designed. This burns out the diodes.

4. Never use a fast charger as a booster for starting the vehicle.

5. Never disconnect the voltage regulator while the engine is running.

6. Do not ground the alternator output terminal.

7. Do not operate the alternator on an open circuit with the field energized.

8. Do not attempt to polarize an alternator.

9. Disconnect the battery cables before using an electric arc welder on the truck.

REMOVAL AND INSTALLATION

1. Disconnect the negative battery terminal.

2. Disconnect the two lead wires and connector from the alternator.

3. Loosen the drive belt adjusting bolt and remove the belt.

4. Unscrew the alternator attaching bolts and remove the alternator from the vehicle.

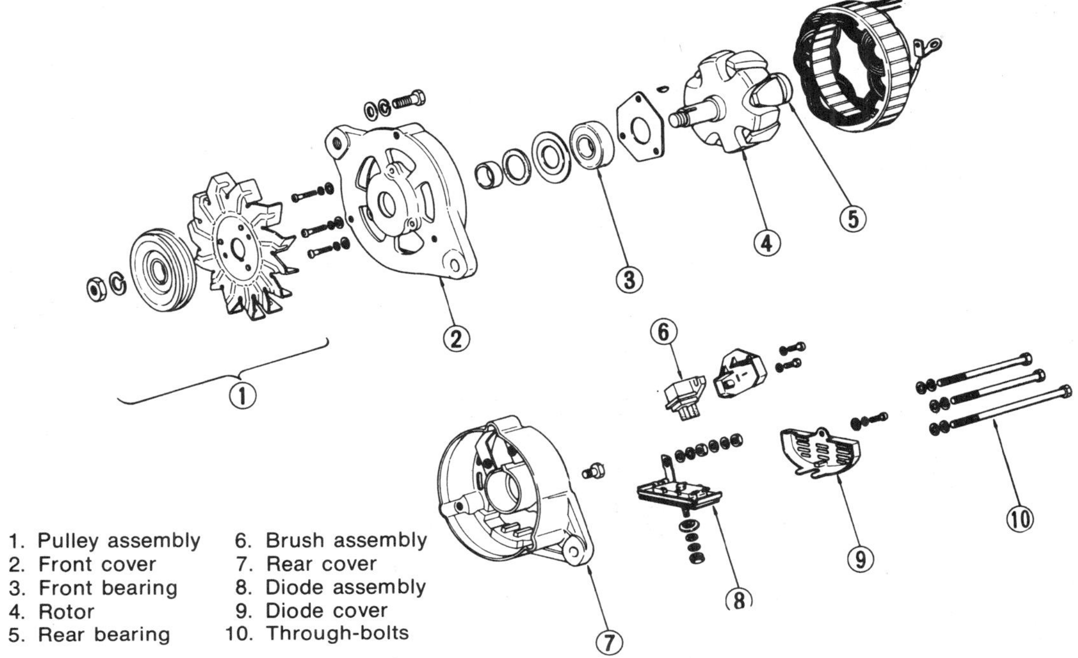

1. Pulley assembly
2. Front cover
3. Front bearing
4. Rotor
5. Rear bearing
6. Brush assembly
7. Rear cover
8. Diode assembly
9. Diode cover
10. Through-bolts

Exploded view of the LT 135-13B alternator. All models through 1977 are similar.

5. Install the alternator in the reverse order of removal.

Regulator

REMOVAL AND INSTALLATION

1970–77

1. Disconnect the negative battery terminal.

2. Disconnect the electrical lead connector of the regulator.

3. Remove the two mounting screws and remove the regulator from the vehicle.

4. Install the regulator in the reverse order of removal.

1978–81

The transistorized regulator is soldered to the brush assembly inside the alternator. It is non-adjustable, and must be replaced together with the brush assembly if faulty.

1. Remove the alternator.

2. Remove the through bolts and separate the front cover from the stator housing.

3. Unsolder the wire connecting the diode plate to the brush at the brush terminal.

4. Remove the bolt retaining the diode plate to the rear cover.

5. Remove the nut securing the battery terminal bolt.

6. Lift the stator slightly, together with the diode plate, to gain access to the diode plate screw. Remove the screw.

7. Separate the stator and diode, and remove the brush and regulator assembly.

8. Assembly is the reverse. Apply soldering heat sparingly, carrying out the operation as quickly as possible, to avoid heat damage to the transistors and diodes. Before assembling the alternator halves, bend a piece of wire in an "L" and slip it through the rear cover next to the brushes. Use the wire to hold the brushes in a retracted position until the case halves are assembled. Remove the wire carefully, to prevent damage to the slip rings.

ADJUSTMENT

1970–72

1. Operate the alternator to raise the temperature of the regulator to normal operating temperature.

2. Adjust the back gap of the voltage regulator coil by loosening the armature set screw and sliding the armature. The gap should be 0.035–0.039 in.

3. Adjust the air gap of the voltage regulator coil by bending the primary contact support to the right or left with a pair of needle-nose pliers. The gap should be 0.032–0.047 in.

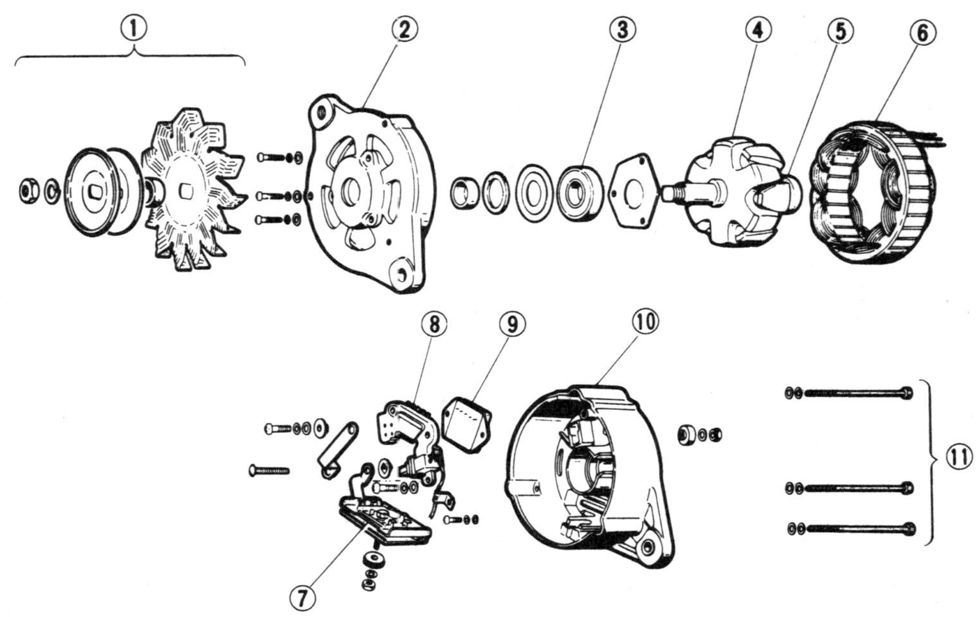

1. Pulley and fan assembly
2. Front cover
3. Front bearing
4. Rotor
5. Rear bearing
6. Stator
7. Diode plate
8. Brush assembly
9. Regulator
10. Rear cover
11. Through bolts

Exploded view of the integral regulator alternator used starting in 1978

4. Adjust the point gap by bending the secondary contact support to the right or left with the pliers. The gap should be 0.016–0.020 in.

5. Adjust the pilot lamp relay coil yoke gap by loosening the armature set screw, adjusting the gap and retightening the set screw. The gap should be 0.008 in.

6. Adjust the pilot lamp relay coil air gap by loosening the contact set screw and inserting a screwdriver in the hole provided and moving the points up or down. The gap should be 0.020–0.024 in.

7. Adjust the pilot lamp relay coil point

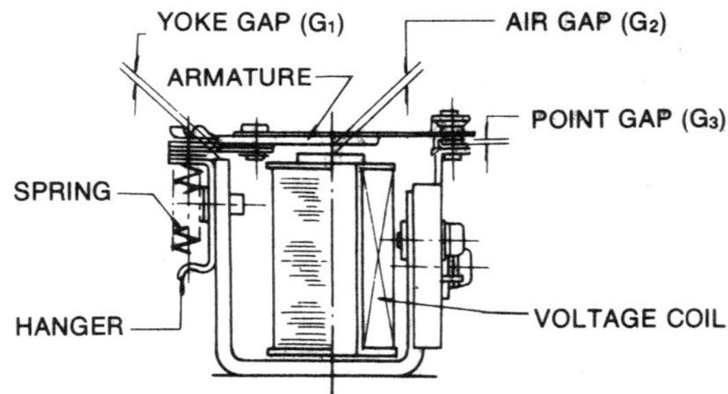

Pilot lamp relay adjustment—1970–72

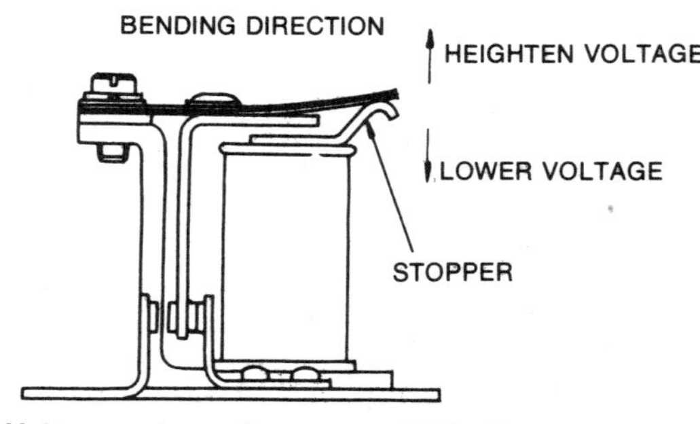

Voltage value adjustment—1970–72

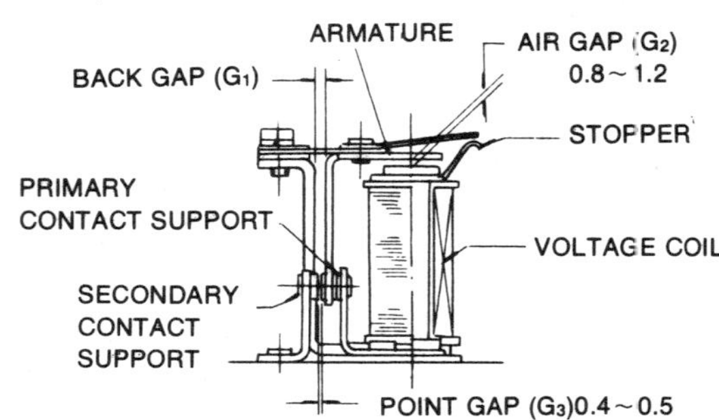

Voltage regulator coil adjustment—1970–72

gap by loosening the contact point retaining screw and inserting a screwdriver in the hole provided and moving the point contact either

Alternator and Regulator Specifications

| | Alternator | | Voltage regulator | | | | | | |
| | | | Charge Indicator Relay | | | Voltage Regulator | | | |
Year	Manufacturer and/or Part Number	Output @ Generator rpm	Back Gap (in.)	Air Gap (in.)	Point Gap (in.)	Back Gap (in.)	Air Gap (in.)	Point Gap (in.)	Regulated Voltage
1970–72	Hitiachi LT130–41	22 Amp @ 2500 (14 volts)	0.007	0.022	0.018	0.037	0.039	0.014	14–15
1973	Hitachi LT135–13B	28 Amp @ 2500 (14 volts)	—	0.035	0.020	—	0.032	0.014	14–15
1974	Hitachi LT135–13B	28 Amp @ 2500 (14 volts)	—	0.035	0.020	—	0.032	0.014	14–15
1975	Hitachi LT135–13B LT135–19B ①	28 Amp @ 2500 (14 volts)	—	0.035	0.020	—	0.032	0.014	14–15
1976	Hitachi LT135–13B LT135–19B ①	28 @ 2500	—	0.035	0.020	—	0.032	0.016	14.3–15.3
1977	Hitachi LT135–36B LT138–01B ①	28 @ 2500 30 @ 2500	—	0.035	0.020	—	0.032	0.016	14.3–15.3
1978–80	Hitachi LR135–44 LR138–01 ①	27.5 @ 2500 30.0 @ 2500	—Transistorized Non-Adjustable Relay—						14.4–15.0
1981	Hitachi LR150–98 LR160–78 LR150–52 LR160–78	50 @ 5000 60 @ 5000 50 @ 5000 60 @ 5000	—Transistorized Non-Adjustable Relay—						14.4–15.0

① With air conditioning   —Not applicable

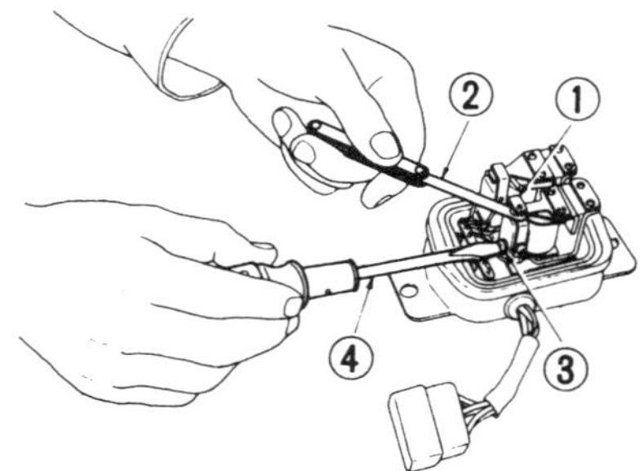

1. Contact set
2. Thickness gauge
3. 4 mm (0.1575 in.) dia. screw
4. Crosshead screwdriver

Voltage regulator core gap adjustment—1973–77

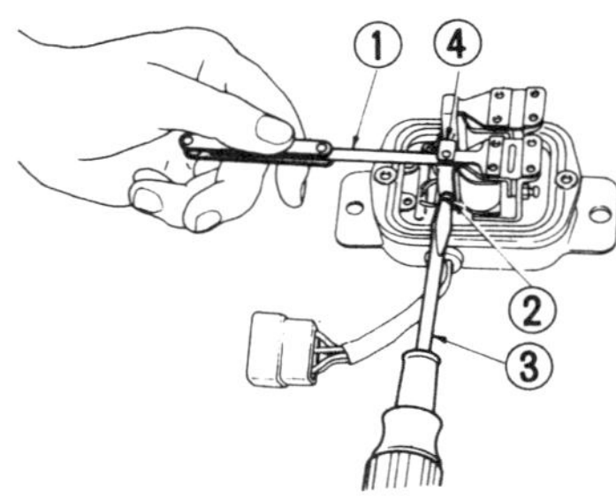

1. Thickness gauge
2. 3 mm (0.1181 in.) dia. screw
3. Crosshead screwdriver
4. Upper contact

Voltage regulator coil point gap adjustment—1973–77

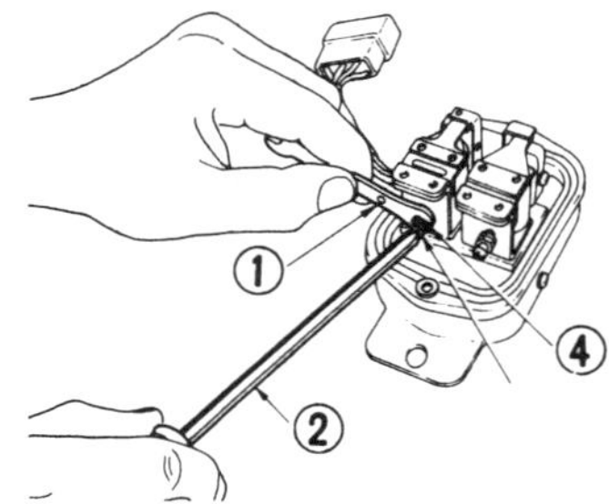

1. Wrench
2. Phillips screwdriver
3. Adjusting screw
4. Locknut

Regulated voltage adjustment—1973–77

up or down. The gap should be between 0.016–0.020 in.

8. Adjust the voltage value by bending the stopper on the voltage regulator coil. The regulated voltage should be between 14–15 volts at 4,000 alternator rpm.

1973–77

1. Adjust the voltage regulator core gap by loosening the screw which is used to secure the contact set on the yoke, and move the contact up or down as necessary. Retighten the screw. The gap should be 0.024–0.039 in.

2. Adjust the point gap of the voltage regulator coil by loosening the screw used to secure the upper contact and move the upper contact up or down. The gap should be 0.012–0.016 in. through 1975, 0.014–0.018 1976–77.

3. The core gap and point gap on the charge relay coil is or are adjusted in the same manner as previously outlined for the voltage regulator coil. The core gap is to be set at 0.032–0.039 in. and the point gap adjusted to 0.016–0.024 in.

4. The regulated voltage is adjusted by loosening the locknut and turning the adjusting screw clockwise to increase, or counterclockwise to decrease the regulated voltage. The voltage should be between 14.3–15.3 volts at 68°F.

Starter

A standard non-reduction gear starting motor is used on all models through 1977, and most 1978–79 models. This motor has its brushes located within the rear cover. A reduction gear starting motor is installed in all 1978–79 Canadian models, and is optional for 1979–81 U.S. trucks. The reduction gear motor brushes are on a plate located just behind the starter drive housing.

REMOVAL AND INSTALLATION

1. Disconnect the negative battery cable from the battery.

2. Disconnect the starter wiring at the starter, taking note of the positions for correct reinstallation.

3. Remove the bolts attaching the starter to the engine and remove the starter from the vehicle.

4. Install the starter in the reverse order of removal.

BRUSH REPLACEMENT

Non-Reduction Gear Type

1. With the starter out of the vehicle, remove the bolts holding the solenoid to the top of the starter and remove the solenoid.

2. To remove the brushes, remove the two thru-bolts, and the two rear cover attaching screws and remove the rear cover.

3. Disconnect the electrical leads and remove the brushes.

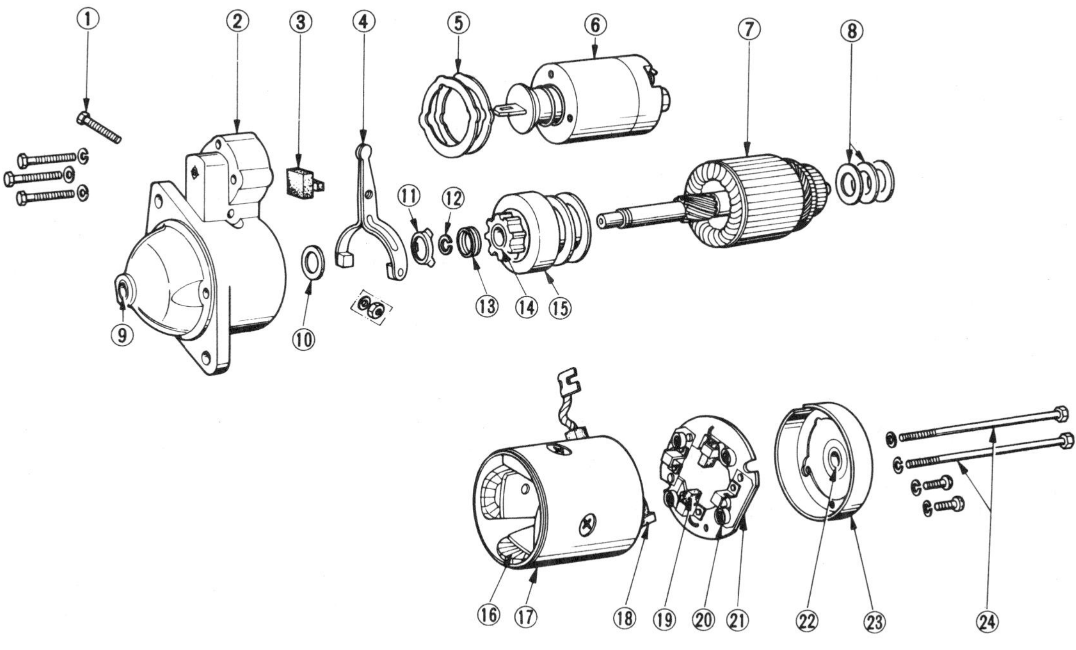

1. Shift lever pin
2. Gear case
3. Dust cover
4. Shift lever
5. Dust cover
6. Magnetic switch
7. Armature
8. Thrust washer
9. Metal
10. Thrust washer
11. Stopper washer
12. Stopper clip
13. Pinion stopper
14. Pinion
15. Overrunning clutch
16. Field coil
17. Yoke
18. Brush (+)
19. Brush (−)
20. Brush spring
21. Brush holder
22. Metal
23. Rear cover
24. Through-bolt

Exploded view of the starter motor through 1974; later years similar

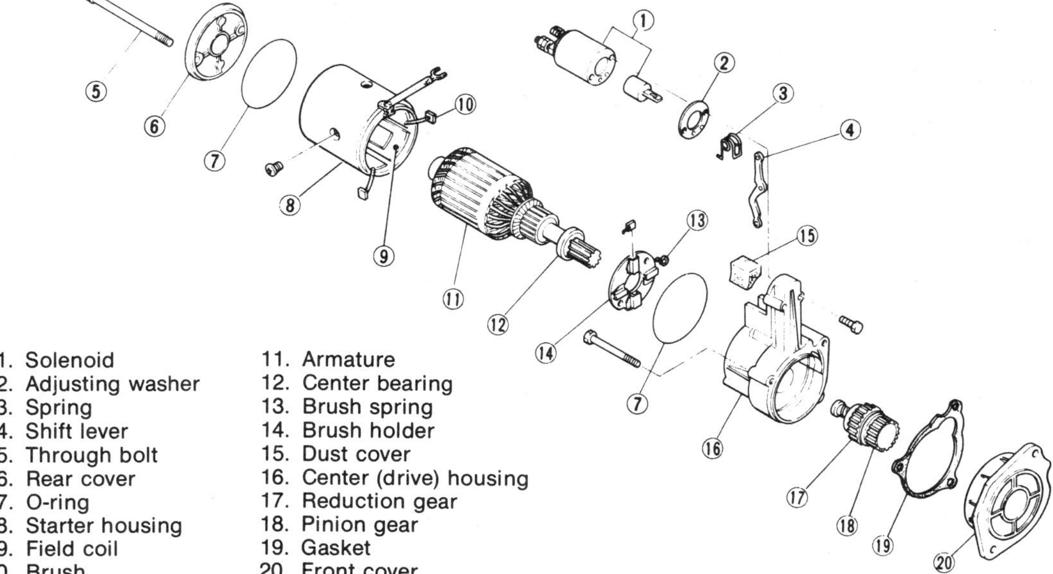

1. Solenoid
2. Adjusting washer
3. Spring
4. Shift lever
5. Through bolt
6. Rear cover
7. O-ring
8. Starter housing
9. Field coil
10. Brush
11. Armature
12. Center bearing
13. Brush spring
14. Brush holder
15. Dust cover
16. Center (drive) housing
17. Reduction gear
18. Pinion gear
19. Gasket
20. Front cover

Reduction gear starter

4. Install the brushes in the reverse order of removal.

Reduction Gear Type

1. Remove the starter. Remove the solenoid.

2. Remove the through bolts and the rear cover. The rear cover can be pried off with a screwdriver, but be careful not to damage the O-ring.

3. Remove the starter housing, armature, and brush holder from the center housing. They can be removed as an assembly.

4. Remove the positive side brush from its holder. The positive brush is insulated from the brush holder, and its lead wire is connected to the field coil.

5. Carefully lift the negative brush from the commutator and remove it from the holder.

6. Installation is the reverse.

STARTER DRIVE REPLACEMENT

Non-Reduction Gear Type

1. With the starter motor removed from the vehicle, remove the solenoid from the starter.

2. Remove the two thru-bolts and separate the gear case from the yoke housing.

3. Remove the pinion stopper clip and the pinion stopper.

4. Slide the starter drive off the armature shaft.

5. Install the starter drive and reassemble the starter in the reverse order of removal.

Reduction Gear Type

1. Remove the starter.

2. Remove the solenoid and the shift lever.

3. Remove the bolts securing the center housing to the front cover and separate the parts.

4. Remove the gears and starter drive.

5. Installation is the reverse.

Battery

Refer to Chapter One for details on battery maintenance.

REMOVAL AND INSTALLATION

1. Disconnect the negative (ground) cable from the terminal, and then the positive ca-

Battery and Starter Specifications

Year	Engine Displacement cc (cu in.)	Battery			Starter							Brush Spring Tension (oz)
		Amp Hour Capacity	Volts	Ground	Lock Test				No Load Test,			
					Amps	Volts	Torque (ft. lbs.)		Amps	Volts	RPM	
1970–72	1595 (97.3)	①	12	Neg	350	10.5	7.95		60	12	7000+②	28
1973	1595 (97.3)	①	12	Neg	Not Recommended				60	12	7000+②	56
1973–74	1770 (108)	①	12	Neg	Not Recommended				60	12	7000+②	56
1975–80	1952 (119)	①	12	Neg	Not Recommended				60	12	7000+②	56
1981	2187 (133.5)	①	12	Neg	Not Recommended				60③	11.5④	6000–7000	56
	2164 (132)	①	12	Neg	800	5.0	21.0		55	12	6500	123.2

① 40, 50, or 60 amp hour batteries were available
② 6000+ if equipped with automatic transmission
③ Canada: 100
④ Canada: 11.0

General Engine Specifications

Year	Engine Displacement cc (cu in.)	Carb Type	Advertised Horsepower @ rpm	Advertised Torque @ rpm (ft. lbs.)	Bore and Stroke (in.)	Advertised Compression Ratio	Oil Pressure (psi/idle)
1970–72	1595 (97.3)	2 bbl	96 @ 5600	100 @ 3600	3.27 x 2.90	8.5 : 1	16
1973	1595 (97.3)	2 bbl	96 @ 5600	100 @ 3600	3.27 x 2.90	8.5 : 1	16
1973–74	1770 (108)	2 bbl	100 @ 5600	100 @ 3600	3.35 x 3.07	8.5 : 1	11–40
1975	1952 (119)	2 bbl	112 @ 5600	108 @ 3600	3.35 x 3.39	8.5 : 1	50–57 ①
1976–80	1952 (119)	2 bbl	97 @ 5600	102 @ 3200	3.35 x 3.39	8.5 : 1	50–57 ①
1981	2164 (132)	Diesel	61 @ 4000	102 @ 1800	3.27 x 3.94	21.6 : 1	60
	2187 (133.5)	2 bbl	98 @ 4000	117 @ 1800	3.43 x 3.62	8.5 : 1	60

① @ 2000 rpm
NA Not available

Crankshaft and Connecting Rod Specifications
(All measurements given in in.)

Year	Engine Displacement cc (cu in.)	Crankshaft				Connecting rod		
		Main Brg Journal Dia	Main Brg Oil Clearance	Shaft End-Play	Thrust on No.	Journal Dia	Oil Clearance	Side Clearance
1970–72	1595 (97.3)	2.1631– 2.1636	0.0008– 0.0028	0.0020– 0.0059	3	1.9670– 1.9675	0.0006– 0.0026	0.0079– 0.0118
1973	1595 (97.3)	2.1631– 2.1636	0.0008– 0.0024	0.0020– 0.0071	3	1.9670– 1.9675	0.0010– 0.0022	0.0079– 0.0118
1973	1770 (108)	2.1631– 2.1636	0.0008– 0.0024	0.0020– 0.0071	3	1.9670– 1.9675	0.0010– 0.0022	0.0079– 0.0118
1974	1770 (108)	2.3599– 2.3600	0.0008– 0.0024	0.0020– 0.0071	3	1.9670– 1.9675	0.0010– 0.0022	0.0079– 0.0118
1975	1952 (119)	2.3599– 2.3600	0.0008– 0.0024	0.0020– 0.0071	3	1.9660– 1.9670	0.0010– 0.0022	0.0079– 0.0118
1976–78	1952 (119)	2.3599– 2.3604	0.0008– 0.0024	0.0020– 0.0071	3	1.9670– 1.9675	0.0010– 0.0020	0.008– 0.012

Crankshaft and Connecting Rod Specifications (cont.)

(All measurements given in in.)

		Crankshaft				Connecting rod		
Year	Engine Displacement cc (cu in.)	Main Brg Journal Dia	Main Brg Oil Clearance	Shaft End-Play	Thrust on No.	Journal Dia	Oil Clearance	Side Clearance
1979–80	1952 (119)	2.3599– 2.3604–	0.0008– 0.0026	0.0020– 0.0071	3	1.9670– 1.9675	0.0009– 0.0026	0.008– 0.012
1981	2164 (132)	2.7918– 2.7988	0.0013– 0.0038	0.0024– 0.0094	3	2.0840– 2.0906	0.0013– 0.0038	0.0039– 0.0079
	2187 (133.5)	2.1631– 2.1636	0.0008– 0.0024	0.0021– 0.0070	3	1.9670– 1.9675	0.0010– 0.0022	0.008– 0.012

ble. Special pullers are avilable to remove the cable clamps.

NOTE: *To avoid sparks, always disconnect the ground cable first, and connect it last.*

2. Remove the battery hold-down clamp.

3. Remove the battery, being careful not to spill the acid.

NOTE: *Spilled acid can be neutralized with a baking soda/water solution. If you somehow get acid into your eyes, flush it out with lots of water and get to a doctor.*

4. Clean the battery posts thoroughly before reinstalling, or when installing a new battery.

5. Clean the cable clamps, using a wire brush, both inside and out.

Piston Clearance

(All measurements given in in.)

Year	Engine Displacement cc (cu in.)	Minimum	Maximum
1970–73	1595 (97.3)	0.0010	0.0018
1973–80	1770 (108), 1952 (119)	0.0010	0.0018
1981	2187 (133.5)	0.0010	0.0018
	2164 (132)	0.0047	0.0075

6. Install the battery and the hold-down clamp or strap. Connect the positive, and then the negative cable. Do not hammer them in place. The terminals should be coated lightly (externally) with grease to prevent corrosion. There are also felt washers impregnated with an anti-corrosion substance which are slipped over the battery posts before installing the cables; these are available in auto parts stores.

CAUTION: *Make absolutely sure that the battery is connected properly before you turn on the ignition switch. Reversed polarity can burn out your alternator and regulator within a matter of seconds.*

ENGINE MECHANICAL

Design

The L16, L18, and L20B gasoline engines used in Datsun pick-ups are water-cooled, 4 cycle, 4 cylinder, overhead camshaft gasoline engines.

The cylinder head is cast aluminum alloy with wedge-type combustion chambers.

The valve system consists of a chain-driven overhead camshaft working directly on the rocker arms which operate valves with dual valve springs.

The crankshaft is fully balanced and is supported by five main bearings.

The intake and exhaust manifolds are mounted on the same side of the engine. The intake manifold is of cast aluminum upon which is mounted a single, two barrel, downdraft carburetor.

The Z22 gasoline engine, new in 1981, is a

Valve Specifications

Year	Engine Displacement cc (cu in.)	Seat Angle (deg)	Face Angle (deg)	Spring Test Pressure (lbs @ in.)		Free Length (in.)		Stem-to-Guide Clearance (in.)		Stem Diameter (in.)	
				Outer	Inner	Outer	Inner	Intake	Exhaust	Intake	Exhaust
1970–72	1595 (97.3)	45	45	64 @ 1.53	27 @ 1.38	2.05	1.77	0.0006–0.0018	0.0016–0.0028	0.3139	0.3131
1973	1595 (97.3)	45	45	47 @ 1.58	27 @ 1.38	1.97	1.77	0.0008–0.0021	0.0016–0.0029	0.3139	0.3131
1973–74	1770 (108)	45	45	47 @ 1.58	27 @ 1.38	1.97	1.77	0.0008–0.0021	0.0016–① 0.0029	0.3139	0.3131
1975–80	1952 (119)	45	45②	47 @ 1.58	27 @ 1.38	1.97	1.77	0.0008–0.0021	0.0016–0.0029	0.3139	0.3131
1981	2164 (132)	45	45	33 @ 1.634	—	1.929	—	0.0006–0.0018	0.0016–0.0028	0.3137–0.3143	0.3137–0.3143
	2187 (133.5)	45.5	45.5	51 @ 1.575	24 @ 1.378	1.959	1.736	0.0008–0.0021	0.0016–0.0029	0.3136–0.3142	0.3128–0.3134

① 0.0002–0.0003 in. for 1974
② 45°30 1978–79

Piston Ring Gaps
(All measurements given in in.)

Year	Engine Displacement cc (cu in.)	Top Compression		Middle Compression		Oil Control	
		Min	Max	Min	Max	Min	Max
1970–72	1595 (97.3)	0.0091	0.0150	0.0059	0.0118	0.0059	0.0118
1973	1595 (97.3)	0.0098	0.0157	0.0059	0.0118	0.0118	0.0354
1973	1770 (108)	0.0138	0.0217	0.0118	0.0197	0.0118	0.0354
1974	1770 (108)	0.0098	0.0157	0.0118	0.0197	0.0118	0.0354
1975–80	1952 (119)	0.0098	0.0157	0.0118	0.0197	0.0118	0.0354
1981	2187 (133.5)	0.0098	0.0157	0.0059	0.0118	0.0118	0.0354
	2164 (132)	0.0118	0.0197	0.0118	0.0197	0.0118	0.0197

Piston Ring Side Clearance
(All measurements given in in.)

Year	Engine Displacement cc (in.)	Top Compression		Middle Compression		Oil Control	
		Min	Max	Min	Max	Min	Max
1970–72	1595 (97.3)	0.0018	0.0031	0.0012	0.0025	0.0010	0.0025
1973	1595 (97.3)	0.0016	0.0031	0.0012	0.0028	Snug	Snug
1973	1770 (108)	0.0018	0.0031	0.0012	0.0028	Snug	Snug
1974–80	1770 (108), 1952 (119)	0.0016	0.0029	0.0012	0.0028	Snug	Snug
1981	2187 (133.5)	0.0016	0.0029	0.0012	0.0025	Snug	Snug
	2164 (132)	0.0024	0.0039	0.0016 ①	0.0032 ①	0.0008	0.0024

① Diesel: 2nd and 3rd rings.

Torque Specifications
(All readings in ft lbs unless noted)

Year	Engine Displacement cc (cu in.)	Cylinder Head Bolts	Rod Bearing Bolts	Main Bearing Bolts	Crankshaft Pulley Bolts	Flywheel to Crankshaft Bolts	Manifolds	
							Intake	Exhaust
1970–72	1595 (97.3)	40①	23	36	122	72	11	11
1973	1595 (97.3)	43①	25	37	102	109	11	11
1973–74	1770 (108)	55②	37	37	102	109	11	11
1975–80	1952 (119)	61③	37	37	102	109	11	11
1981	2187 (133.5)	51–58	33–40	33–40	87–116	101–116	12–15	12–16
	2164 (132)	94 large 36 small	36–40	109–116	217–239	33–36	11–13	11–13

① Tighten in 2 stages—first to 33 ft. lbs.; then to specifications
② Tighten in 3 steps—first to 23 ft. lbs.; then to 40 ft. lbs.; and finally to specifications
③ Tighten in 3 steps—first to 29 ft. lbs.; then to 43 ft. lbs.; and finally to specifications

four cylinder, overhead camshaft engine with a cast iron block and a cross-flow aluminum cylinder head. Hemispherical combustion chambers are employed. This engine has two spark plugs per cylinder which fire simultaneously, effectively burning all of the recirculated exhaust gases and improving idle and performance. The crankshaft employs five main bearings with the thrust on the center bearing.

The SD22 diesel engine is also new for 1981. It has a cast iron block with dry-type, pressed-in cylinder liners. The one-piece cylinder head has swirl-type combustion chambers. Both intake and exhaust valves have cold-fitted, cast iron seats. The camshaft is gear-driven. Also gear-driven is the Bosch-Diesel Kikki injection pump.

REMOVAL AND INSTALLATION

It is much easier to remove the engine and the transmission together as an assembly than to remove only the engine from the engine compartment. After the engine and transmission are removed from the vehicle, the two can be separated.

1. Disconnect the battery ground cable. Remove the battery.

2. Mark the location of the hood hinges on the body in order to facilitate installation and remove the hood.

3. Remove the air cleaner after disconnecting the PCV hose from the rocker cover.

4. Drain the radiator of coolant and the engine crankcase of oil.

5. Disconnect the upper and lower radiator hoses from the engine. Disconnect and plug the automatic transmission cooler lines at the radiator, if so equipped. Use a flare nut wrench if one is available.

6. Remove the four bolts securing the radiator and remove the radiator from the vehicle.

7. Disconnect the engine ground cable at the cylinder head.

8. Disconnect the electrical leads at the starter, alternator, distributor, the high-tension ignition coil cable, and the oil pressure and temperature sending units' wires.

9. Disconnect the fuel line at the fuel pump (or filter on electric pump models), the heater hose at the engine side, and the choke wire and accelerator cable at the carburetor. Disconnect the emission hoses or wires to

the carbon canister, air pump, B.C.D.D. solenoid, and fuel cut solenoid; the vacuum hose to the brake booster (on models so equipped), and any other wires or hoses running to the engine. Tag all wires as they are disconnected for assembly.

10. Remove the transmission control linkage from the transmission; in the case of an automatic transmission, remove the cross-shaft assembly from the transmission. Remove the selector rod from the selector lever on the automatic transmission. On manual transmissions, lift the rubber boot and remove the nut or c-clip from the shift lever and deatch the shift lever from the transmission.

Remove the c-clip and pin on later models for shift lever removal

11. Remove the two bolts securing the clutch slave cylinder. Disconnect the clutch slave cylinder and the flexible tubing as an assembly.

12. Disconnect the speedometer cable and the back-up light wiring (and neutral switch, if equipped) from the rear section of the transmission.

13. Disconnect the exhaust pipe from the exhaust manifold.

14. Disconnect the driveshaft center bearing bracket from the third crossmember of the frame. Disconnect the driveshaft at the differential housing. Remove the driveshaft assembly from the vehicle and plug the rear end of the transmission extension housing to prevent loss of transmission lubricant.

15. Attach a suitable lifting device to the engine and lift the engine slightly.

16. Remove the front engine mount bolts on both sides of the engine.

17. Place a jack under the transmission and lift it slightly.

18. Loosen the two combination engine rear mounting/transmission mounting bolts. On models with a catalytic converter, loosen the two exhaust pipe hanger bolts.

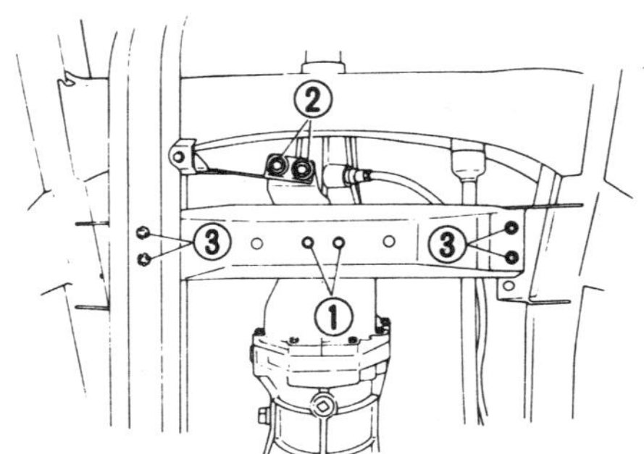

1. Engine mount bolts
2. Exhaust pipe bolts
3. Crossmember bolts

Engine and transmission crossmember removal (1972 and later)

19. On 1972 and later models, remove the four bolts (two on each side) securing the engine rear mounting/transmission support side member and detach the support from the frame.

20. On 1976 and later models, remove the bolts securing the idler arm to the frame, and push down the tie-rod.

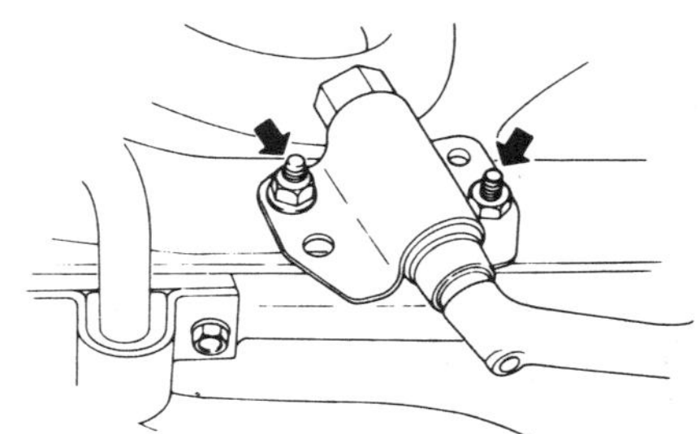

Idler arm removal (1976 and later)

21. Pull the engine toward the front as far as possible and carefully raise the engine with the transmission up and out of the vehicle.

22. Install the engine in the reverse order of removal, taking note of the following:

Do not connect any parts to the engine or transmission until the engine and transmission are in place on the engine/transmission mounts and secured by the mounting bolts. Secure the rear support first and then the front engine mounts, using the upper bolt hole as a guide.

Cylinder Head

REMOVAL AND INSTALLATION
L16, L18 and L20B Engines

1. Crank the engine until the No. 1 piston is at TDC of the compression stroke and

disconnect the negative battery cable, drain the cooling system, and remove the air cleaner and attending hoses.

2. Remove the alternator.

3. If equipped with air conditioning, unbolt the compressor and move it aside onto the fender. Do not detach any of the compressor lines; the escaping refrigerant will freeze any surface it contacts, including your skin.

4. Disconnect the carburetor throttle linkage, the fuel line and any other vacuum lines or electrical leads, and remove the carburetor.

5. Disconnect the exhaust pipe from the exhaust manifold.

6. Remove the fan and fan pulley.

7. Remove the spark plugs to protect them from damage. Lay the spark plugs aside and out of the way.

8. Remove the rocker cover.

9. Remove the water pump.

10. Remove the fuel pump from the head, on models without the electric pump.

11. Remove the fuel pump drive cam.

12. Mark the relationship of the camshaft sprocket to the timing chain with paint or chalk. If this is done, it will not be necessary to locate the factory timing marks. Before removing the camshaft sprocket, it will be necessary to wedge the chain in place so that it will not fall down into the front cover. The factory procedure is to wedge the timing chain in place with the wooden wedge shown here. The problem with this is that it may allow the chain tensioner to move out far enough to cock itself against the chain. If this happens, you'll find that the chain won't go back over the sprocket after you've put the sprocket back on. In this case, you'll have to remove the front cover and push the tensioner back.

After installing the wedge, unbolt and remove the camshaft sprocket.

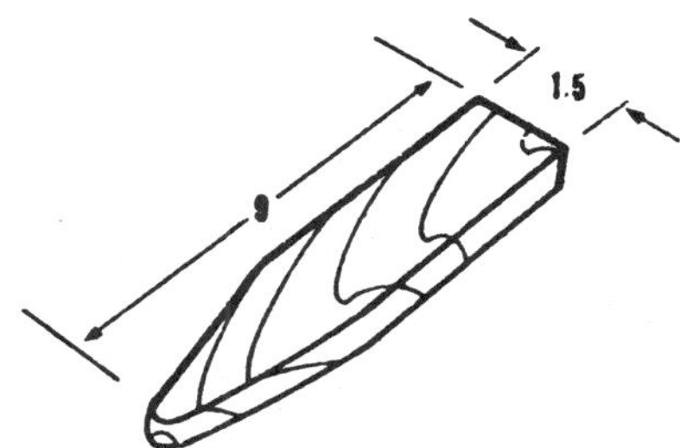

Dimensions for fabricating the wooden wedge used to support the timing chain

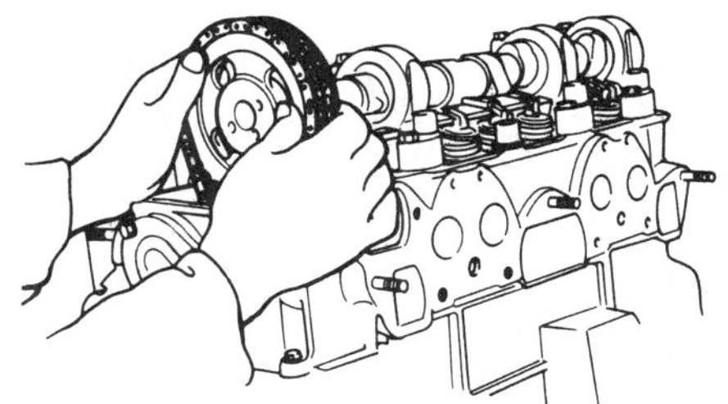

Removing the camshaft sprocket and chain

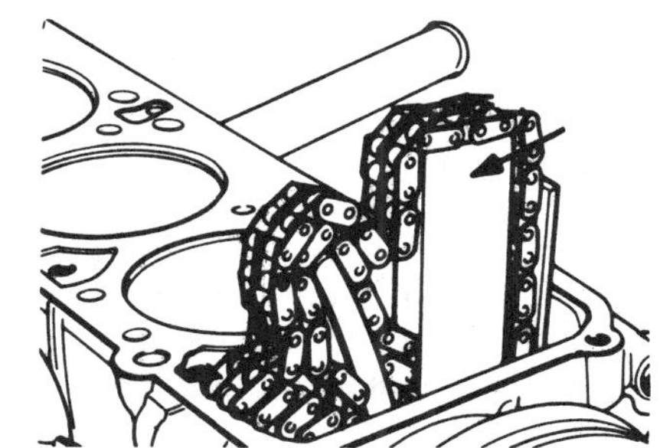

Support the timing chain with a wedge

13. Loosen and remove the cylinder head bolts. You will need a 10 mm Allen wrench to remove the head bolts. Keep the bolts in order, because they are different sizes. Lift the cylinder head assembly from the engine. Remove the intake and exhaust manifolds as necessary.

14. Thoroughly clean the cylinder block and head mating surfaces. Check the block and head for flatness before installing the head. See the Engine Rebuilding section at the end of this Chapter for details on how to do this. Install a new cylinder head gasket. Do not use sealer on the cylinder head gasket of 1973 and later models. Use sealer on all other models.

15. With the crankshaft turned so that the No. 1 piston is at TDC of the compression stroke (if not already done so as mentioned in Step 1), make sure that the camshaft sprocket timing mark and the oblong groove in the camshaft retaining plate are aligned.

16. Place the cylinder head in position on the cylinder block, being careful not to allow any of the valves to come in contact with any of the pistons. Do not rotate the crankshaft or camshaft separately because of possible damage which might occur to the valves.

17. Temporarily tighten the two center right and left cylinder head bolts to 14.5 ft. lbs.

18. Install the camshaft sprocket together with the timing chain to the camshaft. Make

Installing the camshaft sprocket

sure that the marks you made earlier line up. If the chain will not stretch over the sprocket, the problem lies in the tensioner. See "Timing Chain Removal and Installation" for timing procedure, if necessary.

19. Install the cylinder head bolts. Note that there are two sizes of bolts used; the longer bolts are installed on the driver side of the engine with a smaller bolt in the center position. The remaining small bolts are installed on the opposite side of the cylinder head.

20. Tighten the cylinder head bolts in three stages on 1973 and later models: first to 29 ft. lbs, second to 43 ft. lbs, and lastly to 47–62 ft. lbs.

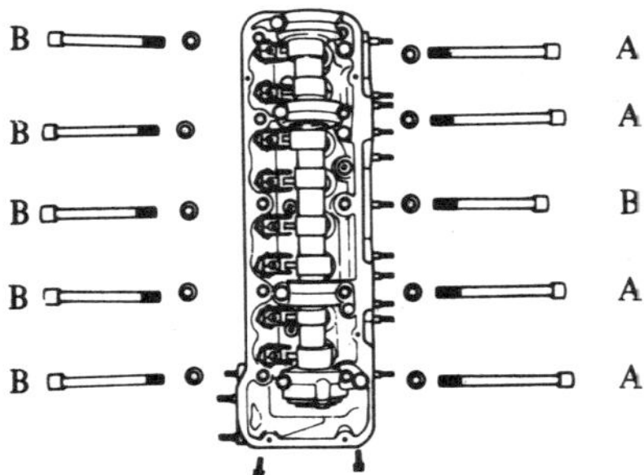

Different size cylinder head bolts

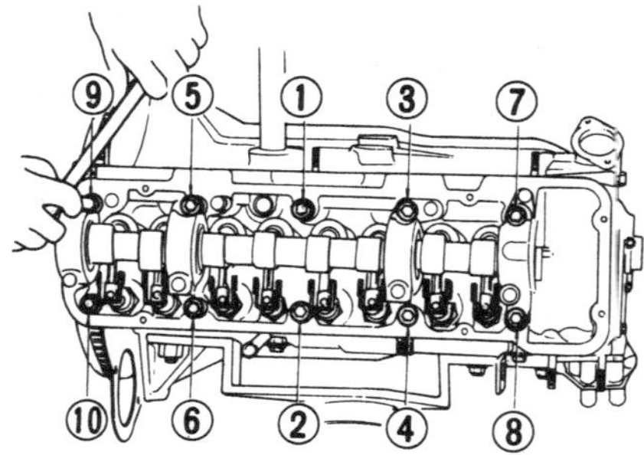

Cylinder head bolt tightening sequence

On 1970–72 models, tighten the cylinder head bolts to 36–43 ft. lbs. in three progressive steps.

Tighten the cylinder head bolts on all models in the proper sequence.

21. Install and assemble the remaining components of the engine in the reverse order of removal.

Z22, Overhead Camshaft Engine

1. Complete steps 1 through 5 under L20 Overhead Camshaft Engine. Observe the following note for step 5.

NOTE: *The spark plug leads should be marked, however it would be wise to mark them yourself, especially the dual spark plug California models.*

2. Disconnect the throttle linkage, the air cleaner or its intake hose assembly (fuel injection). Disconnect the fuel line, the return fuel line and any other vacuum lines or electrical leads. Remove the carburetor to avoid damaging it while removing the head.

NOTE: *A good rule of thumb when disconnecting the rather complex engine wiring of today's automobiles is to put a piece of masking tape on the wire or hose and on the connection you removed the wire or hose from, then mark both pieces of tape 1, 2, 3, etc. When replacing wiring, simply match the pieces of tape.*

3. Remove the E.G.R. tube from around the rear of the engine.

4. Remove the exhaust air induction tubes from around the front of the engine and from the exhaust manifold.

5. Unbolt the exhaust manifold from the exhaust pipe. Remove the fuel pump.

6. Remove the intake manifold supports from under the manifold. Remove the P.C.V. valve from around the rear of the engine if necessary.

7. Remove the spark plugs to protect them from damage. Remove the valve cover.

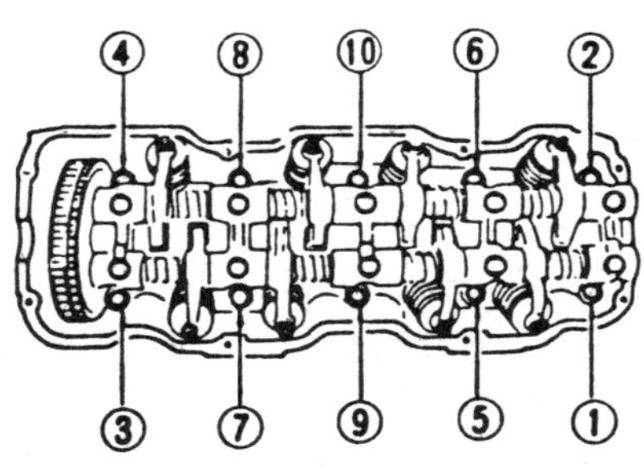

Z22 engine cylinder head loosening sequence

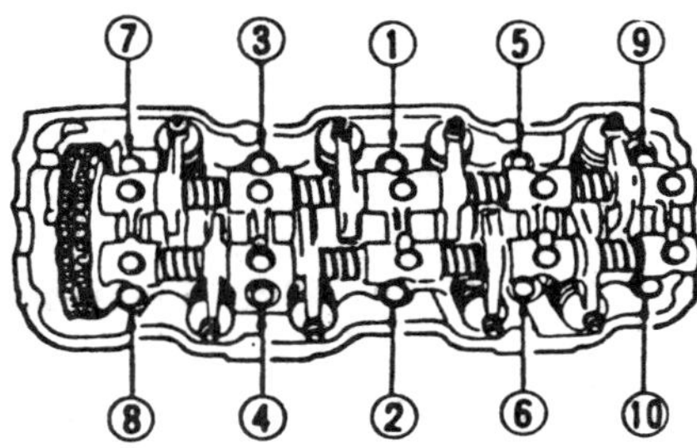

Z22 engine head tightening sequence

8. Mark the relationship of the camshaft sprocket to the timing chain with paint or chalk. If this is done, it will not be necessary to locate the factory timing marks. Before removing the camshaft sprocket, it will be necessary to wedge the chain in place so that it will not fall down into the front cover. The factory procedure is to wedge the timing chain in place with the wooden wedge shown here. The problem with this procedure is that it may allow the chain tensioner to move out far enough to cock itself against the chain. If this happens, you'll find that the chain won't go back over the sprocket after you've put the sprocket back on. In this case, you'll have to remove the front cover and push the tensioner back. After you've wedged the chain, unbolt the camshaft sprocket and remove it.

9. Working from both ends in, loosen the cylinder head bolts and remove them. Remove the bolts securing the cylinder head to the front cover assembly.

10. Lift the cylinder head off the engine block. It may be necessary to tap the head lightly with a copper or brass mallet to loosen it.

To install the cylinder head:

11. Thoroughly clean the cylinder block and head surfaces and check both for warpage. See A12, etc. Overhead Valve Engines cylinder head removal section for procedure. (Step 1 of assembly process).

12. Fit the new head gasket. Don't use sealant. Make sure that no open valves are in the way of raised pistons, and do not rotate the crankshaft or camshaft separately because of possible damage which might occur to the valves.

13. Temporarily tighten the two center right and left cylinder head bolts to 14 ft. lbs.

14. Install the camshaft sprocket together with the timing chain to the camshaft. Make sure the marks you made earlier line up with each other. If you get into trouble, see "Tim-

ing Chain Removal and Installation" for timing procedures.

15. Install the cylinder head bolts and torque them to 20 ft. lbs., then 40 ft. lbs., then 58 ft. lbs. in the order shown in the illustration.

16. Assemble the rest of the components in the reverse order of disassembly.

NOTE: *It is always wise to drain the crankcase oil after the cylinder head has been installed to avoid coolant contamination.*

Overhaul

Cylinder head overhaul should be referred to a competent automotive machine shop. Valve guides and seats are removable and oversizes are available from Datsun. For procedures, see overhaul section at the end of this chapter.

SD22 Diesel

1. Remove the air cleaner.
2. Remove the crankcase vent hose and remove the intake and exhaust manifolds. These are bolted together.
3. Remove the alternator, bracket and belts.
4. Disconnect the coolant hose between the head and the oil cooler.
5. Remove the fuel filter assembly.
6. Disconnect the injection lines from the pump and the injectors. Cap all openings at once.
7. Remove the bypass hoses between the coolant pump and the thermostat housing.
8. Remove the fan.
9. Remove the rocker arm cover.
10. Remove the rocker arm shaft assembly.
11. Remove the pushrods and keep them in order.
12. Remove the fuel return lines.
13. Remove the nozzles from the head.
14. Remove the cylinder head bolts in the sequence shown.

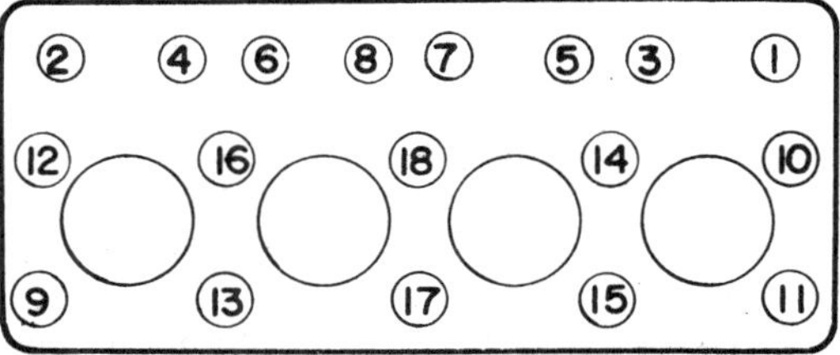

SD22 diesel cylinder head bolt loosening sequence

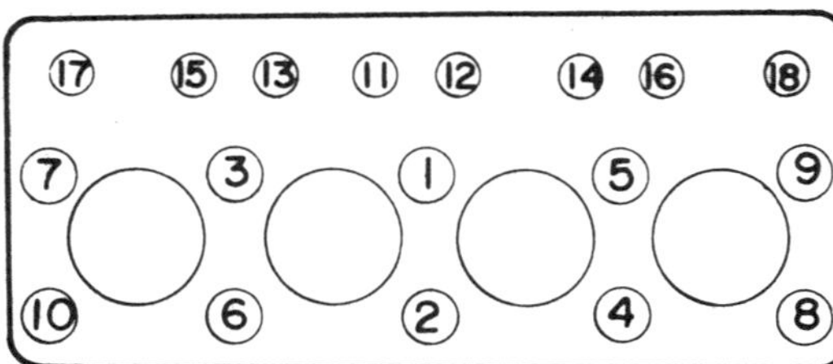

SD22 diesel cylinder head bolt tightening sequence

15. Attach a hoist to the head and lift it clear of the block. On occasion, the precombustion chambers may fall out, especially if the head is bumped or handled roughly. Take care that they are returned to their original positions if this occurs.

16. Remove the head gasket and O-rings.

17. Clean and inspect all parts.

18. Check the head with a straight-edge. Maximum warpage is .0079". Do not remove more than .011" from the head.

19. Place a new cylinder head gasket on the block with the stainless steel inset side facing up.

20. Install the O-rings around the water and oil passages.

21. Position the head on the block.

22. Coat the head bolts with clean engine oil and torque them in sequence, in stages as follows:

Large: 43, 94
Small: 21, 36

23. Install the pushrods, pressing down and turning them to be sure of proper seating.

24. Install the rocker arm shaft assembly, torquing the bolts to 18 ft. lb. in sequence from the center to each end.

25. Install the injection nozzles.

26. Install all other parts in reverse order of removal.

VALVE GUIDE REMOVAL AND INSTALLATION

Gasoline Engines

1. With the cylinder head removed from the engine, and the valves removed from the head, use a drift and a hammer or press. Drive the valve guides out from the combustion chamber side toward the rocker cover side. A heated cylinder head will facilitate the operation.

2. Ream the cylinder head side guide hole at room temperature. The guide hole should be 0.4719–0.4723 in. for standard valves and

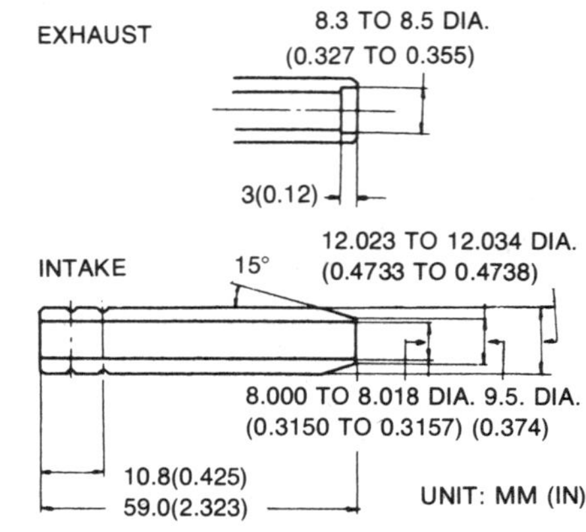

Valve guide dimensions

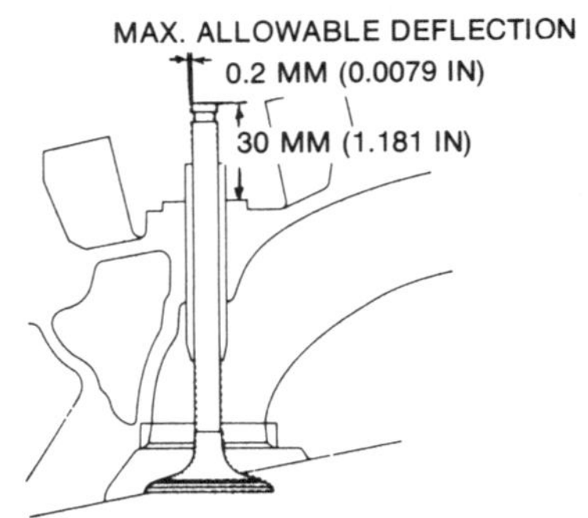

Measuring the valve stem-to-guide clearance

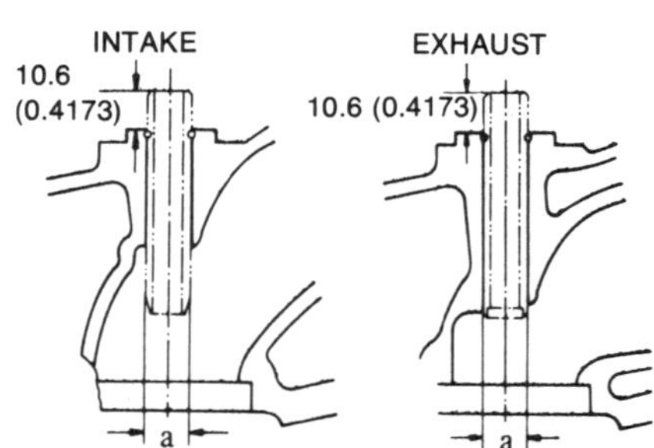

Gasoline engine valve guide installation

0.4797–0.4802 in. for 0.0079 in. oversize valves which are available for service.

3. After heating the cylinder head to 302–392°F, press the new valve guide carefully into the cylinder head. The top of the valve guide should protrude out the top of the guide hole 0.4173 in.

4. Ream the bore of the valve guide with the valve guide pressed into the cylinder head. The standard valve guide bore size is 0.3150–0.3157 in.

5. Assemble the cylinder head and install it on the engine in the reverse order of removal.

Diesel Engine

The guides are not replaceable.

VALVE SEAT REMOVAL AND INSTALLATION

Gasoline Engines

1. With the cylinder head removed from the engine and the valves removed from the cylinder head, old valve seat inserts can be removed by boring them out until they collapse. Be careful that the boring doesn't continue beyond the bottom face of the insert recess in the cylinder head.

2. Select the suitable valve seat insert and check its outside diameter.

3. Machine the cylinder head recess using the center of the valve guide as the center of the valve seat insert so that the insert will have the correct fit.

4. Ream the cylinder head recess at room temperature.

5. Heat the cylinder head to 302–392°F.

6. Fit the insert, making sure that it seats fully in the recess in the cylinder head. Peen the insert with a punch in at least four places equally spaced around its circumference.

7. Grind the valve seats to the proper angle.

8. Lap the valves with lapping compound to each seat to which they are to be mated. Thoroughly clean both the valve and the seat of all lapping compound before installing the valves.

Diesel Engines

1. The seats may be removed by cracking with a cold chisel.

2. Immerse the head in water at 175°F while at the same time cool the valve seats in dry ice. The processes should take about 5–10 minutes.

3. Install the seats and reface and lap according to specifications.

Valve Rockers and Rocker Pivots

REMOVAL AND INSTALLATION

L20B

1. Loosen the rocker pivot locknut, lower the pivot by screwing it down into the cylinder head, and remove the rocker arm by pressing down on the valve spring.

2. To remove the rocker pivots, loosen the locknut, then unscrew the pivot from the cylinder head.

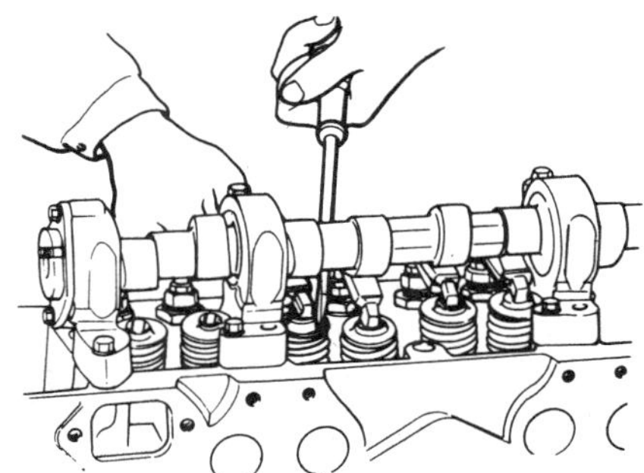

Removing the rocker arms on L20B engines

3. Install the pivots and rockers and assemble the engine in the reverse order of removal.

Rocker Shaft Assembly

REMOVAL AND INSTALLATION

Z22

1. The rocker shaft assembly is removed by simply removing the retaining bolts.

NOTE: *When removing the bolts, DO NOT REMOVE THE NO. 1 AND NO. 5 BRACKET BOLTS SINCE THE ROCKER SHAFT BRACKET AND ROCKER ARM WILL SPRING OUT!*

2. Installation is the reverse of removal. Torque the bolts evenly from the ends toward the center to 11–18 ft. lb.

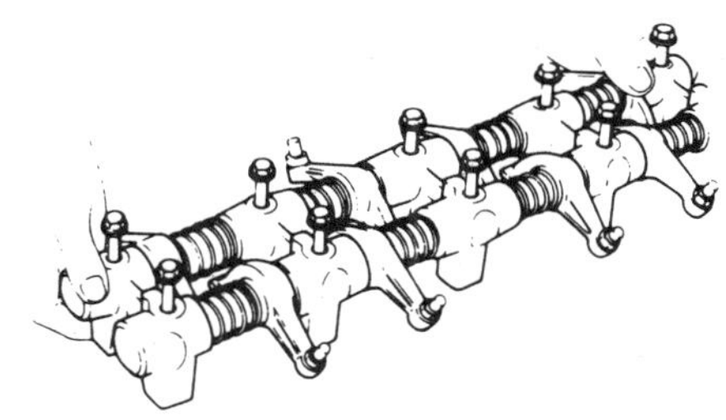

Rocker arm assembly on the Z22 engine

SD22 Diesel

1. Remove the shaft retaining bolts evenly, from the center towards the ends.

2. Lift the shaft assembly off the head.

3. If you are disassembling the shaft and rocker arms, it may be necessary to immerse the assembly in water heated to 160°F for a few minutes to free the rocker arms. NEVER HAMMER THEM OFF!

4. Installation is the reverse of removal. Torque the retaining bolts evenly from the ends toward the center to 14–18 ft. lb.

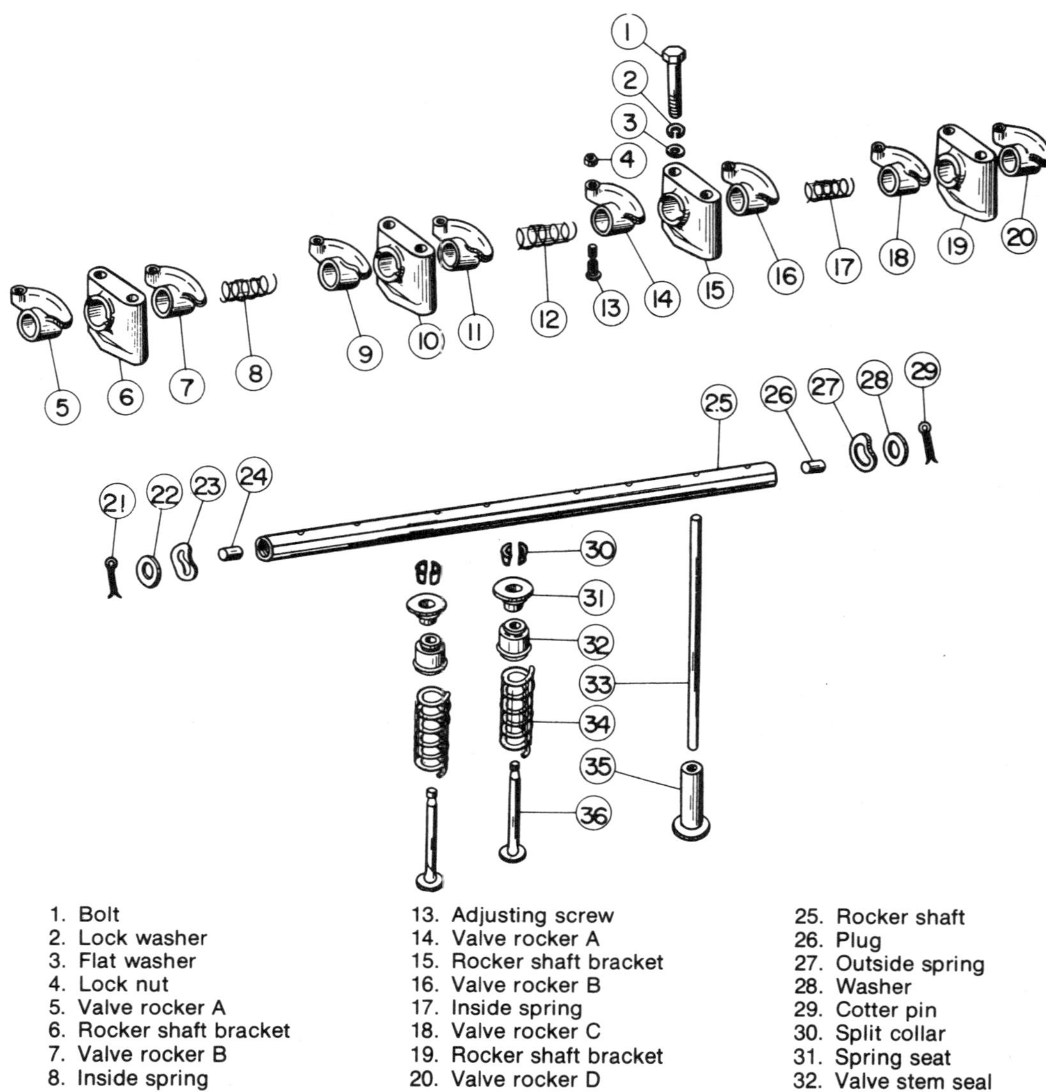

1. Bolt	13. Adjusting screw	25. Rocker shaft
2. Lock washer	14. Valve rocker A	26. Plug
3. Flat washer	15. Rocker shaft bracket	27. Outside spring
4. Lock nut	16. Valve rocker B	28. Washer
5. Valve rocker A	17. Inside spring	29. Cotter pin
6. Rocker shaft bracket	18. Valve rocker C	30. Split collar
7. Valve rocker B	19. Rocker shaft bracket	31. Spring seat
8. Inside spring	20. Valve rocker D	32. Valve stem seal
9. Valve rocker C	21. Cotter pin	33. Push rod
10. Rocker shaft bracket	22. Washer	34. Valve spring
11. Valve rocker D	23. Outside spring	35. Valve lifter
12. Inside spring	24. Plug	36. Valve

SD22 diesel rocker arm assembly

Intake Manifold

REMOVAL AND INSTALLATION

All Engines

1. Remove the air cleaner assembly together with all of the attending hoses. Remove the EGR tube on 1974 and later gasoline models.

NOTE: *It is important to replace the gasket whenever the intake manifold is removed. Because the intake and exhaust manifolds share a common gasket, whenever the intake manifold is removed, the exhaust manifold must also be removed, so that the gasket can be replaced.*

2. Disconnect the throttle linkage, fuel, and vacuum lines from the carburetor. Label all wires and hoses as they are removed to simplify installation.

3. The carburetor can be removed from the manifold at this point or can be removed as an assembly with the intake manifold.

4. Loosen the intake manifold attaching nuts, working from the two ends toward the center, and then remove them.

Intake manifold and gasket

5. Remove the intake manifold from the engine.

6. Install the intake manifold in the reverse order of removal. Always use a new gasket when installing the manifold; air leaks will cause burnt valves. Tighten the manifold bolts from the center outwards, in two progressive steps, to 9–12 ft. lbs.

Exhaust Manifold

REMOVAL AND INSTALLATION

All Engines

1. Remove the air cleaner assembly.

2. Disconnect the exhaust pipe from the exhaust manifold.

NOTE: *It is not absolutely necessary to replace the gasket when only the exhaust manifold is removed, unless the gasket is damaged, or leaks develop.*

3. Loosen and remove the exhaust manifold attaching nuts and remove the manifold from the engine.

4. Install the exhaust manifold in the reverse order of removal. Use new gaskets at the cylinder head (if necessary) and exhaust pipe. Tighten the mounting bolts in a circular pattern, working from the center to the ends, in two progressive steps to the figures in the torque chart.

NOTE: *On 1978–81 gasoline models, install the stud bolt into the center of the outermost guide hole (no. 4 cylinder) of the manifold.*

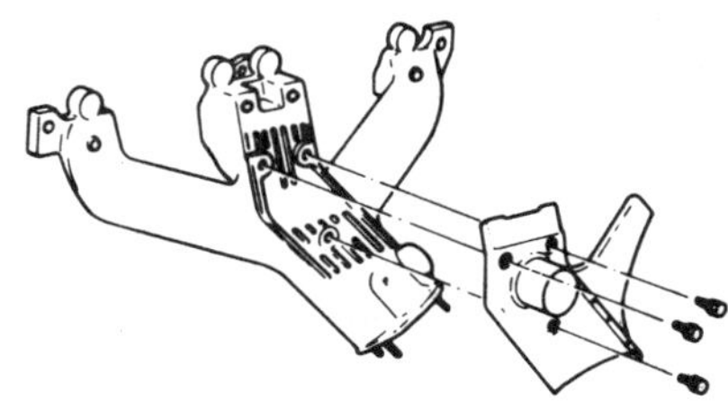

Exhaust manifold and heat stove

Timing Gear Cover

REMOVAL AND INSTALLATION

Gasoline Engines

1. Disconnect the negative battery cable from the battery, drain the cooling system, and remove the radiator together with the upper and lower radiator hoses.

2. Loosen the alternator drive belt adjusting screw and remove the drive belt. Remove the bolts which attach the alternator bracket to the engine and set the alternator aside out of the way.

3. Remove the distributor.

4. Remove the oil pump attaching screws, and take out the pump and its drive spindle.

5. Remove the cooling fan and the fan pulley together with the drive belt.

6. Remove the water pump.

7. Remove the crankshaft pulley bolt and remove the crankshaft pulley.

8. Remove the bolts holding the front cover to the front of the cylinder block, the four bolts which retain the front of the oil pan to the bottom of the front cover and the two bolts which are screwed down through the front of the cylinder head and into the top of the front cover.

9. Carefully pry the front cover off the front of the engine.

10. Cut the exposed front section of the oil

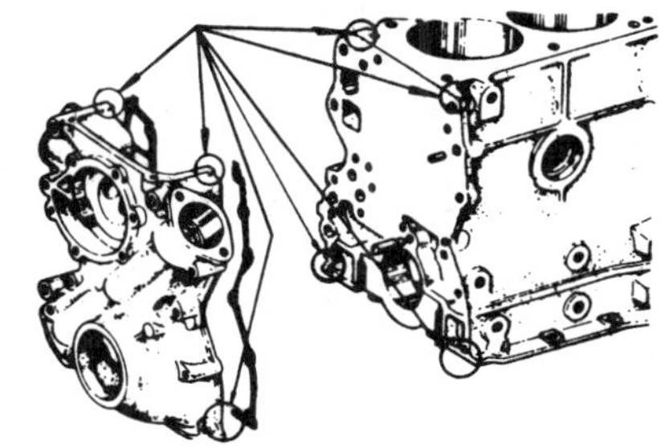

Gasoline engine front cover installation

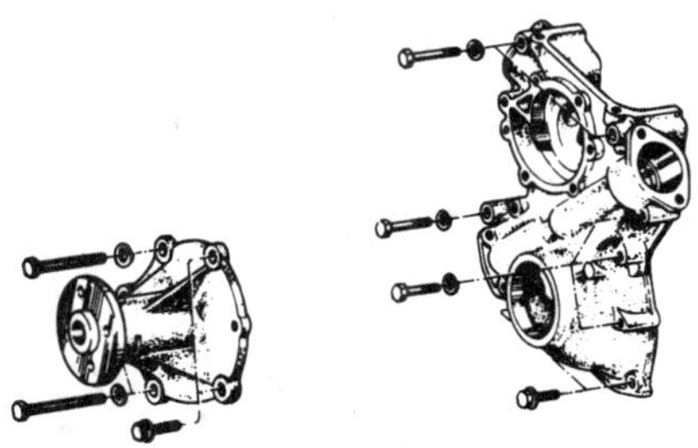

Gasoline engine front cover bolts

pan gasket away from the oil pan. Do the same to the gasket at the top of the front cover. Remove the two side gaskets and clean all of the mating surfaces.

11. Cut the portions needed from a new oil pan gasket and top front cover gasket.

12. Apply sealer to all of the gaskets and position them on the engine in their proper places.

13. Apply a light coating of oil to the crankshaft oil seal and carefully mount the front cover to the front of the engine and install all of the mounting bolts.

Tighten the 8 mm bolts to 7–12 ft. lbs. and the 6 mm bolts to 3–6 ft. lbs. Tighten the oil pan attaching bolts to 4–7 ft. lbs.

14. Before installing the oil pump, place the gasket over the shaft and make sure that the mark on the drive spindle faces (aligned with) the oil pump hole. Install the oil pump so that the projection on the top of the shaft is located in the exact position as when it was removed or is in the 11:25 o'clock position with the piston in the No. 1 cylinder is placed at TDC on the compression stroke, if the engine was disturbed since disassembly. Tighten the oil pump attaching screws to 8–10 ft. lbs. See "Oil Pump Removal and Installation."

SD22 Diesel

NOTE: *A 41mm (1.614 in.) socket is needed for this procedure.*

1. Remove the fan and pulley.

2. Remove the water pump bypass hose and allow the cooling system to drain below the level of the water pump.

3. Remove the three bolts and lift the water pump and gasket off the block. Discard the gasket.

Pulley nut removal

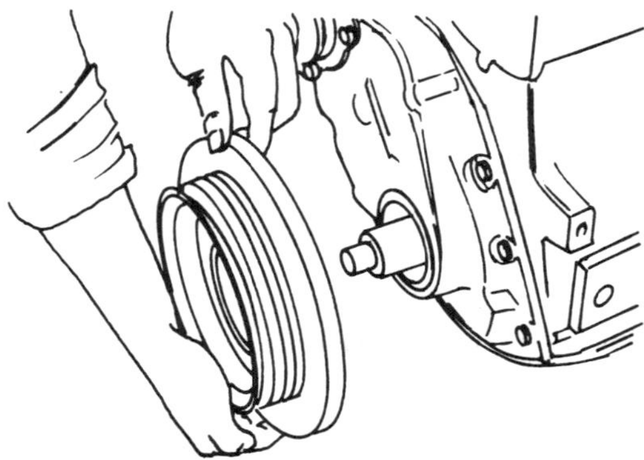

Lifting pulley assembly

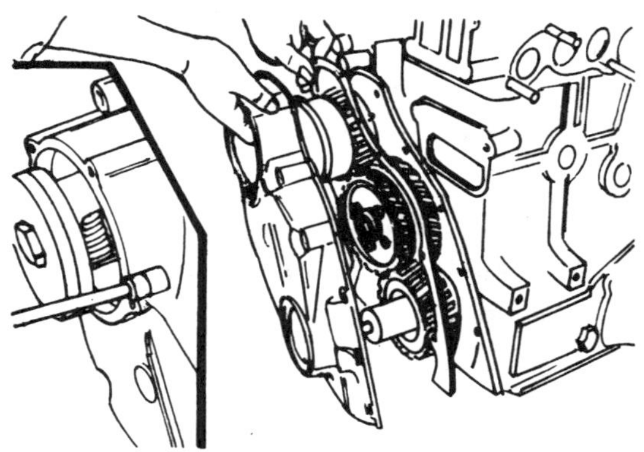

Timing gear case removal

4. Remove the crankshaft pulley nut with a 41mm socket.

5. Drive the pulley from the crankshaft with a wooden or plastic mallet.

6. Remove the five bolts and lift the timing gear cover from the case.

7. Installation is the reverse of removal. Always replace the cover oil seal and use a new cover gasket. Torque the cover bolts to 8 ft. lb., the crankshaft pulley nut to 238 ft. lb., and the water pump bolts to 8 ft. lb. for the 8mm bolt and 16 ft. lb. for the 10mm bolts.

NOTE: *Do not tighten the water pump bolts until the belt adjuster is installed when installing the alternator.*

Oil Seal

1. Remove the front cover.

2. Pry the old seal from the cover with a pointed piece of plastic or wood. Do not use a screwdriver to avoid scratching the seal surface.

3. Oil the lip of the new seal. Do not use grease. Press it into place, making sure the

flat side faces forward and the lip faces the engine.

4. Install the front cover.

Timing Gears and Case

The following requires use of special tools.

SD22 Diesel

1. Remove the timing gear cover.
2. Remove the timing gear round nut.
3. Using timer extractor 57926-581, thread the tool into the timer weight holder. Remove the timing gear assembly by threading in the extractor tool bolt.

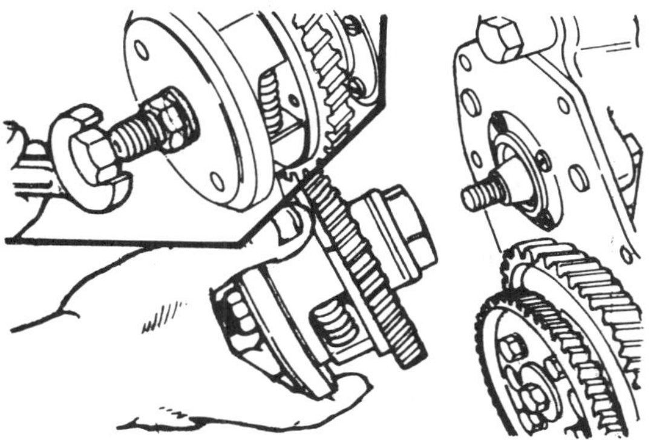

Timer removal and installation

4. Unbolt and remove the camshaft gear set.
5. Remove the oil slinger. Unbolt the crankshaft gear and remove it with a gear puller.
6. Install the camshaft gear.
7. Install the crankshaft gear and oil slinger while carefully aligning the timing marks as shown. Measure the gear backlash. Backlash should be 0.0028–0.0079 in.

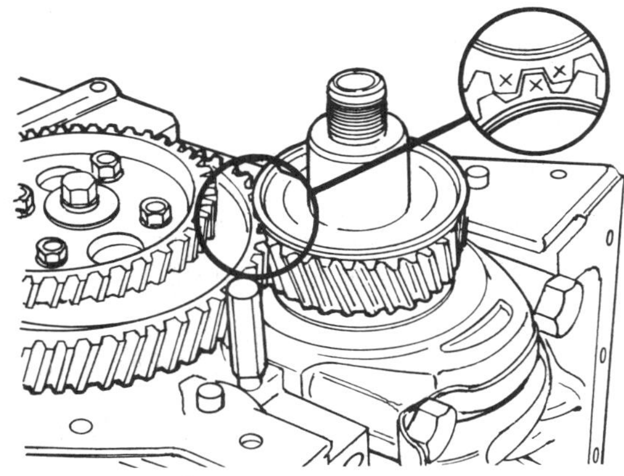

Timing mark alignment

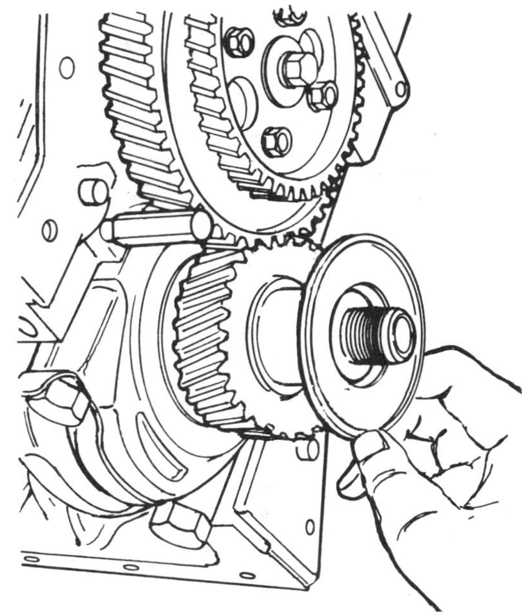

8. With the #1 piston at TDC, mesh the timing gear and idler gear at the "Y" marks. After aligning the gear with the keyway, secure the timer assembly with a lock washer and the round nut. Torque the nut to 50–58 ft. lb.

9. Install the cover.

NOTE: *If the gear case oil jet was removed, install in the relationship as shown.*

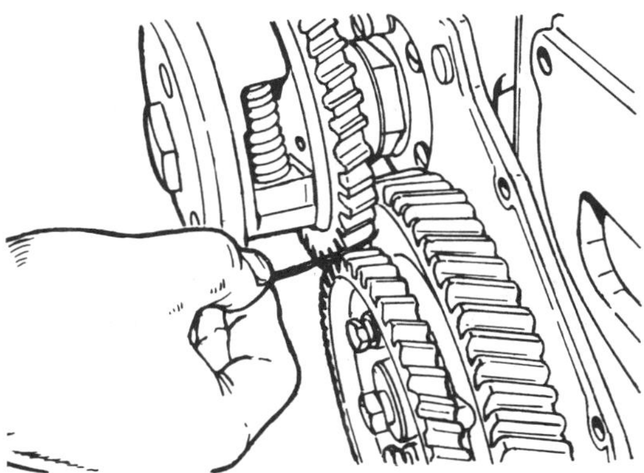

Measuring backlash

Timing Chain and Tensioner
REMOVAL AND INSTALLATION
Gasoline Engines

1. Before beginning any disassembly procedures, position the no. 1 piston at TDC on the compression stroke.

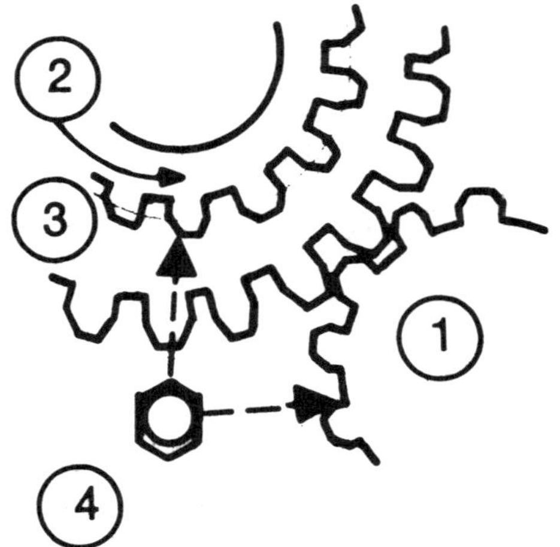

1. Crankshaft gear
2. Idle gear
3. Camshaft gear
4. Oil jet

Oil jet orientation

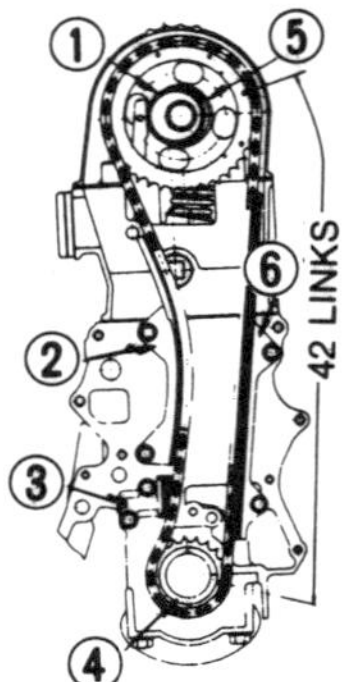

1. Fuel pump drive cam
2. Chain guide
3. Chain tensioner
4. Crank sprocket
5. Cam sprocket
6. Chain guide

Installing the timing chain. The number of "links" refers to pre-1974 models

2. Remove the front cover. Remove the camshaft cover.

3. With the no. 1 piston at TDC, the timing marks on the camshaft sprocket and the timing chain should be visible. Mark both of them with paint. Also mark the relationship of the camshaft sprocket to the camshaft. At this point you will see that there are three sets of timing marks and locating holes in the sprocket. They are for making adjustments to compensate for timing chain stretch. See the "Timing Chain Adjustment" section following for details.

4. With the timing marks on the cam sprocket clearly marked, locate and mark the timing marks on the crankshaft sprocket. Also mark the chain timing mark. Of course, if the chain is not to be reused, marking it is useless.

5. Unbolt the camshaft sprocket and remove the sprocket along with the chain. As you remove the chain, hold it where the chain tensioner contacts it. When the chain is removed, the tensioner is going to come apart. Hold on to it and you won't lose any of the parts.

The crankshaft sprocket can be removed with a puller, if necessary. There is no need to remove the chain guide unless it is being replaced.

6. Install the timing chain and the camshaft sprocket together after first positioning the chain over the crankshaft sprocket. Position the sprocket so that the marks made earlier line up. This is assuming that the engine has not been disturbed. The camshaft and the crankshaft keys should both be pointing upward. If a new chain and/or gear is being installed, position the sprocket so that the timing marks on the chain align with the marks on the sprocket (with both keys pointing up). The marks are on the right-hand side of the sprockets as you face the engine. Engines up to 1973 have 42 pins between the mating marks of the chain and sprockets when the chain is installed correctly. 1974 and later engines have 44 pins. The factory refers to the pins as links, but in American terminology this is incorrect. Count the pins. There are two pins per chain link. This is an important step. If you do not get the exact

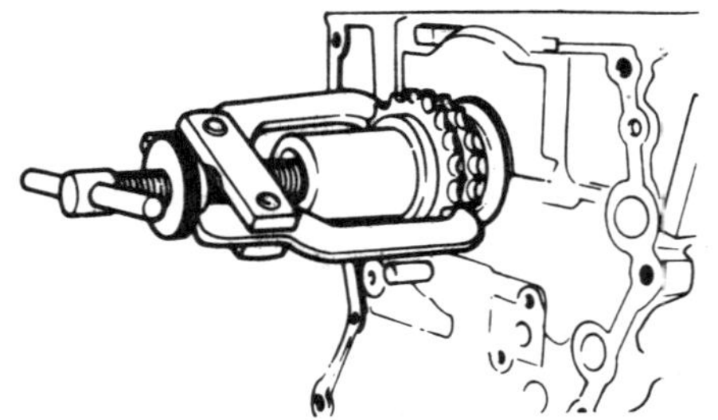

The gasoline engine crankshaft sprocket removal

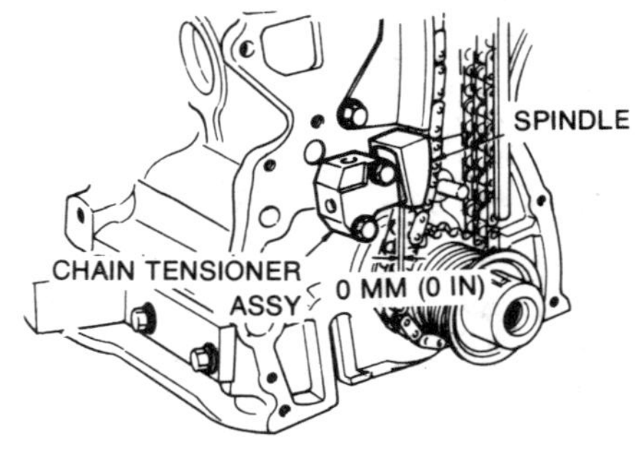

Installing the timing chain tensioner

number of pins between the timing marks, valve timing will be incorrect, and the engine will either not run at all or run very badly.

7. Install the chain tensioner. Adjust the protrusion of the chain tensioner spindle to zero clearance.

8. With a new seal installed in the front cover and a light coat of oil applied to the seal, assemble the remaining components of the engine in the reverse order of disassembly.

TIMING CHAIN ADJUSTMENT

When the timing chain stretches excessively, the valve timing will be adversely affected. There are two camshaft sprocket locating holes provided to correct the valve timing. Actually there are three sets of holes and timing marks on the camshaft sprocket; the third hole and timing mark are for 6 cylinder Datsun engines and in the case of the Datsun pick-up 4 cylinder engines, are obviously ignored.

If the stretch of the chain roller links is excessive, adjust the camshaft sprocket location by transferring the camshaft set position of the camshaft sprocket from the factory position of No. 1 to No. 2 as follows:

1. Turn the crankshaft until the No. 1 piston is at TDC on its compression stroke. Exxamine whether the camshaft sprocket location notch is to the left of the oblong groove on the camshaft retaining plate. If the notch in the sprocket is to the left of the groove in the retaining plate, then the chain is stretched and needs adjusting.

2. Remove the camshaft sprocket together with the chain and reinstall the sprocket and chain with the locating dowel on the camshaft inserted into the No. 2 hole of the sprocket and the timing mark on the timing chain aligned with the No. 2 mark on the sprocket. The amount of modification is 4° of the crankshaft rotation.

3. Recheck the valve timing as outlined in Step 1. The notch in the sprocket should be to the right of the groove in the camshaft retaining plate.

4. If and when the notch cannot be brought to the right of the groove with the sprocket installed in the No. 2 hole, the timing chain must be replaced to gain the proper valve timing.

Camshaft
REMOVAL AND INSTALLATION
Gasoline Engines

1. Removal of the cylinder head from the engine is optional. Remove the camshaft sprocket from the camshaft together with the timing chain.

2. Loosen the valve rocker pivot locknut and remove the rocker arm by pressing down on the valve spring. Remove all of the rocker arms in this manner.

3. Remove the two retaining nuts on the camshaft retainer plate at the front of the cylinder head and carefully slide the camshaft out of the camshaft carrier.

4. Lightly coat the camshaft bearings with clean motor oil and carefully slide the camshaft in place in the camshaft carrier.

5. Install the camshaft retainer plate with the oblong groove in the face of the plate facing toward the front of the engine.

6. Check the valve timing as outlined under "Timing Chain Removal and Installation" and install the timing sprocket on the camshaft, tightening the bolt together with the fuel pump cam to 86–116 ft. lbs.

7. Install the rocker arms by pressing

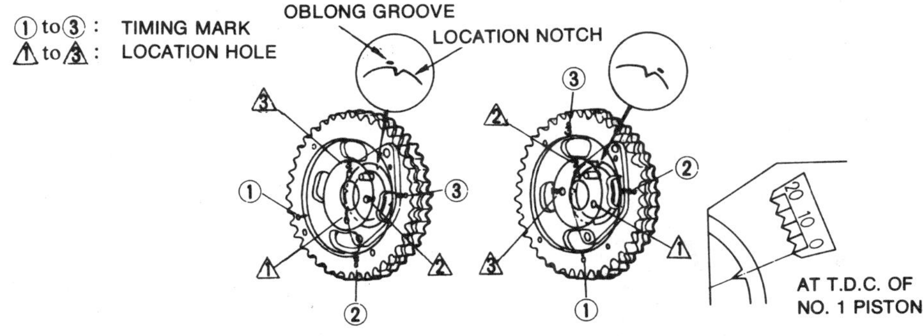

Adjusting the camshaft sprocket location in order to obtain proper valve timing due to a worn timing chain

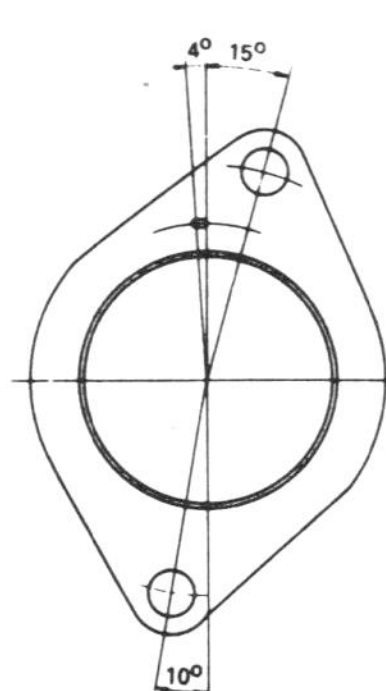

The camshaft retaining plate

down the valve springs with a screwdriver and install the valve rocker springs.

8. Install the cylinder head, if it was removed, and assemble the rest of the engine in the reverse order of removal.

SD22 Diesel

1. Remove the head.

2. Remove the lifters and mark them for reassembly.

3. Remove the front case and cover.

4. Remove the tachometer drive support nuts.

5. Remove the timer round nut.

6. Thread the timer extractor, ST 57926-581, into the timer weight holder. Remove the timer assembly by tightening the extractor bolt.

7. Remove the oil pump drive spindle.

8. Remove the camshaft locating plate bolts and carefully slide the camshaft from the engine.

9. Coat the camshaft with clean engine oil and carefully slide it into the block. Install the locating plate.

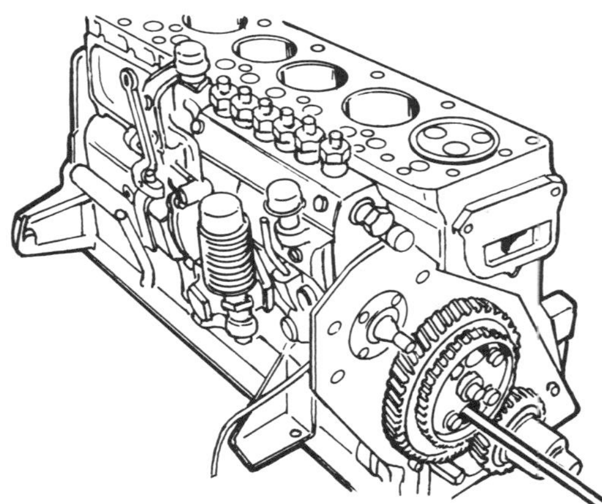

Removing the SD22 camshaft locating bolts

10. Install the oil pump drive spindle by aligning the oil pump drive shaft groove and the camshaft oil pump drive gear with the spindle.

11. Install all other parts in reverse order.

Pistons and Connecting Rods
REMOVAL AND INSTALLATION

See the "Engine Rebuilding" section for general procedures.

1. Remove the cylinder head.

2. Remove the oil pan.

3. Remove any carbon buildup from the cylinder wall at the top end of the piston travel with a ridge reamer tool.

4. Position the piston to be removed at the bottom of its stroke so that the connecting rod bearing cap can be reached easily from under the engine.

5. Unscrew the connecting rod bearing cap and remove the cap and lower half of the bearing.

6. Push the piston and connecting rod up and out of the cylinder block with a length of wood. Use care not to scratch the cylinder wall with the connecting rod or the wooden tool.

7. Keep all of the components from each cylinder together and install them in the cylinder from which they were removed.

8. Coat the bearing face of the connecting rod and the outer face of the pistons with engine oil.

9. Turn the top compression ring to bring its gap to about the 1:30 o'clock position. Set the remaining rings so that their gaps are positioned 180° apart around the piston. The oil ring gap will be directly under the top compression ring gap.

10. Turn the crankshaft until the rod journal of the particular cylinder you are working on is brought to the TDC position.

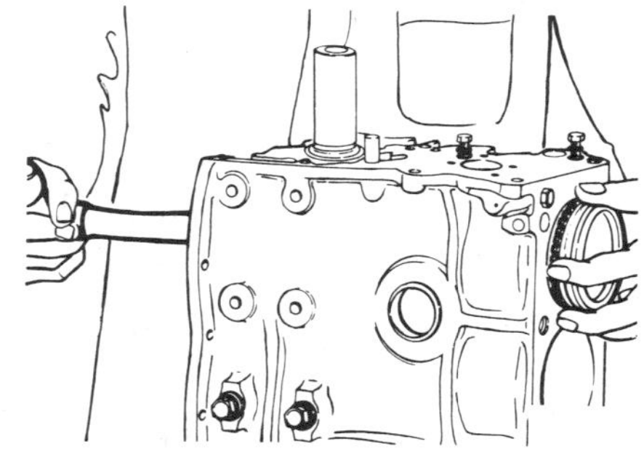

Removing the piston and connecting rod assembly from the cylinder block

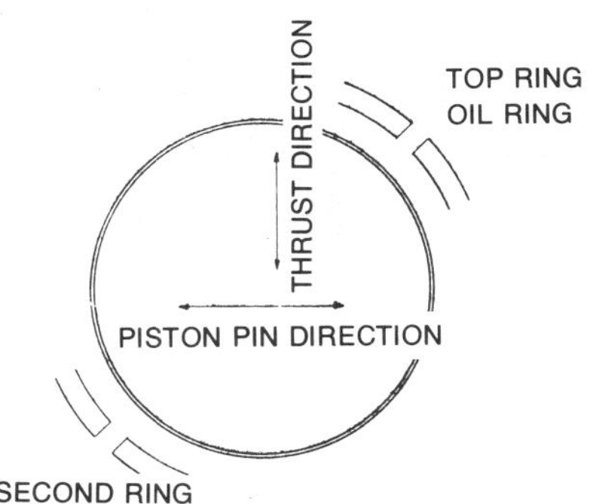

Arrangement of the piston ring gaps around the piston

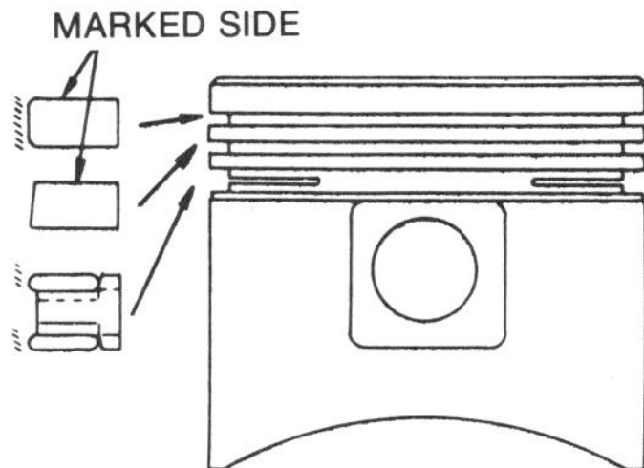

Piston ring installation

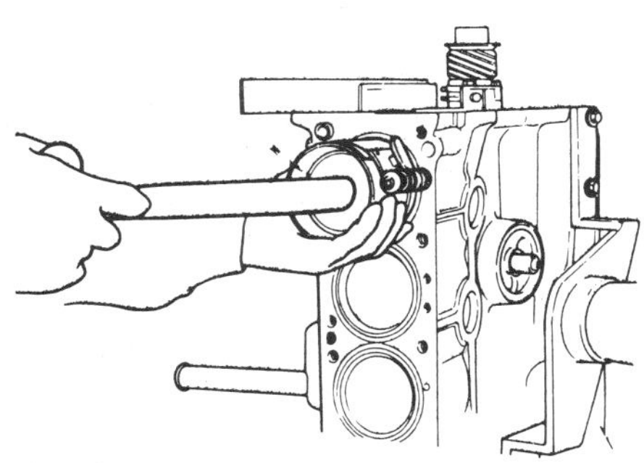

Installing the piston and connecting rod

11. With the piston and rings clamped in a ring compressor, the notched mark on the head of the piston toward the front of the engine, and the oil hole side of the connecting rod toward the right side of the engine, push the piston and connecting rod assembly into the cylinder bore until the big bearing end of the connecting rod contacts and is seated on the rod journal of the crankshaft. Use care not to scratch the cylinder wall with the connecting rod.

12. Push down farther on the piston and turn the crankshaft while the connecting rod rides around on the crankshaft rod journal. Turn the crankshaft until the crankshaft rod journal is at BDC (bottom dead center).

13. Align the mark on the connecting rod bearing cap with that on the connecting rod and tighten the bearing cap bolts to the specified torque.

14. Install all of the piston/connecting rod assemblies in the manner outlined above and assemble the oil pan and cylinder head to the engine in the reverse order of removal.

PISTON AND CONNECTING ROD IDENTIFICATION AND POSITIONING

The pistons are marked with a notch in the piston head. When installed in the engine, the notch markings are to be facing toward the front of the engine.

The connecting rods are installed in the engine with the oil hole facing toward the fuel pump side (right) of the engine.

NOTE: *It is advisable to number the pistons, connecting rods, and bearing caps in some manner so that they can be reinstalled in the same cylinder, facing the same direction from which they are removed.*

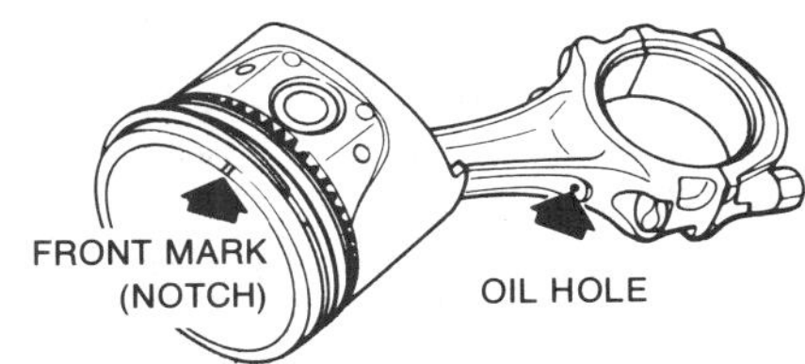

Piston and connecting rod identification and positioning

ENGINE LUBRICATION

Oil Pan

REMOVAL AND INSTALLATION

To remove the oil pan it will be necessary to unbolt the motor mounts and jack the engine to gain clearance. Drain the oil and remove the attaching screws and remove the oil pan and gasket. Install the oil pan in the reverse order with a new gasket. Apply a thin bead of silicone seal to the engine block at the junction of the block and front cover, and the junction of the block and main bearing cap. Then apply a thin coat of silicone seal to the new oil pan gasket, install the gasket to the block, and install the pan. Tighten the pan bolts in a circular pattern from the center to the ends, to 4–7 ft. lbs. Overtightening will distort the pan lip, causing leakage.

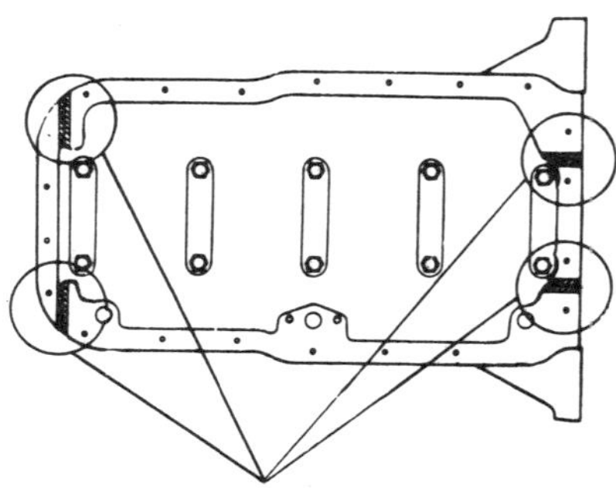

Apply a thin bead of silicone seal to these areas before installing the oil pan gasket on the block

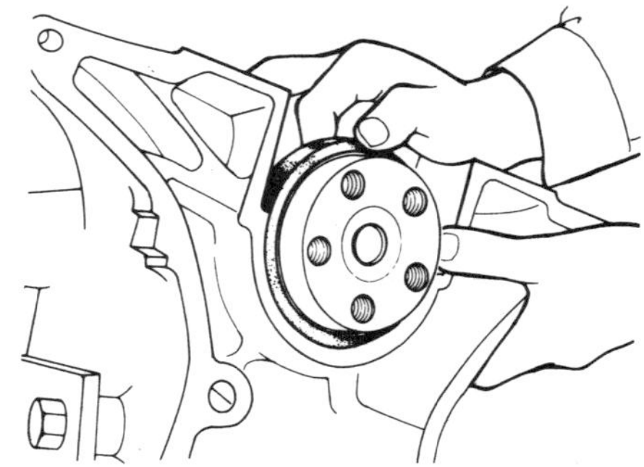

Removing the rear main seal

Rear Main Oil Seal

REPLACEMENT

In order to replace the rear main oil sea, the rear main bearing cap must be removed. Removal of the rear main bearing cap requires the use of a special rear main bearing cap puller. Also, the oil seal is installed with a special crankshaft rear oil seal drift. Unless these or similar tools are available to you, it is recommended that the oil seal be replaced by a Datsun service center.

1. Remove the engine and transmission assembly from the vehicle.
2. Remove the transmission from the engine.
3. Remove the clutch from the flywheel.
4. Remove the flywheel from the crankshaft.
5. Remove the rear main bearing cap together with the bearing cap side seals.
6. Remove the rear main oil seal from around the crankshaft.
7. Apply oil to the sealing lip of the oil seal and install the seal around the crankshaft using a suitable tool.
8. Apply sealer to the rear main bearing cap as indicated and install the rear main

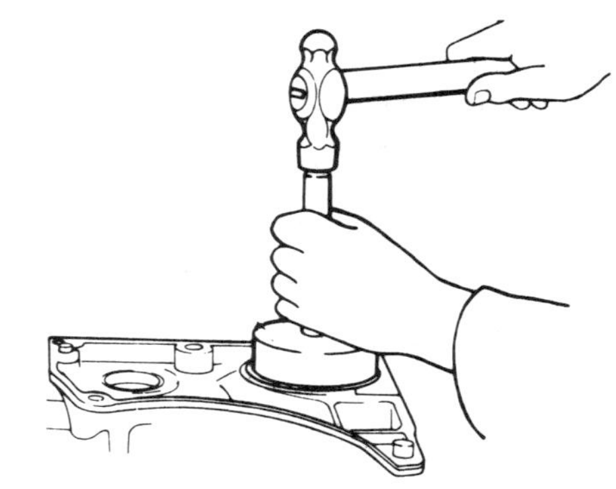

Installing the rear main seal

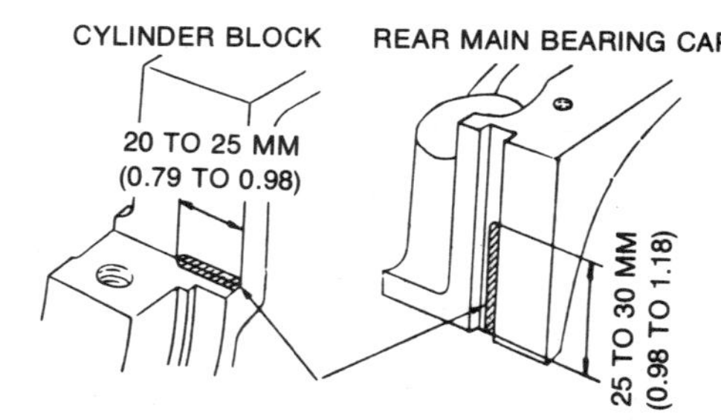

Application of sealer to the rear main bearing cap

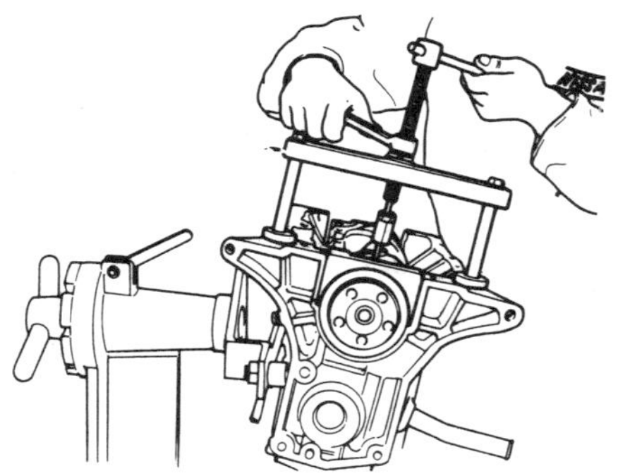

Removing the rear main bearing cap with a puller

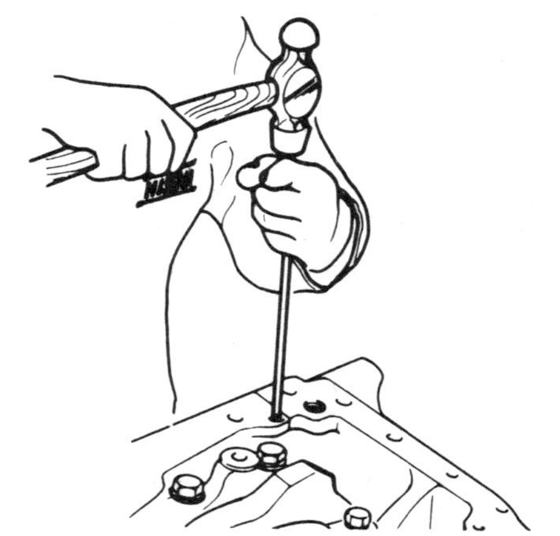

Installing the rear main bearing cap side seals

bearing cap and tighten the cap bolts to 33–40 ft. lbs.

9. Apply sealant to the rear main bearing cap side seals and install the side seals, driving the seals into place with a suitable drift.

10. Assemble the engine and install it in the vehicle in the reverse order of removal.

Oil Pump

REMOVAL AND INSTALLATION

Gasoline Engines

The oil pump is mounted externally on the engine, eliminating the need to remove the oil pan in order to remove the oil pump.

1. Remove the distributor.
2. Drain the engine oil.
3. Remove the front stabilizer.
4. Remove the splash shield board.
5. Remove the oil pump body with the drive spindle assembly.
6. Before installing the oil pump in the engine, turn the crankshaft so that the No. 1 piston is at TDC of the compression stroke.
7. Fill the pump housing with engine oil, then align the punch mark on the spindle with the hole in the oil pump.

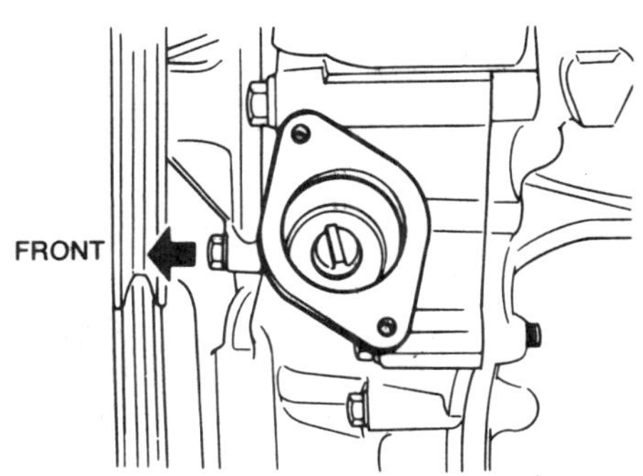

The projection on top of the oil pump drive spindle located in the 11:25 o'clock position. The smaller crescent formed by the notch faces forward.

8. With a new gasket placed over the drive spindle, install the oil pump and drive spindle assembly so that the projection on the top of the drive spindle is located in the 11:25 o'clock position.

9. Install the distributor with the metal tip of the rotor pointing toward the No. 1 spark plug tower of the distributor cap.

10. Assemble the remaining components in the reverse order of removal.

SD22 Diesel

1. Remove the oil pump drive spindle.
2. Remove the oil pan.
3. Unbolt and remove the oil pump. Discard the gasket.
4. Using a new gasket, install the oil pump. Torque the bolts to 7–9 ft. lb.
5. Install the drive spindle by aligning it with the oil pump drive shaft groove in the cylinder block and the camshaft oil pump drive gear.
6. Place a new O-ring on the spindle support and bolt it to the block.

Removing the gasoline engine oil pump

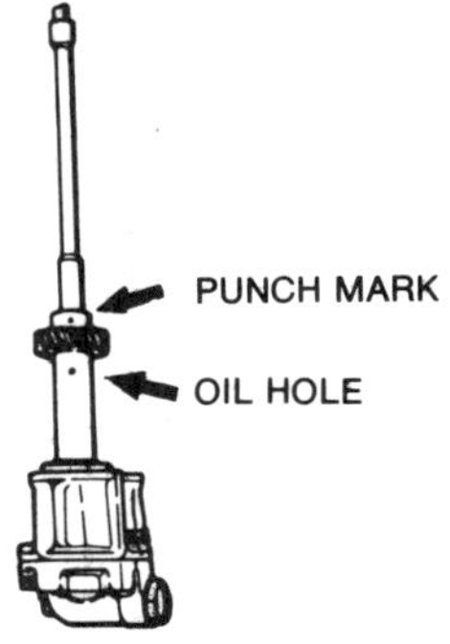

Aligning the punch mark on the spindle with the hole in the oil pump on gasoline engines

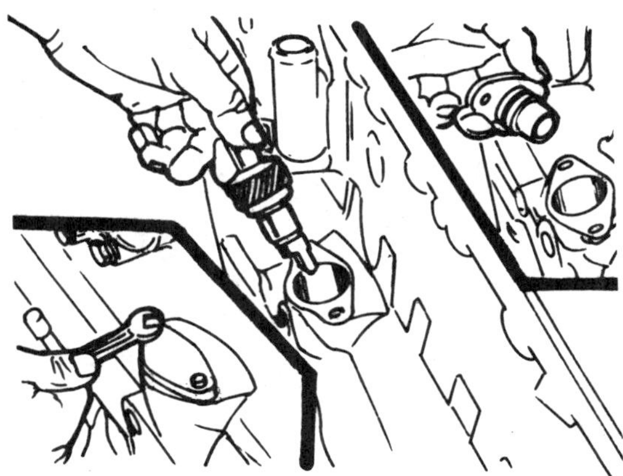

Oil pump drive spindle removal, diesel engine

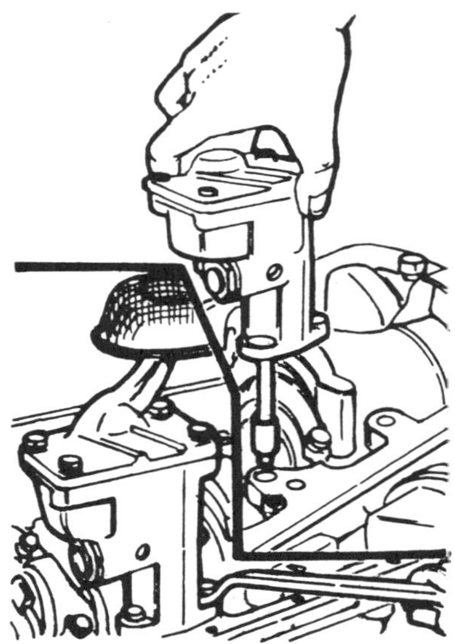

Removing the diesel engine oil pump

Oil Filter Canister Assembly
REMOVAL AND INSTALLATION
SD22 Diesel

1. Remove the bolts at the oil filter end of the oil inlet and outlet lines.

2. Remove the four filter assembly mounting bolts and separate the filter from the block.

NOTE: *Have a drip pan ready, since some oil will drain out.*

3. Installation is the reverse of removal. Torque the bolts to 14–18 ft. lb.

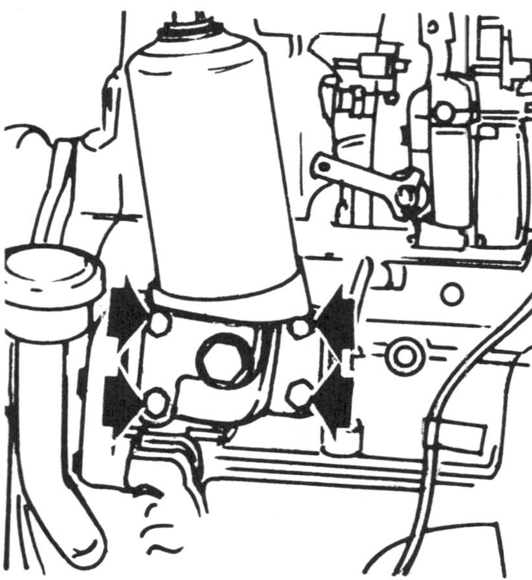

Oil filter showing the four mounting bolts

Oil Cooler
REMOVAL AND INSTALLATION
SD22 Diesel

1. Remove the water hose from the cooler.

2. Remove the eight mounting bolts and lift the cooler from the block.

NOTE: *Have a drip pan ready, since some oil will drain out.*

3. Installation is the reverse of removal. Torque bolts to 14–18 ft. lb.

Oil cooler removal

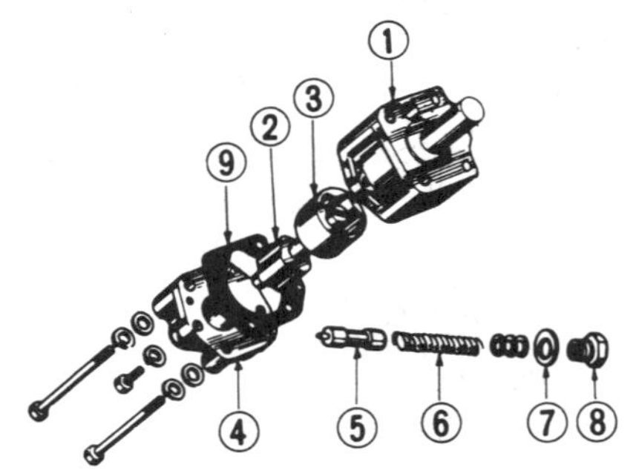

1. Oil pump body	6. Regulator spring
2. Inner rotor and shaft	7. Washer
3. Outer rotor	8. Regulator cap
4. Oil pump cover	9. Cover gasket
5. Regulator valve	

Exploded view of the gasoline engine oil pump

ENGINE COOLING

Coolant Changes
DRAINING, FLUSHING AND REFILLING

Once every 24 months or 24,000 miles, the cooling system should be drained, thor-

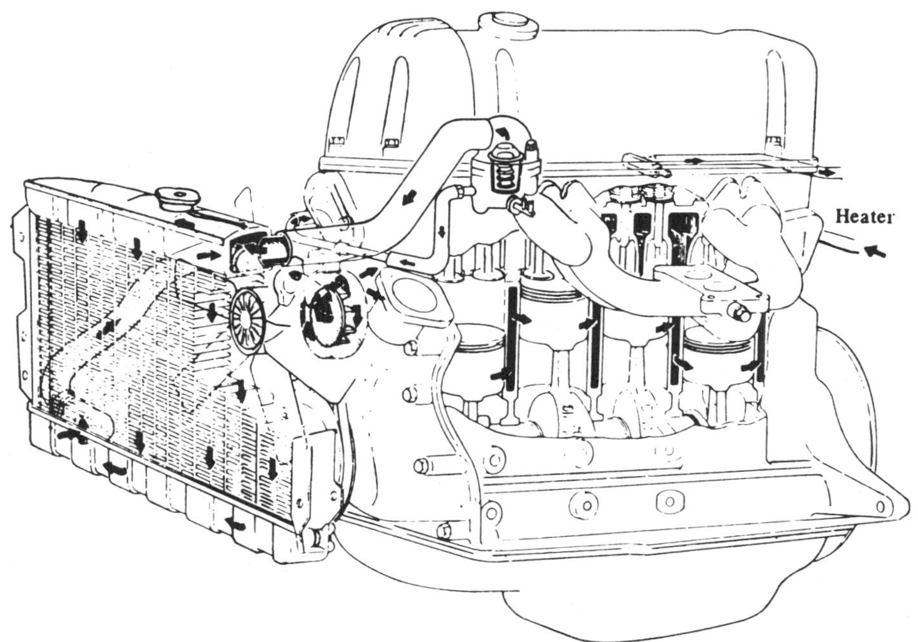

Cutaway of the gasoline engine cooling system

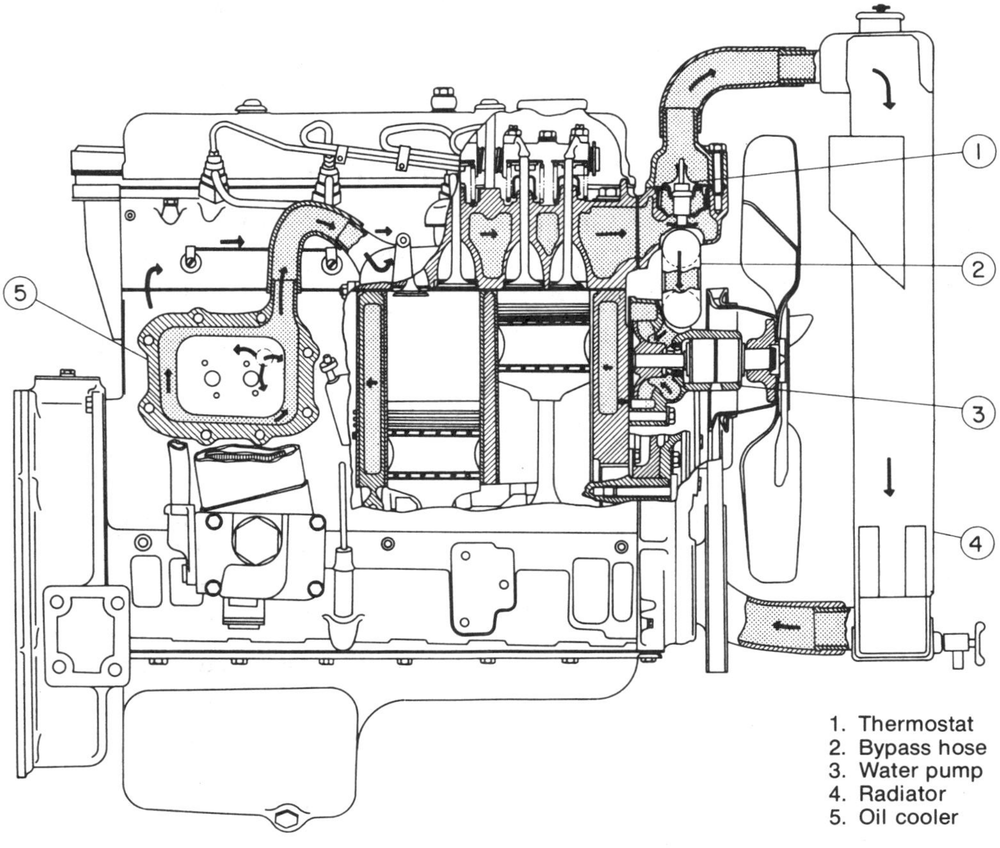

Cutaway of the SD22 diesel engine cooling system

oughly flushed, and refilled. This should be done with the engine cold.

1. Remove the radiator cap.

2. There are two drain plugs in the cooling system; one at the bottom of the radiator and one at the rear of the driver's side of the engine. Both should be loosened to allow the coolant to drain.

3. Turn on the heater inside the truck to its hottest position. This ensures that the heater core is flushed out completely. Flush out the system thoroughly by refilling it with clean water through the radiator opening as it escapes from the two drain cocks. Continue until the water running out is clear. Be sure to clean out the coolant recovery tank as well if your truck has one.

4. If the system is badly contaminated with rust or scale, you can use a commercial flushing solution to clear it out. Follow the manufacturer's instructions. Some causes of rust are air in the system, caused by a leaky radiator cap or an insufficiently filled or leaking system, failure to change the coolant regularly, use of excessively hard or soft water, and failure to use a proper mix of antifreeze and water.

5. When the system is clear, allow all the water to drain, then close the drain plugs. Fill the system through the radiator with a 50/50 mix of ethylene glycol type antifreeze and water.

6. Start the engine and top off the radiator with water. If your truck has a coolant recovery tank, fill it half full with the coolant mix.

7. Replace the radiator and coolant tank caps, and check for leaks. When the engine has reached normal operating temperature, shut it off, allow it to cool, then top off the radiator or coolant tank as necessary.

Radiator

REMOVAL AND INSTALLATION

1. Drain the engine coolant into a clean container.

2. Remove the front grille.

3. Disconnect the upper and lower radiator hoses. On a truck with an automatic transmission, disconnect the fluid cooler inlet and outlet lines from the radiator. Plug the lines to prevent the loss of transmission fluid and the entrance of dirt. Remove the fan shroud, if equipped.

4. Remove the bolts retaining the radiator from the radiator side supports and remove the radiator upward.

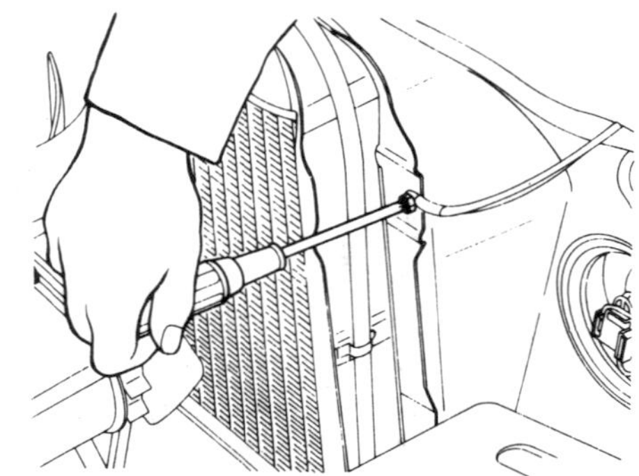

Removing the radiator securing bolts

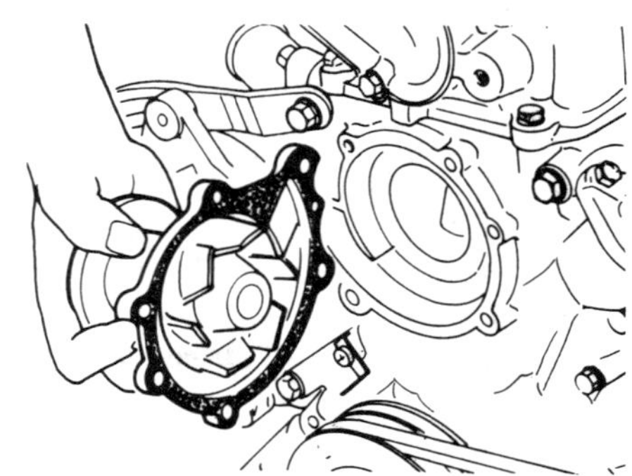

Removing the water pump

5. Install the radiator in the reverse order of removal.

Water Pump

REMOVAL AND INSTALLATION

1. Drain the engine coolant into a clean container. On the SD22, remove the bypass hose from the pump.

2. Loosen the four bolts retaining the fan shroud to the radiator and remove the shroud.

3. Loosen the belt, then remove the fan and pulley from the water pump hub.

4. Remove the bolts (3 on the SD22 and 5 on the L20B and Z22) retaining the pump and remove the pump together with the gasket from the front cover.

5. Remove all traces of gasket material and install the water pump in the reverse order with a new gasket and sealer. Tighten the bolts uniformly.

Thermostat

The factory-installed thermostat opening temperature is 180°F for trucks sold in the U.S., 190°F for trucks sold in Canada.

REMOVAL AND INSTALLATION

1. Drain the engine coolant into a clean container so that the level is below the thermostat housing.

2. Disconnect the upper radiator hose at the water outlet.

3. Loosen the two securing nuts and remove the water outlet, gasket, and the thermostat from the thermostat housing.

4. Install the thermostat in the reverse order of removal, using a new gasket with sealer and with the thermostat spring toward the inside of the engine.

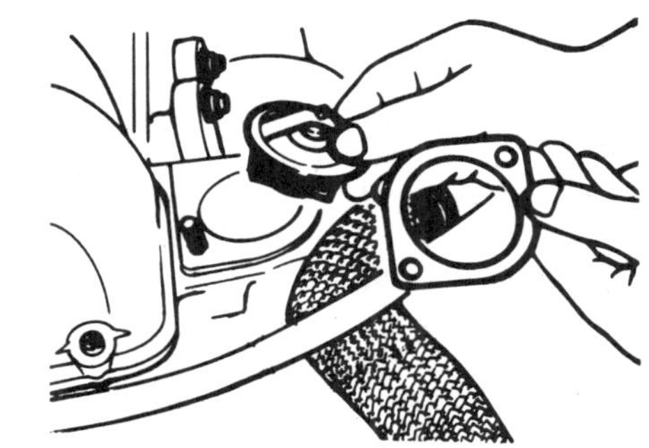

Removing the thermostat

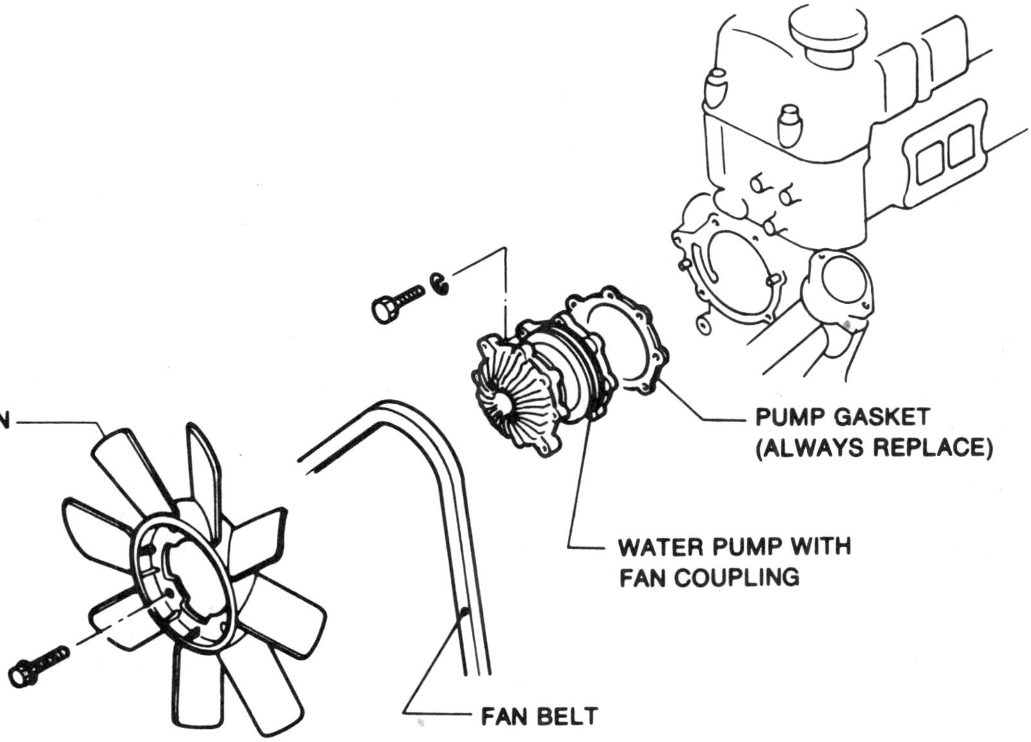

Water pump and fan installation details

ENGINE REBUILDING

Most procedures involved in rebuilding an engine are fairly standard, regardless of the type of engine involved. This section is a guide to accepted rebuilding procedures. Examples of standard rebuilding practices are illustrated and should be used along with specific details concerning your particular engine, found earlier in this chapter.

The procedures given here are those used by any competent rebuilder. Obviously some of the procedures cannot be performed by the do-it-yourself mechanic, but are provided so that you will be familiar with the services that should be offered by rebuilding or machine shops. As an example, in most instances, it is more profitable for the home mechanic to remove the cylinder heads, buy the necessary parts (new valves, seals, keepers, keys, etc.) and deliver these to a machine shop for the necessary work. In this way you will save the money to remove and install the cylinder head and the mark-up on parts.

On the other hand, most of the work involved in rebuilding the lower end is well within the scope of the do-it-yourself mechanic. Only work such as hot-tanking, actually boring the block or Magnafluxing (invisible crack detection) need be sent to a machine shop.

Tools

The tools required for basic engine rebuilding should, with a few exceptions, be those included in a mechanic's tool kit. An accurate torque wrench, and a dial indicator (reading in thousandths) mounted on a universal base should be available. Special tools, where required, are available from the major tool suppliers. The services of a competent automotive machine shop must also be readily available.

Precautions

Aluminum has become increasingly popular for use in engines, due to its low weight and excellent heat transfer characteristics. The following precautions must be observed when handling aluminum (or any other) engine parts:

—Never hot-tank aluminum parts.
—Remove all aluminum parts (identification tags, etc.) from engine parts before hot-tanking (otherwise they will be removed during the process).

—Always coat threads lightly with engine oil or anti-seize compounds before installation, to prevent seizure.
—Never over-torque bolts or spark plugs in aluminum threads. Should stripping occur, threads can be restored using any of a number of thread repair kits available (see next section).

Inspection Techniques

Magnaflux and Zyglo are inspection techniques used to locate material flaws, such as stress cracks. Magnaflux is a magnetic process, applicable only to ferrous materials. The Zyglo process coats the matrial with a fluorescent dye penetrant, and any material may be tested using Zyglo. Specific checks of suspected surface cracks may be made at lower cost and more readily using spot check dye. The dye is sprayed onto the suspected area, wiped off, and the area is then sprayed with a developer. Cracks then will show up brightly.

Overhaul

The section is divided into two parts. The first, Cylinder Head Reconditioning, assumes that the cylinder head is removed from the engine, all manifolds are removed, and the cylinder head is on a workbench. The camshaft should be removed from overhead cam cylinder heads. The second section, Cylinder Block Reconditioning, covers the block, pistons, connecting rods and crankshaft. It is assumed that the engine is mounted on a work stand, and the cylinder head and all accessories are removed.

Procedures are identified as follows:
Unmarked—Basic procedures that must be performed in order to successfully complete the rebuilding process.
Starred (*)—Procedures that should be performed to ensure maximum performance and engine life.
Double starred (**)—Procedures that may be performed to increase engine performance and reliability.

When assembling the engine, any parts that will be in frictional contact must be pre-lubricated, to provide protection on initial start-up. Any product specifically formulated for this purpose may be used. NOTE: *Do not use engine oil. Where semi-permanent* (locked but removable) installation of bolts or nuts is desired, threads should be cleaned and located with Loctite ® or a similar product (non-hardening).

Repairing Damaged Threads

Several methods of repairing damaged threads are available. Heli-Coil® (shown here), Keenserts® and Microdot® are among the most widely used. All involve basically the same principle—drilling out stripped threads, tapping the hole and installing a pre-wound insert—making welding, plugging and oversize fasteners unnecessary.

Two types of thread repair inserts are usually supplied—a standard type for most Inch Coarse, Inch Fine, Metric Coarse and Metric Fine thread sizes and a spark plug type to fit most spark plug port sizes. Consult the individual manufacturer's catalog to determine exact applications. Typical thread repair kits will contain a selection of pre-wound threaded inserts, a tap (corresponding to the outside diameter threads of the insert) and an installation tool. Spark plug inserts usually differ because they require a tap equipped with pilot threads and a combined reamer/tap section. Most manufacturers also supply blister-packed thread repair inserts separately in addition to a master kit containing a variety of taps and inserts plus installation tools.

Before effecting a repair to a threaded hole, remove any snapped, broken or damaged bolts or studs. Penetrating oil can be used to free frozen threads; the offending item can be removed with locking pliers or with a screw or stud extractor. After the hole is clear, the thread can be repaired, as follows:

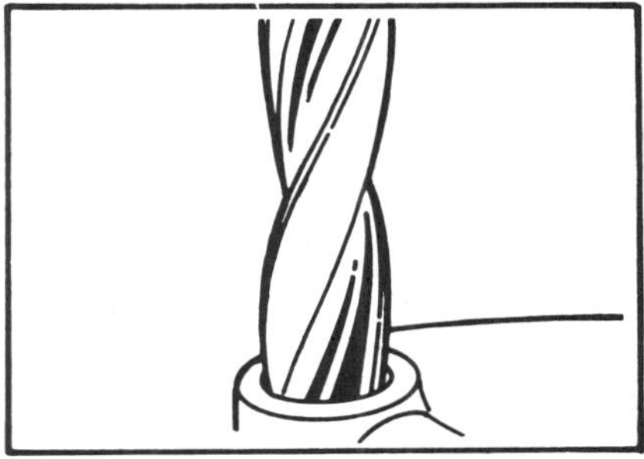

Drill out the damaged threads with specified drill. Drill completely through the hole or to the bottom of a blind hole

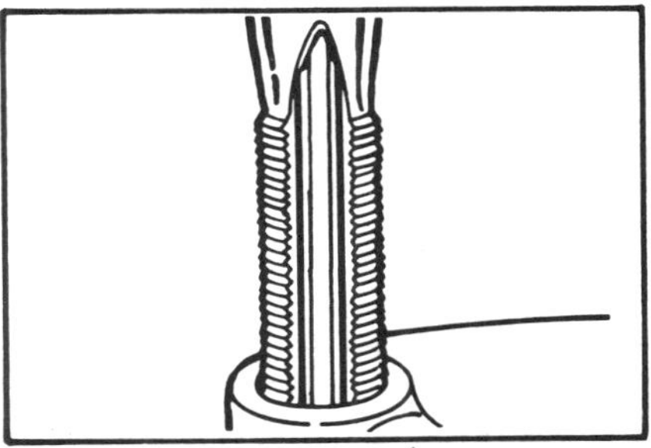

With the tap supplied, tap the hole to receive the thread insert. Keep the tap well oiled and back it out frequently to avoid clogging the threads

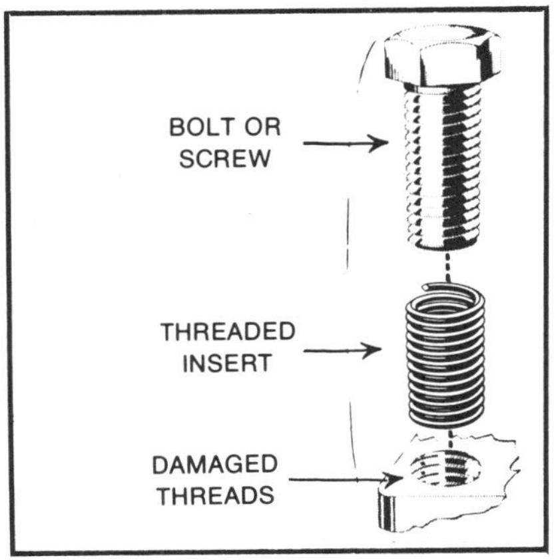

Damaged bolt holes can be repaired with thread repair inserts

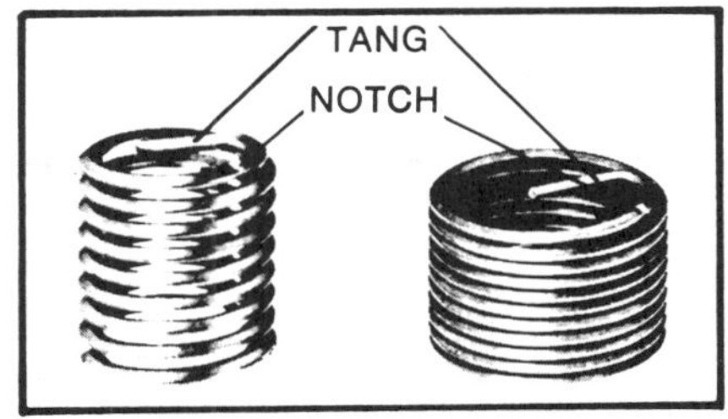

Standard thread repair insert (left) and spark plug thread insert (right)

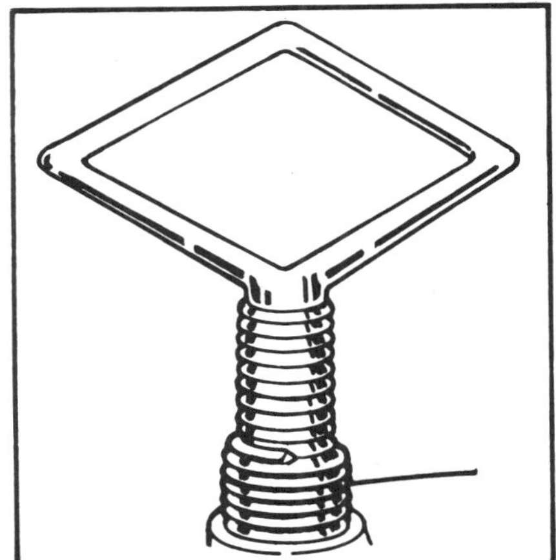

Screw the threaded insert onto the installation tool until the tang engages the slot. Screw the insert into the tapped hole until it is ¼–½ turn below the top surface. After installation break off the tang with a hammer and punch

Standard Torque Specifications and Fastener Markings

The Newton-metre has been designated the world standard for measuring torque and will gradually replace the foot-pound and kilogram-meter. In the absence of specific torques, the following chart can be used as a guide to the maximum safe torque of a particular size/grade of fastener.

- There is no torque difference for fine or coarse threads.
- Torque values are based on clean, dry threads. Reduce the value by 10% if threads are oiled prior to assembly.
- The torque required for aluminum components or fasteners is considerably less.

U. S. BOLTS

SAE Grade Number	1 or 2			5			6 or 7		
Bolt Markings Manufacturer's marks may vary—number of lines always 2 less than the grade number.									
Usage	*Frequent*			*Frequent*			*Infrequent*		
Bolt Size (inches)—(Thread)	*Maximum Torque*			*Maximum Torque*			*Maximum Torque*		
	Ft-Lb	*kgm*	*Nm*	*Ft-Lb*	*kgm*	*Nm*	*Ft-Lb*	*kgm*	*Nm*
¼—20	5	0.7	6.8	8	1.1	10.8	10	1.4	13.5
—28	6	0.8	8.1	10	1.4	13.6			
⁵⁄₁₆—18	11	1.5	14.9	17	2.3	23.0	19	2.6	25.8
—24	13	1.8	17.6	19	2.6	25.7			
⅜—16	18	2.5	24.4	31	4.3	42.0	34	4.7	46.0
—24	20	2.75	27.1	35	4.8	47.5			
⁷⁄₁₆—14	28	3.8	37.0	49	6.8	66.4	55	7.6	74.5
—20	30	4.2	40.7	55	7.6	74.5			
½—13	39	5.4	52.8	75	10.4	101.7	85	11.75	115.2
—20	41	5.7	55.6	85	11.7	115.2			
⁹⁄₁₆—12	51	7.0	69.2	110	15.2	149.1	120	16.6	162.7
—18	55	7.6	74.5	120	16.6	162.7			
⅝—11	83	11.5	112.5	150	20.7	203.3	167	23.0	226.5
—18	95	13.1	128.8	170	23.5	230.5			
¾—10	105	14.5	142.3	270	37.3	366.0	280	38.7	379.6
—16	115	15.9	155.9	295	40.8	400.0			
⅞— 9	160	22.1	216.9	395	54.6	535.5	440	60.9	596.5
—14	175	24.2	237.2	435	60.1	589.7			
1— 8	236	32.5	318.6	590	81.6	799.9	660	91.3	894.8
—14	250	34.6	338.9	660	91.3	849.8			

METRIC BOLTS

NOTE: *Metric bolts are marked with a number indicating the relative strength of the bolt. These numbers have nothing to do with size.*

Description	Torque ft-lbs (Nm)			
Thread size x pitch (mm)	Head mark—4		Head mark—7	
6 x 1.0	2.2–2.9	(3.0–3.9)	3.6–5.8	(4.9–7.8)
8 x 1.25	5.8–8.7	(7.9–12)	9.4–14	(13–19)
10 x 1.25	12–17	(16–23)	20–29	(27–39)
12 x 1.25	21–32	(29–43)	35–53	(47–72)
14 x 1.5	35–52	(48–70)	57–85	(77–110)
16 x 1.5	51–77	(67–100)	90–120	(130–160)
18 x 1.5	74–110	(100–150)	130–170	(180–230)
20 x 1.5	110–140	(150–190)	190–240	(160–320)
22 x 1.5	150–190	(200–260)	250–320	(340–430)
24 x 1.5	190–240	(260–320)	310–410	(420–550)

NOTE: *This engine rebuilding section is a guide to accepted rebuilding procedures. Typical examples of standard rebuilding procedures are illustrated. Use these procedures along with the detailed instructions earlier in this chapter, concerning your particular engine.*

Cylinder Head Reconditioning

Procedure	Method
Remove the cylinder head:	See the engine service procedures earlier in this chapter for details concerning specific engines.
Identify the valves:	Invert the cylinder head, and number the valve faces front to rear, using a permanent felt-tip marker.
Remove the rocker arms (OHV engines only):	Remove the rocker arms with shaft(s) or balls and nuts. Wire the sets of rockers, balls and nuts together, and identify according to the corresponding valve.
Remove the camshaft (OHC engines only):	See the engine service procedures earlier in this chapter for details concerning specific engines.
Remove the valves and springs:	Using an appropriate valve spring compressor (depending on the configuration of the cylinder head), compress the valve springs. Lift out the keepers with needlenose pliers, release the compressor, and remove the valve, spring, and spring retainer. See the engine service procedures earlier in this chapter for details concerning specific engines.

Cylinder Head Reconditioning

Procedure	Method

Check the valve stem-to-guide clearance:

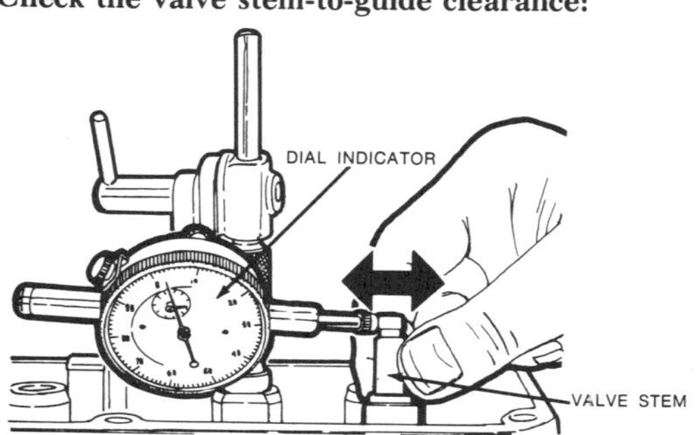

Check the valve stem-to-guide clearance

Clean the valve stem with lacquer thinner or a similar solvent to remove all gum and varnish. Clean the valve guides using solvent and an expanding wire-type valve guide cleaner. Mount a dial indicator so that the stem is at 90° to the valve stem, as close to the valve guide as possible. Move the valve off its seat, and measure the valve guide-to-stem clearance by rocking the stem back and forth to actuate the dial indicator. Measure the valve stems using a micrometer, and compare to specifications, to determine whether stem or guide wear is responsible for excessive clearance.
NOTE: *Consult the Specifications tables earlier in this chapter.*

De-carbon the cylinder head and valves:

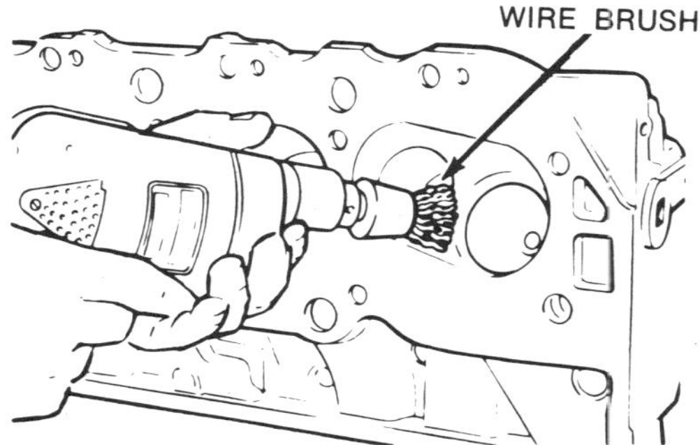

Remove the carbon from the cylinder head with a wire brush and electric drill

Chip carbon away from the valve heads, combustion chambers, and ports, using a chisel made of hardwood. Remove the remaining deposits with a stiff wire brush.
NOTE: *Be sure that the deposits are actually removed, rather than burnished.*

Hot-tank the cylinder head (cast iron heads only):
CAUTION: *Do not hot-tank aluminum parts.*

Have the cylinder head hot-tanked to remove grease, corrosion, and scale from the water passages.
NOTE: *In the case of overhead cam cylinder heads, consult the operator to determine whether the camshaft bearings will be damaged by the caustic solution.*

Degrease the remaining cylinder head parts:

Clean the remaining cylinder head parts in an engine cleaning solvent. Do not remove the protective coating from the springs.

Check the cylinder head for warpage:

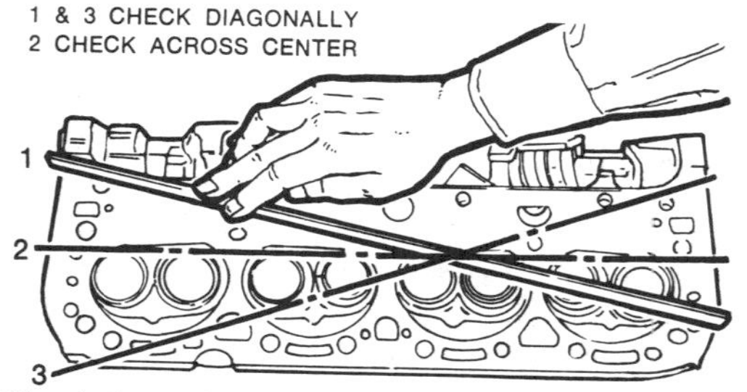

Check the cylinder head for warpage

Place a straight-edge across the gasket surface of the cylinder head. Using feeler gauges, determine the clearance at the center of the straight-edge. If warpage exceeds .003″ in a 6″ span, or .006″ over the total length, the cylinder head must be resurfaced.
NOTE: *If warpage exceeds the manufacturer's maximum tolerance for material removal, the cylinder head must be replaced.* When milling the cylinder heads of V-type engines, the intake manifold mounting position is altered, and must be corrected by milling the manifold flange a proportionate amount.

Cylinder Head Reconditioning

Procedure	Method

***Knurl the valve guides:**

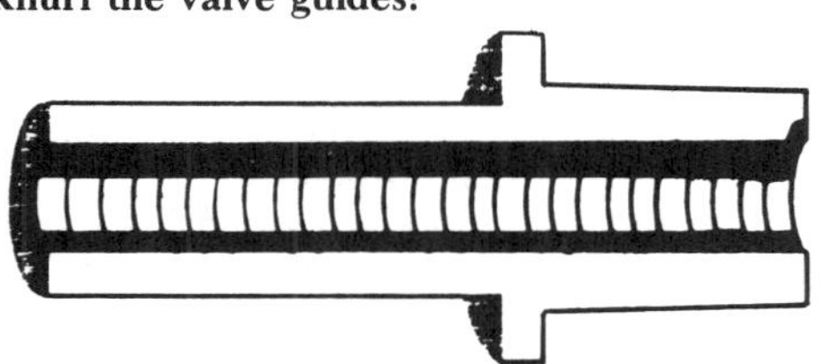

Cut-away view of a knurled valve guide

*Valve guides which are not excessively worn or distorted may, in some cases, be knurled rather than replaced. Knurling is a process in which metal is displaced and raised, thereby reducing clearance. Knurling also provides excellent oil control. The possibility of knurling rather than replacing valve guides should be discussed with a machinist.

Replace the valve guides:
NOTE: *Valve guides should only be replaced if damaged or if an oversize valve stem is not available.*

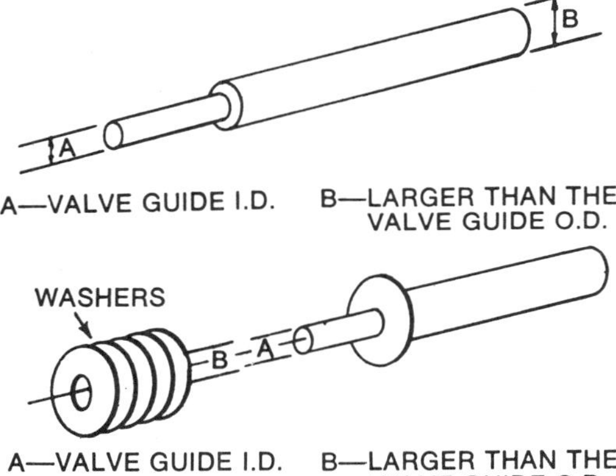

Valve guide installation tool using washers for installation

See the engine service procedures earlier in this chapter for details concerning specific engines. Depending on the type of cylinder head, valve guides may be pressed, hammered, or shrunk in. In cases where the guides are shrunk into the head, replacement should be left to an equipped machine shop. In other cases, the guides are replaced using a stepped drift (see illustration). Determine the height above the boss that the guide must extend, and obtain a stack of washers, their I.D. similar to the guide's O.D., of that height. Place the stack of washers on the guide, and insert the guide into the boss.
NOTE: *Valve guides are often tapered or beveled for installation.* Using the stepped installation tool (see illustration), press or tap the guides into position. Ream the guides according to the size of the valve stem.

Replace valve seat inserts:

Replacement of valve seat inserts which are worn beyond resurfacing or broken, if feasible, must be done by a machine shop.

Resurface (grind) the valve face:

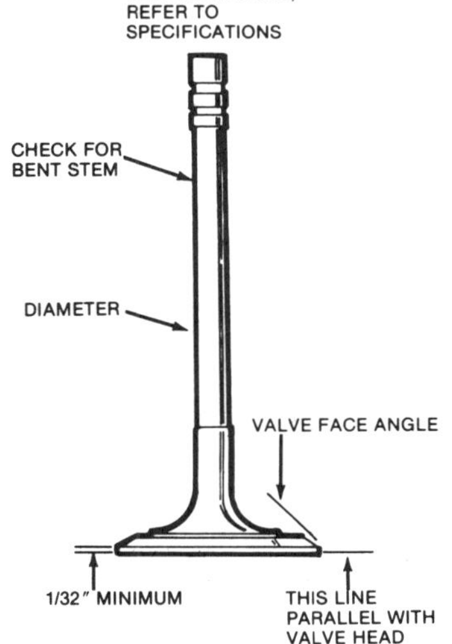

Critical valve dimensions

Using a valve grinder, resurface the valves according to specifications given earlier in this chapter.
CAUTION: *Valve face angle is not always identical to valve seat angle.* A minimum margin of

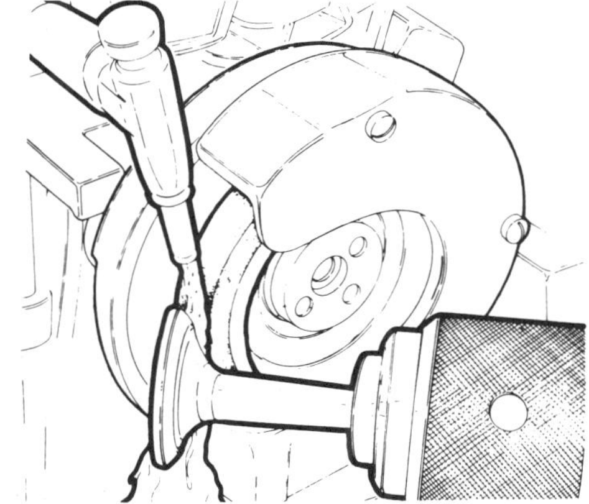

Valve grinding by machine

Cylinder Head Reconditioning

Procedure	Method

$^1/_{32}$" should remain after grinding the valve. The valve stem top should also be squared and resurfaced, by placing the stem in the V-block of the grinder, and turning it while pressing lightly against the grinding wheel.

NOTE: *Do not grind sodium filled exhaust valves on a machine. These should be hand lapped.*

Resurface the valve seats using reamers or grinder:

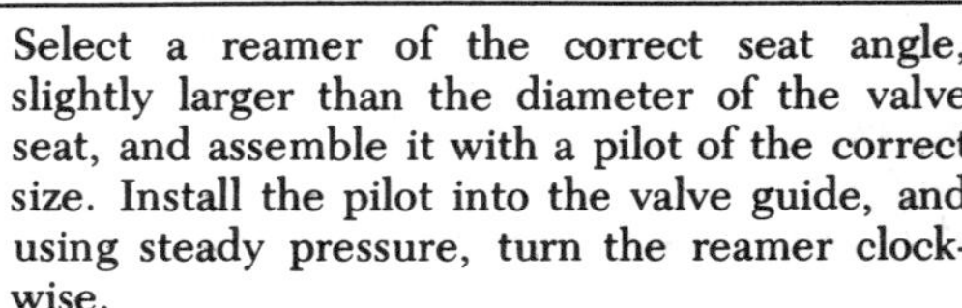
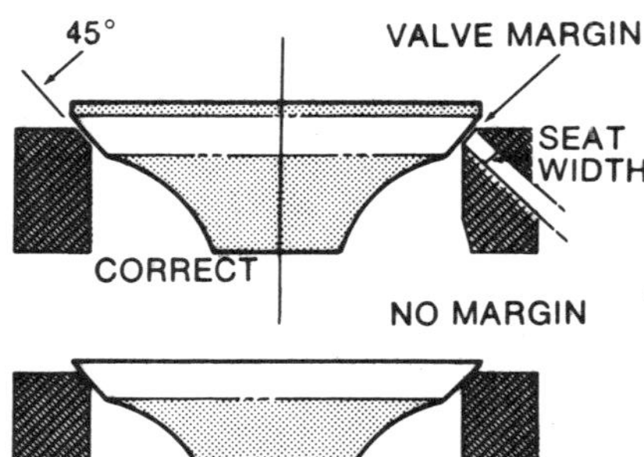

Valve seat width and centering

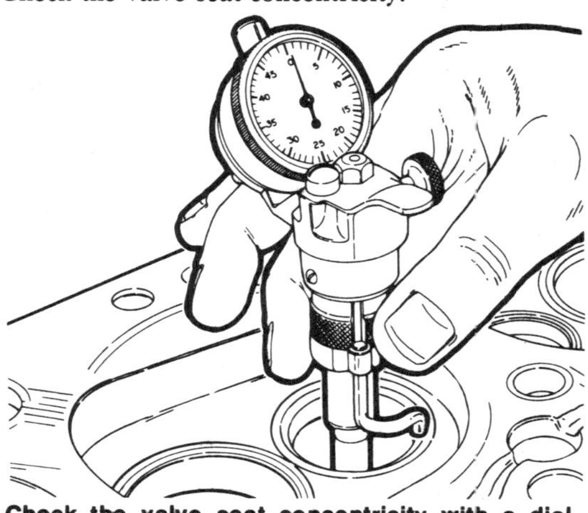

Reaming the valve seat with a hand reamer

Select a reamer of the correct seat angle, slightly larger than the diameter of the valve seat, and assemble it with a pilot of the correct size. Install the pilot into the valve guide, and using steady pressure, turn the reamer clockwise.

CAUTION: *Do not turn the reamer counterclockwise.* Remove only as much material as necessary to clean the seat. Check the concentricity of the seat (following). If the dye method is not used, coat the valve face with Prussian blue dye, install and rotate it on the valve seat. Using the dye marked area as a centering guide, center and narrow the valve seat to specifications with correction cutters.

NOTE: *When no specifications are available, minimum seat width for exhaust valves should be $^5/_{64}$", intake valves $^1/_{16}$".*

After making correction cuts, check the position of the valve seat on the valve face using Prussian blue dye.

To resurface the seat with a power grinder, select a pilot of the correct size and coarse stone of the proper angle. Lubricate the pilot and move the stone on and off the valve seat at 2 cycles per second, until all flaws are gone. Finish the seat with a fine stone. If necessary the seat can be corrected or narrowed using correction stones.

Check the valve seat concentricity:

Coat the valve face with Prussian blue dye, install the valve, and rotate it on the valve seat. If the entire seat becomes coated, and the valve is known to be concentric, the seat is concentric.

---

* Install the dial gauge pilot into the guide, and rest of the arm on the valve seat. Zero the gauge, and rotate the arm around the seat. Run-out should not exceed .002".

Check the valve seat concentricity with a dial gauge

Cylinder Head Reconditioning

Procedure	Method
***Lap the valves:** NOTE: *Valve lapping is done to ensure efficient sealing of resurfaced valves and seats.*	*Invert the cylinder head, lightly lubricate the valve stems, and install the valves in the head as numbered. Coat valve seats with fine grinding compound, and attach the lapping tool suction cup to a valve head. NOTE: *Moisten the suction cup.* Rotate the tool between the palms, changing position and lifting the tool often to prevent grooving. Lap the valve until a smooth, polished seat is evident. Remove the valve and tool, and rinse away all traces of grinding compound.

Lapping the valves by hand

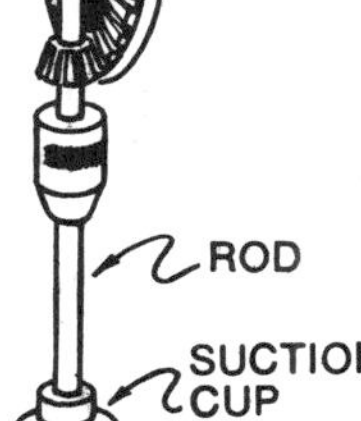

Home-made valve lapping tool

**Fasten a suction cup to a piece of drill rod, and mount the rod in a hand drill. Proceed as above, using the hand drill as a lapping tool.

CAUTION: *Due to the higher speeds involved when using the hand drill, care must be exercised to avoid grooving the seat.* Lift the tool and change direction of rotation often.

Check the valve springs:

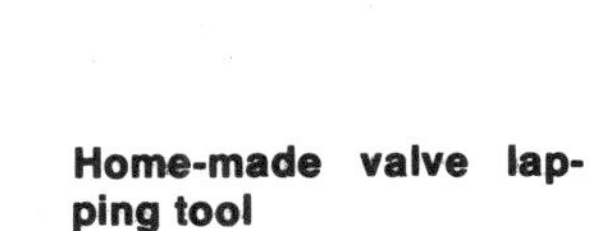

Check the valve spring free length and squareness

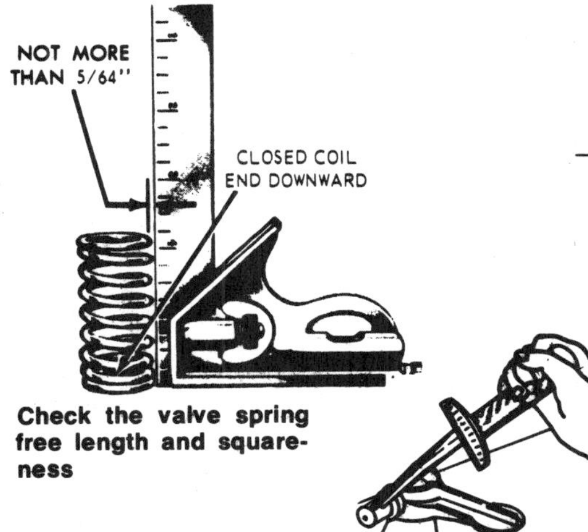

Check the valve spring test pressure

Place the spring on a flat surface next to a square. Measure the height of the spring, and rotate it against the edge of the square to measure distortion. If spring height varies (by comparison) by more than $1/16''$ or if distortion exceeds $1/16''$, replace the spring.

**In addition to evaluating the spring as above, test the spring pressure at the installed and compressed (installed height minus valve lift) height using a valve spring tester. Springs used on small displacement engines (up to 3 liters) should be ∓ 1 lb of all other springs in either position. A tolerance of ∓ 5 lbs is permissible on larger engines.

Cylinder Head Reconditioning

Procedure	Method

***Install valve stem seals:**

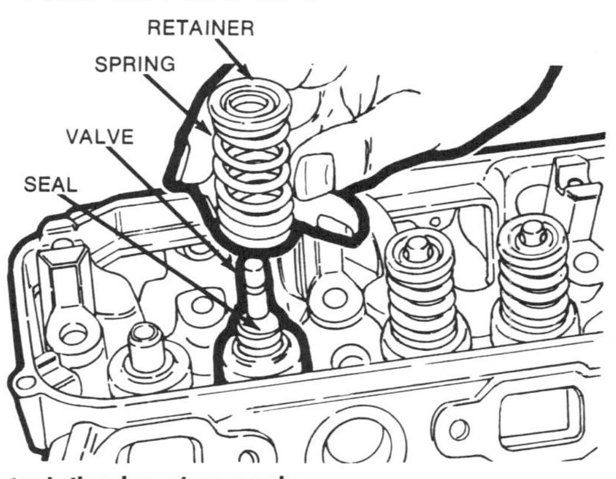

Install valve stem seals

*Due to the pressure differential that exists at the ends of the intake valve guides (atmospheric pressure above, manifold vacuum below), oil is drawn through the valve guides into the intake port. This has been alleviated somewhat since the addition of positive crankcase ventilation, which lowers the pressure above the guides. Several types of valve stem seals are available to rocker arms and balls, and install them on the the stem and guide boss, while others require that the boss be machined. Recently, Teflon guide seals have become popular. Consult a parts supplier or machinist concerning availability and suggested usages.

NOTE: *When installing seals, ensure that a small amount of oil is able to pass the seal to lubricate the valve guides; otherwise, excessive wear may result.*

Install the valves:

See the engine service procedures earlier in this chapter for details concerning specific engines.

Lubricate the valve stems, and install the valves in the cylinder head as numbered. Lubricate and position the seals (if used) and the valve springs. Install the spring retainers, compress the springs, and insert the keys using needle-nose pliers or a tool designed for this purpose.

NOTE: *Retain the keys with wheel bearing grease during installation.*

Check valve spring installed height:

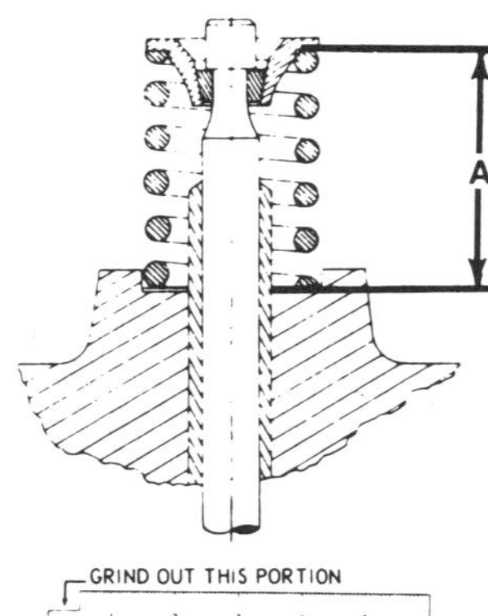

Measure the valve spring installed height (A) with a modified steel rule

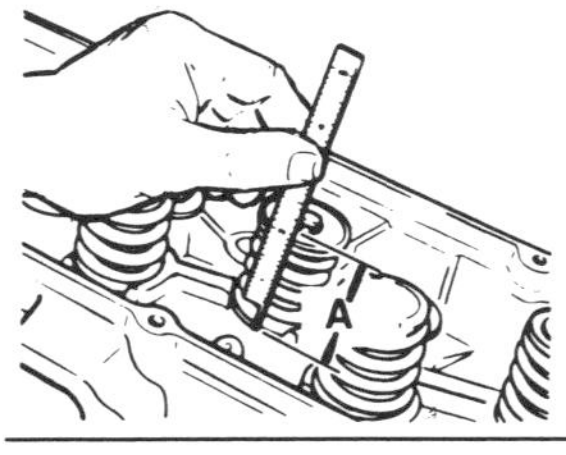

Valve spring installed height (A)

Measure the distance between the spring pad and the lower edge of the spring retainer, and compare to specifications. If the installed height is incorrect, add shim washers between the spring pad and the spring.

CAUTION: *Use only washers designed for this purpose.*

Install the camshaft (OHC engines only) and check end-play:

See the engine service procedures earlier in this chapter for details concerning specific engines.

Cylinder Head Reconditioning

Procedure	Method

Inspect the rocker arms, balls, studs, and nuts (OHV engines only):

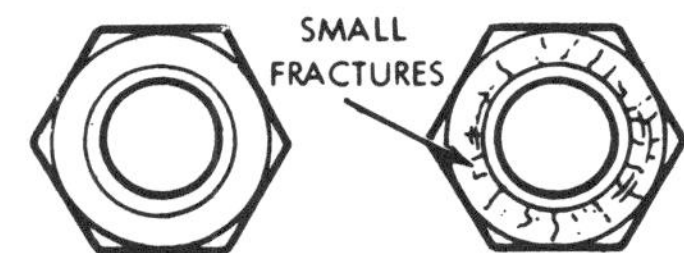

Stress cracks in the rocker nuts

Visually inspect the rocker arms, balls, studs, and nuts for cracks, galling, burning, scoring, or wear. If all parts are intact, liberally lubricate the rocker arms and balls, and install them on the cylinder head. If wear is noted on a rocker arm at the point of valve contact, grind it smooth and square, removing as little material as possible. Replace the rocker arm if excessively worn. If a rocker stud shows signs of wear, it must be replaced (see below). If a rocker nut shows stress cracks, replace it. If an exhaust ball is galled or burned, substitute the intake ball from the same cylinder (if it is intact), and install a new intake ball.

NOTE: *Avoid using new rocker balls on exhaust valves.*

Replacing rocker studs (OHV engines only):

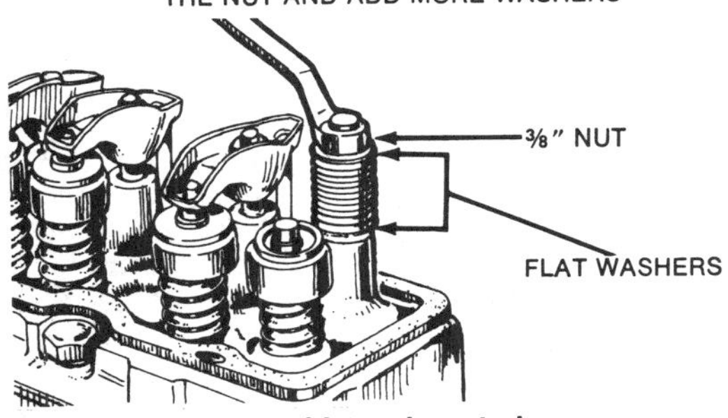

Extracting a pressed-in rocker stud

In order to remove a threaded stud, lock two nuts on the stud, and unscrew the stud using the lower nut. Coat the lower threads of the new stud with Loctite, and install.

Two alternative methods are available for replacing pressed in studs. Remove the damaged stud using a stack of washers and a nut (see illustration). In the first, the boss is reamed .005–.006″ oversize, and an oversize stud pressed in. Control the stud extension over the boss using washers, in the same manner as valve guides. Before installing the stud, coat it with white lead and grease. To retain the stud more positively drill a hole through the stud and boss, and install a roll pin. In the second method, the boss is tapped, and a threaded stud installed.

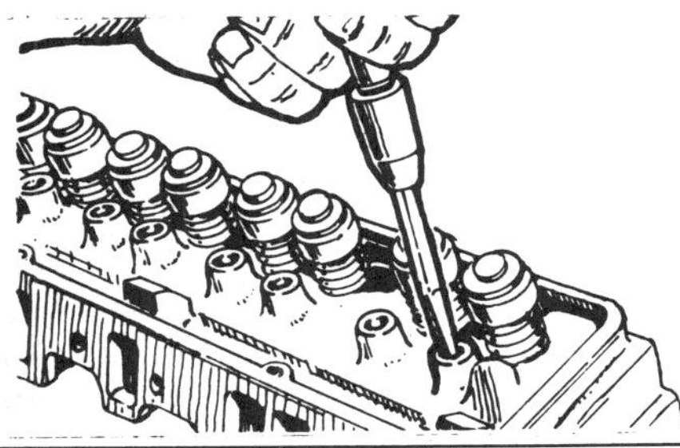

Ream the stud bore for oversize rocker studs

Inspect the rocker shaft(s) and rocker arms (OHV engines only)

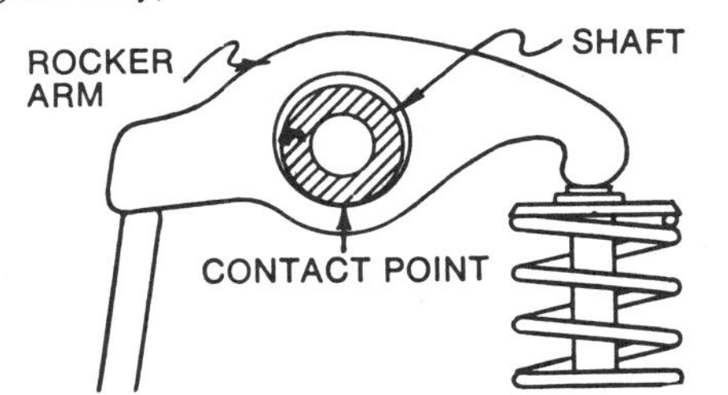

Check the rocker arm-to-rocker shaft contact area

Remove rocker arms, springs and washers from rocker shaft.

NOTE: *Lay out parts in the order as they are removed.* Inspect rocker arms for pitting or wear on the valve contact point, or excessive bushing wear. Bushings need only be replaced if wear is excessive, because the rocker arm normally contacts the shaft at one point only. Grind the valve contact point of rocker arm smooth if necessary, removing as little material as possible. If excessive material must be removed to smooth and square the arm, it should be replaced. Clean out all oil holes and passages in rocker shaft. If shaft is grooved or worn, replace it. Lubricate and assemble the rocker shaft.

Cylinder Head Reconditioning

Procedure	Method
Inspect the pushrods (OHV engines only):	Remove the pushrods, and, if hollow, clean out the oil passages using fine wire. Roll each pushrod over a piece of clean glass. If a distinct clicking sound is heard as the pushrod rolls, the rod is bent, and must be replaced.
	*The length of all pushrods must be equal. Measure the length of the pushrods, compare to specifications, and replace as necessary.
Inspect the valve lifters (OHV engines only): 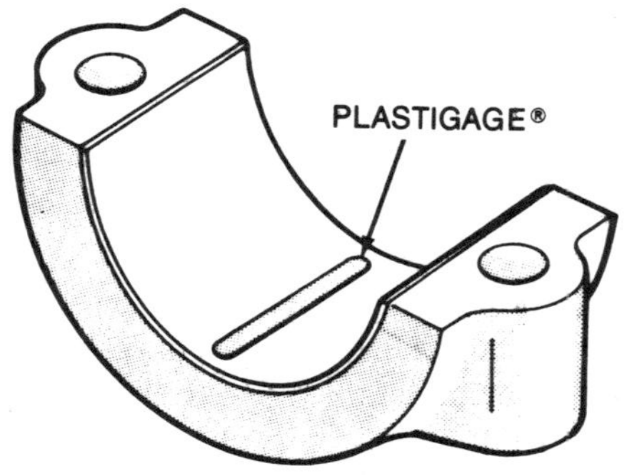 **Check the lifter face for squareness**	Remove lifters from their bores, and remove gum and varnish, using solvent. Clean walls of lifter bores. Check lifters for concave wear as illustrated. If face is worn concave, replace lifter, and carefully inspect the camshaft. Lightly lubricate lifter and insert it into its bore. If play is excessive, an oversize lifter must be installed (where possible). Consult a machinist concerning feasibility. If play is satisfactory, remove, lubricate, and reinstall the lifter.
*Testing hydraulic lifter leak down (OHV engines only):	Submerge lifter in a container of kerosene. Chuck a used pushrod or its equivalent into a drill press. Position container of kerosene so pushrod acts on the lifter plunger. Pump lifter with the drill press, until resistance increases. Pump several more times to bleed any air out of lifter. Apply very firm, constant pressure to the lifter, and observe rate at which fluid bleeds out of lifter. If the fluid bleeds very quickly (less than 15 seconds), lifter is defective. If the time exceeds 60 seconds, lifter is sticking. In either case, recondition or replace lifter. If lifter is operating properly (leak down time 15–60 seconds), lubricate and install it.

Cylinder Block Reconditioning

Procedure	Method
Checking the main bearing clearance: **Plastigage® installed on the lower bearing shell**	Invert engine, and remove cap from the bearing to be checked. Using a clean, dry rag, thoroughly clean all oil from crankshaft journal and bearing insert. NOTE: *Plastigage® is soluble in oil; therefore, oil on the journal or bearing could result in erroneous readings.* Place a piece of Plastigage along the full length of journal, reinstall cap, and torque to specifications. NOTE: *Specifications are given in the engine specifications earlier in this chapter.* Remove bearing cap, and determine bearing clearance by comparing width of Plastigage to the scale on Plastigage envelope. Journal taper is determined by comparing width of the Plastigage strip near its ends. Rotate crankshaft 90° and retest, to determine journal eccentricity. NOTE: *Do not rotate crankshaft with Plastigage*

Cylinder Block Reconditioning

Procedure	Method

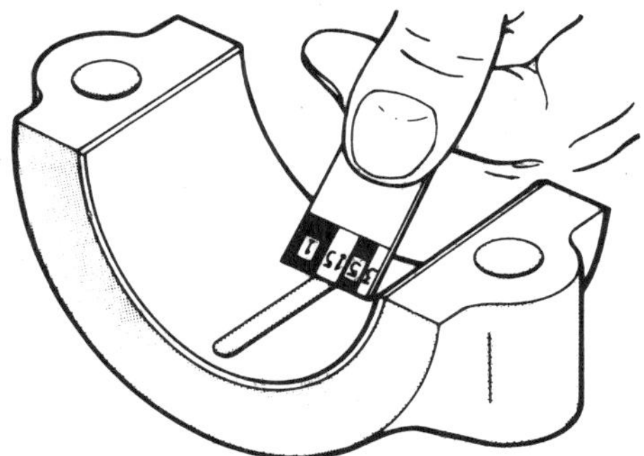

Measure Plastigage® to determine main bearing clearance

installed. If bearing insert and journal appear intact, and are within tolerances, no further main bearing service is required. If bearing or journal appear defective, cause of failure should be determined before replacement.

* Remove crankshaft from block (see below). Measure the main bearing journals at each end tiwce (90° apart) using a micrometer, to determine diameter, journal taper and eccentricity. If journals are within tolerances, reinstall bearing caps at their specified torque. Using a telescope gauge and micrometer, measure bearing I.D. parallel to piston axis and at 30° on each side of piston axis. Subtract journal O.D. from bearing I.D. to determine oil clearance. If crankshaft journals appear defective, or do not meet tolerances, there is no need to measure bearings; for the crankshaft will require grinding and/or undersize bearings will be required. If bearing appears defective, cause for failure should be determined prior to replacement.

Check the connecting rod bearing clearance:

Connecting rod bearing clearance is checked in the same manner as main bearing clearance, using Plastigage. Before removing the crankshaft, connecting rod side clearance also should be measured and recorded.

* Checking connecting rod bearing clearance, using a micrometer, is identical to checking main bearing clearance. If no other service is required, the piston and rod assemblies need not be removed.

Remove the crankshaft:

Using a punch, mark the corresponding main bearing caps and saddles according to position (i.e., one punch on the front main cap and saddle, two on the second, three on the third, etc.). Using number stamps, identify the corresponding connecting rods and caps, according to cylinder (if no numbers are present). Remove the main and connecting rod caps, and place sleeves of plastic tubing or vacuum hose over the connecting rod bolts, to protect the journals as the crankshaft is removed. Lift the crankshaft out of the block.

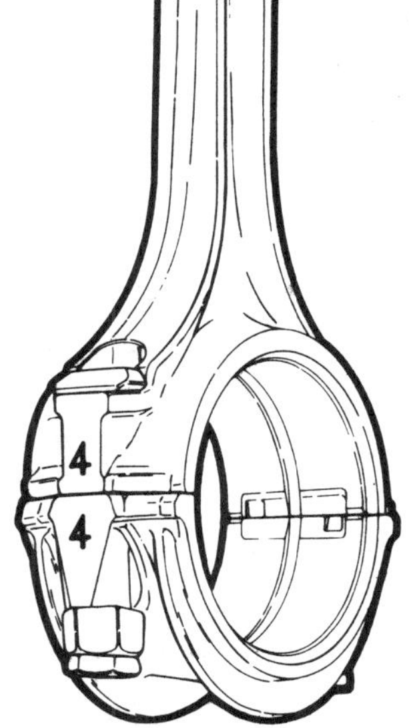

Match the connecting rod to the cylinder with a number stamp

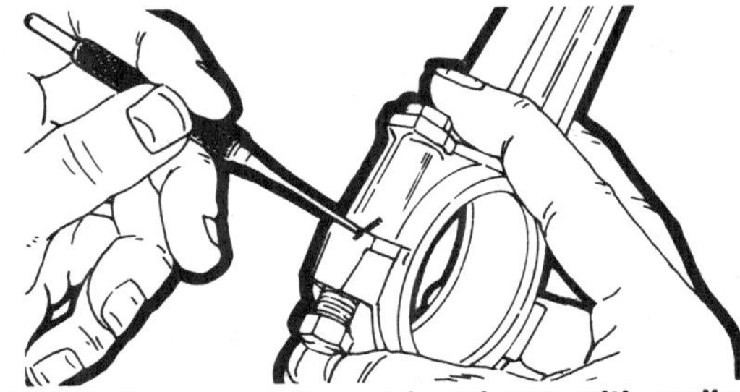

Match the connecting rod and cap with scribe marks

Cylinder Block Reconditioning

Procedure	Method
Remove the ridge from the top of the cylinder: 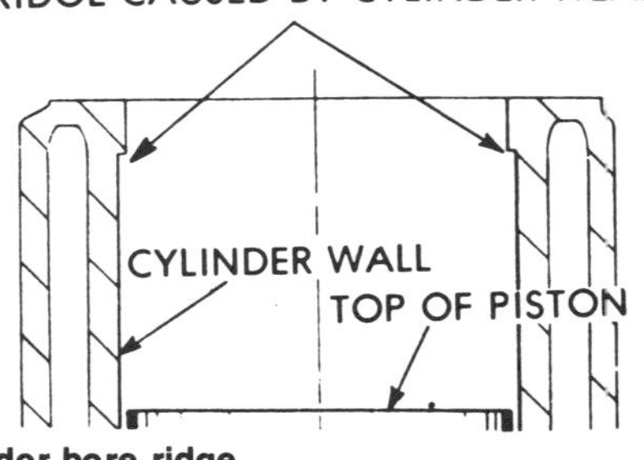 **Cylinder bore ridge**	In order to facilitate removal of the piston and connecting rod, the ridge at the top of the cylinder (unworn area; see illustration) must be removed. Place the piston at the bottom of the bore, and cover it with a rag. Cut the ridge away using a ridge reamer, exercising extreme care to avoid cutting too deeply. Remove the rag, and remove cuttings that remain on the piston. **CAUTION:** *If the ridge is not removed, and new rings are installed, damage to rings will result.*
Remove the piston and connecting rod: 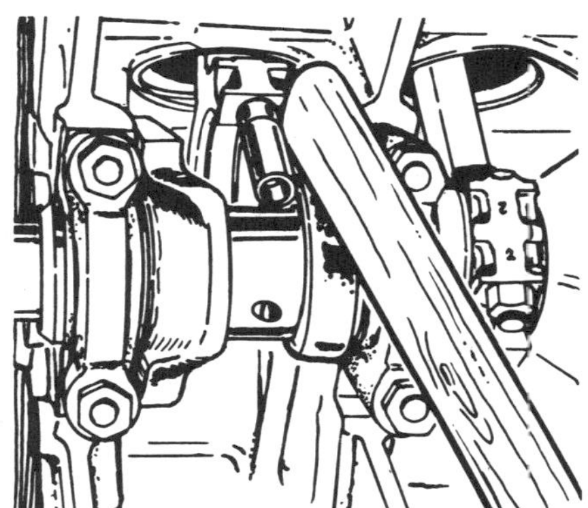 **Push the piston out with a hammer handle**	Invert the engine, and push the pistons and connecting rods out of the cylinders. If necessary, tap the connecting rod boss with a wooden hammer handle, to force the piston out. **CAUTION:** *Do not attempt to force the piston past the cylinder ridge* (see above).
Service the crankshaft:	Ensure that all oil holes and passages in the crankshaft are open and free of sludge. If necessary, have the crankshaft ground to the largest possible undersize.
	** Have the crankshaft Magnafluxed, to locate stress cracks. Consult a machinist concerning additional service procedures, such as surface hardening (e.g., nitriding, Tuftriding) to improve wear characteristics, cross drilling and chamfering the oil holes to improve lubrication, and balancing.
Removing freeze plugs:	Drill a small hole in the middle of the freeze plugs. Thread a large sheet metal screw into the hole and remove the plug with a slide hammer.
Remove the oil gallery plugs:	Threaded plugs should be removed using an appropriate (usually square) wrench. To remove soft, pressed in plugs, drill a hole in the plug, and thread in a sheet metal screw. Pull the plug out by the screw using pliers.
Hot-tank the block: **NOTE:** *Do not hot-tank aluminum parts.*	Have the block hot-tanked to remove grease, corrosion, and scale from the water jackets. **NOTE:** *Consult the operator to determine whether the camshaft bearings will be damaged during the hot-tank process.*

Cylinder Block Reconditioning

Procedure	Method
Check the block for cracks:	Visually inspect the block for cracks or chips. The most common locations are as follows: Adjacent to freeze plugs. Between the cylinders and water jackets. Adjacent to the main bearing saddles. At the extreme bottom of the cylinders. Check only suspected cracks using spot check dye (see introduction). If a crack is located, consult a machinist concerning possible repairs.
	** Magnaflux the block to locate hidden cracks. If cracks are located, consult a machinist about feasibility of repair.
Install the oil gallery plugs and freeze plugs:	Coat freeze plugs with sealer and tap into position using a piece of pipe, slightly smaller than the plug, as a driver. To ensure retention, stake the edges of the plugs. Coat threaded oil gallery plugs with sealer and install. Drive replacement soft plugs into block using a large drift as driver.
	* Rather than reinstalling lead plugs, drill and tap the holes, and install threaded plugs.
Check the bore diameter and surface: 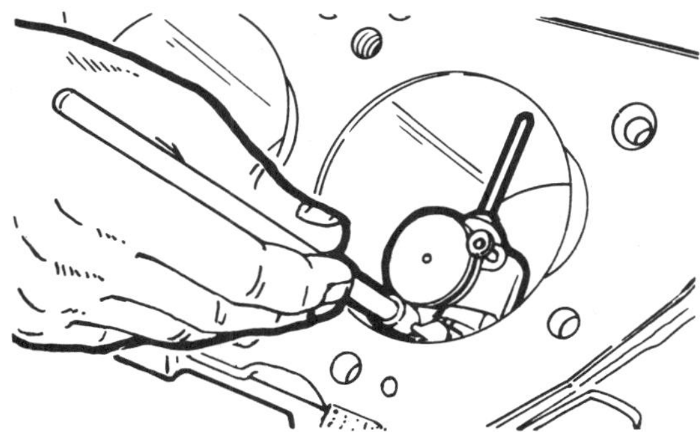 **Measure the cylinder bore with a dial gauge**	Visually inspect the cylinder bores for roughness, scoring, or scuffing. If evident, the cylinder bore must be bored or honed oversize to eliminate imperfections, and the smallest possible oversize piston used. The new pistons should be given to the machinist with the block, so that the cylinders can be bored or honed exactly to the piston size (plus clearance). If no flaws are evident, measure the bore diameter using a telescope gauge and micrometer, or dial gauge, parallel and perpendicular to the engine centerline, at the top (below the ridge) and bottom of the bore. Subtract the bottom measurements from the top to determine taper, and the parallel to the centerline measurements from the perpendicular measurements to determine eccentricity. If the measurements are not within specifications, the cylinder must be bored or honed, and an oversize piston installed. If the measurements are within specifications the cylinder may be used as is, with only finish honing (see below).

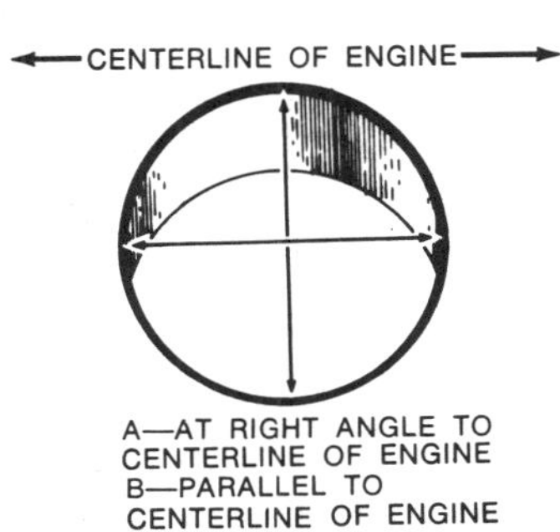

Cylinder bore measuring points

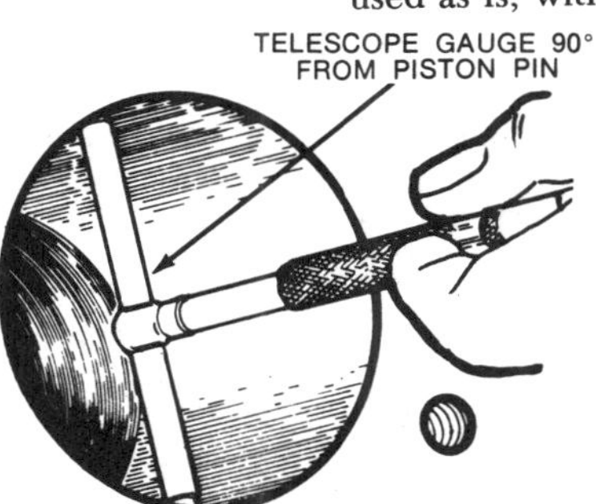

Measure the cylinder bore with a telescope gauge

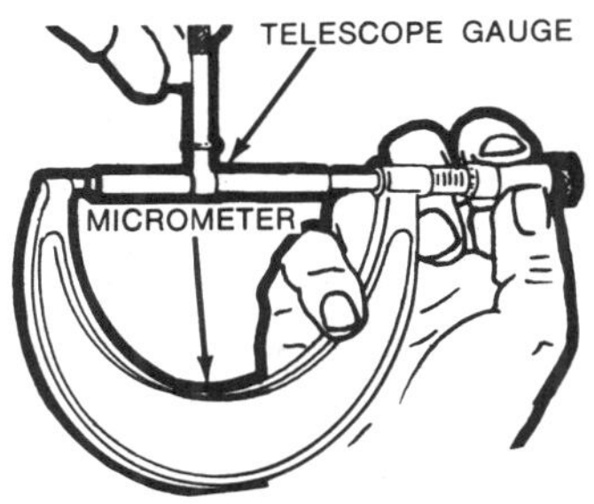

Measure the telescope gauge with a micrometer to determine the cylinder bore

Cylinder Block Reconditioning

Procedure	Method
	NOTE: *Prior to submitting the block for boring, perform the following operation(s).*
Check the cylinder block bearing alignment:	Remove the upper bearing inserts. Place a straightedge in the bearing saddles along the centerline of the crankshaft. If clearance exists between the straightedge and the center saddle, the block must be alignbored.

Check the main bearing saddle alignment

*Check the deck height:	The deck height is the distance from the crankshaft centerline to the block deck. To measure, invert the engine, and install the crankshaft, retaining it with the center main cap. Measure the distance from the crankshaft journal to the block deck, parallel to the cylinder centerline. Measure the diameter of the end (front and rear) main journals, parallel to the centerline of the cylinders, divide the diameter in half, and subtract it from the previous measurement. The results of the front and rear measurements should be identical. If the difference exceeds .005", the deck height should be corrected. NOTE: *Block deck height and warpage should be corrected at the same time.*
Check the block deck for warpage:	Using a straightedge and feeler gauges, check the block deck for warpage in the same manner that the cylinder head is checked (see Cylinder Head Reconditioning). If warpage exceeds specifications, have the deck resurfaced. NOTE: *In certain cases a specification for total material removal (Cylinder head and block deck) is provided. This specification must not be exceeded.*
Clean and inspect the pistons and connecting rods:	Using a ring expander, remove the rings from the piston. Remove the retaining rings (if so equipped) and remove piston pin. NOTE: *If the piston pin must be pressed out, determine the proper method and use the proper tools; otherwise the piston will distort.* Clean the ring grooves using an appropriate tool, exercising care to avoid cutting too deeply. Thoroughly clean all carbon and varnish from the piston with solvent. CAUTION: *Do not use a wire brush or caustic solvent on pistons.* Inspect the pistons for scuffing, scoring, cracks, pitting, or excessive ring groove wear. If wear is evident, the piston must be replaced. Check the connecting rod length by measuring the rod from the inside of the large end to the

Remove the piston rings

Cylinder Block Reconditioning

Procedure	Method

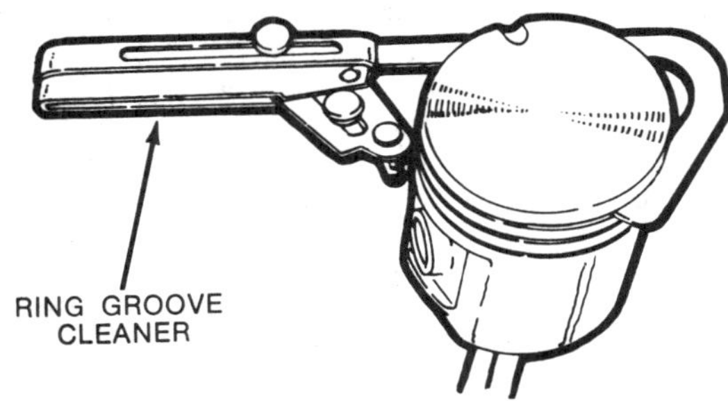

Clean the piston ring grooves

inside of the small end using calipers (see illustration). All connecting rods should be equal length. Replace any rod that differs from the others in the engine.

* Have the connecting rod alignment checked in an alignment fixture by a machinist. Replace any twisted or bent rods.

* Magnaflux the connecting rods to locate stress cracks. If cracks are found, replace the connecting rod.

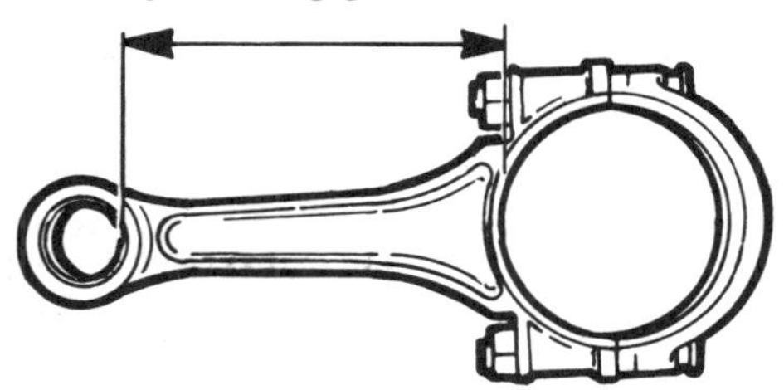

Check the connecting rod length (arrow)

Fit the pistons to the cylinders:

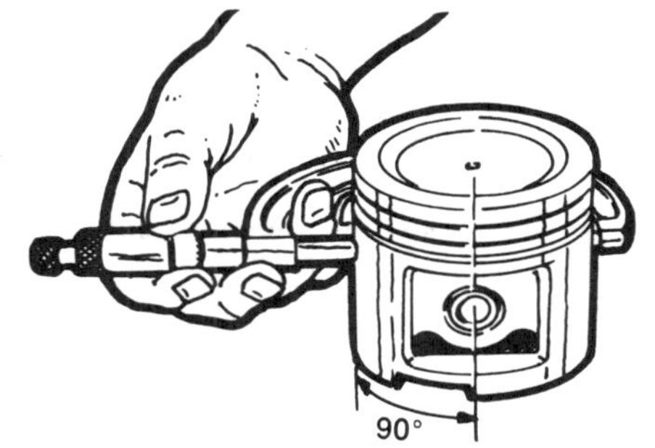

Measure the piston prior to fitting

Using a telescope gauge and micrometer, or a dial gauge, measure the cylinder bore diameter perpendicular to the piston pin, 2½" below the deck. Measure the piston perpendicular to its pin on the skirt. The difference between the two measurements is the piston clearance. If the clearance is within specifications or slightly below (after boring or honing), finish honing is all that is required. If the clearance is excessive, try to obtain a slightly larger piston to bring clearance within specifications. Where this is not possible, obtain the first oversize piston, and hone (or if necessary, bore) the cylinder to size.

Assemble the pistons and connecting rods:

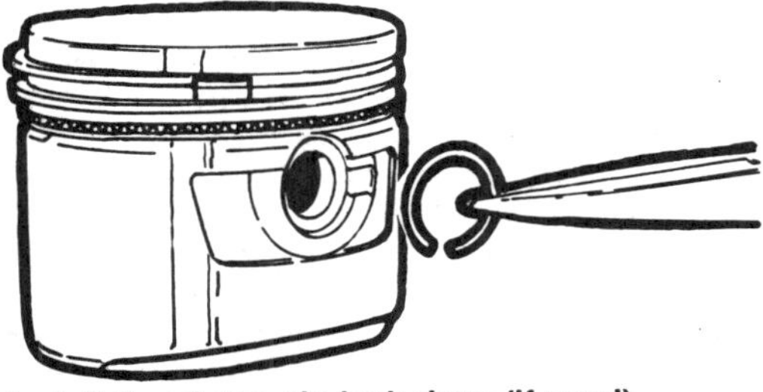

Install the piston pin lock-rings (if used)

Inspect piston pin, connecting rod small end bushing, and piston bore for galling, scoring, or excessive wear. If evident, replace defective part(s). Measure the I.D. of the piston boss and connecting rod small end, and the O.D. of the piston pin. If within specifications, assemble piston pin and rod.
CAUTION: *If piston pin must be pressed in, determine the proper method and use the proper tools; otherwise the piston will distort.*
Install the lock rings; ensure that they seat properly. If the parts are not within specifications, determine the service method for the type of engine. In some cases, piston and pin are serviced as an assembly when either is defective. Others specify reaming the piston and connecting rods for an oversize pin. If the connecting rod bushing is worn, it may in many cases be replaced. Reaming the piston and replacing the rod bushing are machine shop operations.

Cylinder Block Reconditioning

Procedure	Method

Clean and inspect the camshaft:

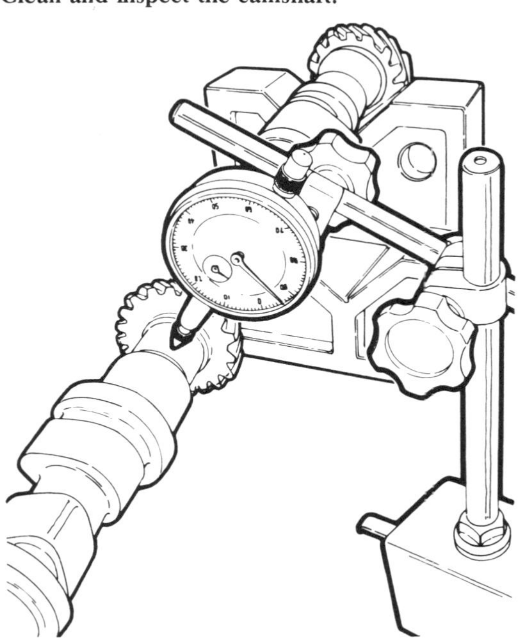

Check the camshaft for straightness

Degrease the camshaft, using solvent, and clean out all oil holes. Visually inspect cam lobes and bearing journals for excessive wear. If a lobe is questionable, check all lobes as indicated below. If a journal or lobe is worn, the camshaft must be reground or replaced.

NOTE: *If a journal is worn, there is a good chance that the bushings are worn.* If lobes and journals appear intact, place the front and rear journals in V-blocks, and rest a dial indicator on the center journal. Rotate the camshaft to check straightness. If deviation exceeds .001", replace the camshaft.

---

*Check the camshaft lobes with a micrometer, by measuring the lobes from the nose to base and again at 90° (see illustration). The lift is determined by subtracting the second measurement from the first. If all exhaust lobes and all intake lobes are not identical, the camshaft must be reground or replaced.

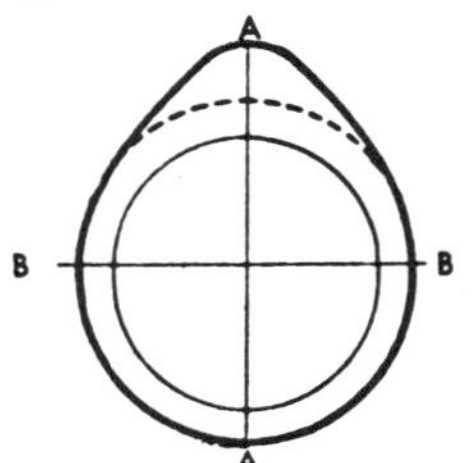

Camshaft lobe measurement

Replace the camshaft bearings (OHV engines only):

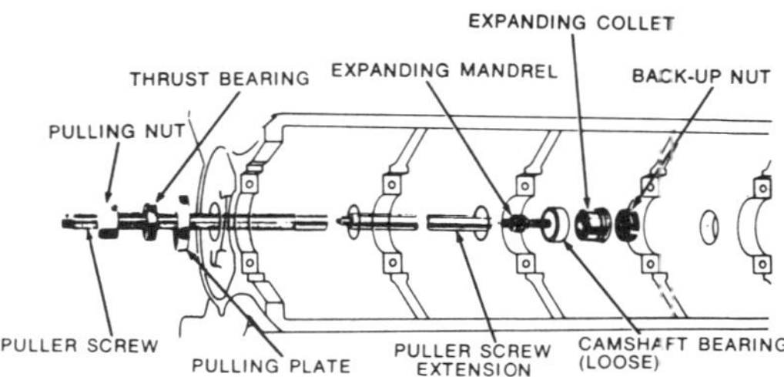

Camshaft bearing removal and installation tool (OHV engines only)

If excessive wear is indicated, or if the engine is being completely rebuilt, camshaft bearings should be replaced as follows: Drive the camshaft rear plug from the block. Assemble the removal puller with its shoulder on the bearing to be removed. Gradually tighten the puller nut until bearing is removed. Remove remaining bearings, leaving the front and rear for last. To remove front and rear bearings, reverse position of the tool, so as to pull the bearings in toward the center of the block. Leave the tool in this position, pilot the new front and rear bearings on the installer, and pull them into position: Return the tool to its original position and pull remaining bearings into position.

NOTE: *Ensure that oil holes align when installing bearings.* Replace camshaft rear plug, and stake it into position to aid retention.

Finish hone the cylinders:

Chuck a flexible drive hone into a power drill, and insert it into the cylinder. Start the hone, and move it up and down in the cylinder at a rate which will produce approximately a 60° cross-hatch pattern.

NOTE: *Do not extend the hone below the cylin-*

Cylinder Block Reconditioning

Procedure	Method

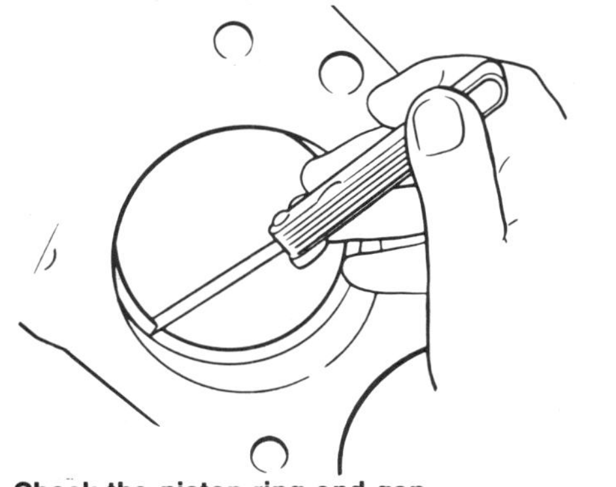

Cylinder bore after honing

der bore. After developing the pattern, remove the hone and recheck piston fit. Wash the cylinders with a detergent and water solution to remove abrasive dust, dry, and wipe several times with a rag soaked in engine oil.

Check piston ring end-gap:

Check the piston ring end gap

Compress the piston rings to be used in a cylinder, one at a time, into that cylinder, and press them approximately 1″ below the deck with an inverted piston. Using feeler gauges, measure the ring end-gap, and compare to specifications. Pull the ring out of the cylinder and file the ends with a fine file to obtain proper clearance.

CAUTION: *If inadequate ring end-gap is utilized, ring breakage will result.*

Install the piston rings:

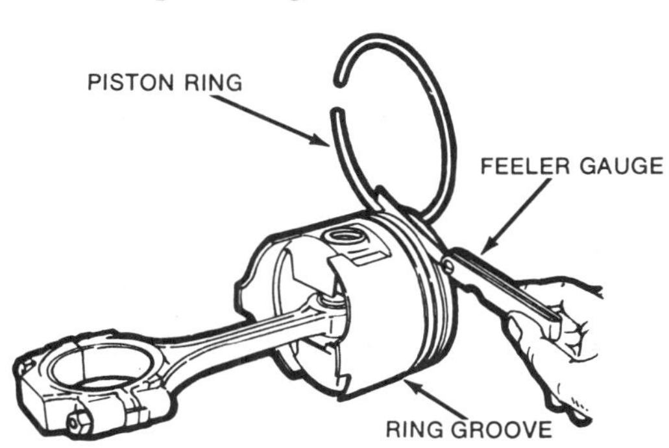

Check the piston ring side clearance

Inspect the ring grooves in the piston for excessive wear or taper. If necessary, recut the grooves(s) for use with an overwidth ring or a standard ring and spacer. If the groove is worn uniformly, overwidth rings, or standard rings and spacers may be installed without recutting. Roll the outside of the ring around the groove to check for burrs or deposits. If any are found, remove with a fine file. Hold the ring in the groove, and measure side clearance. If necessary, correct as indicated above.

NOTE: *Always install any additional spacers above the piston ring.*

The ring groove must be deep enough to allow the ring to seat below the lands (see illustration). In many cases, a "go-no-go" depth gauge will be provided with the piston rings. Shallow grooves may be corrected by recutting, while deep grooves require some type of filler or expander behind the piston. Consult the piston ring sup-

Cylinder Block Reconditioning

Procedure	Method
	plier concerning the suggested method. Install the rings on the piston, lowest ring first, using a ring expander. **NOTE:** *Position the ring as specified by the manufacturer.* Consult the engine service procedures earlier in this chapter for details concerning specific engines.
Install the camshaft (OHV engines only):	Liberally lubricate the camshaft lobes and journals, and install the camshaft. **CAUTION:** *Exercise extreme care to avoid damaging the bearings when inserting the camshaft.* Install and tighten the camshaft thrust plate retaining bolts. See the engine service procedures earlier in this chapter for details concerning specific engines.

Check camshaft end-play (OHV engines only):

Check the camshaft end-play with a feeler gauge

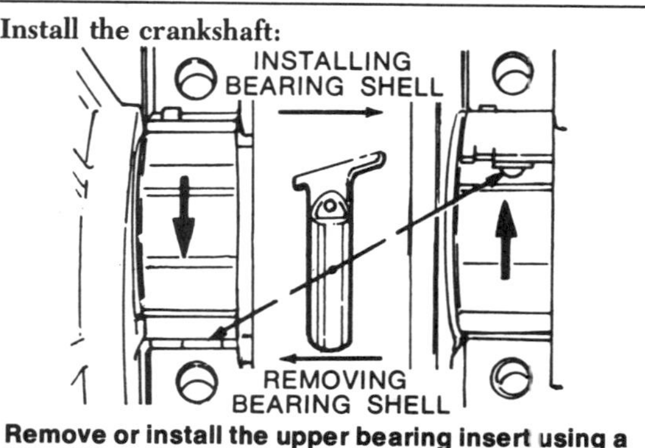

Check the camshaft end-play with a dial indicator

Using feeler gauges, determine whether the clearance between the camshaft boss (or gear) and backing plate is within specifications. Install shims behind the thrust plate, or reposition the camshaft gear and retest endplay. In some cases, adjustment is by replacing the thrust plate.
See the engine service procedures earlier in this chapter for details concerning specific engines.

*Mount a dial indicator stand so that the stem of the dial indicator rests on the nose of the camshaft, parallel to the camshaft axis. Push the camshaft as far in as possible and zero the gauge. Move the camshaft outward to determine the amount of camshaft endplay. If the endplay is not within tolerance, install shims behind the thrust plate, or reposition the camshaft gear and retest.
See the engine service procedures earlier in this chapter for details concerning specific engines.

Install the rear main seal:

See the engine service procedures earlier in this chapter for details concerning specific engines.

Install the crankshaft:

Remove or install the upper bearing insert using a roll-out pin

Thoroughly clean the main bearing saddles and caps. Place the upper halves of the bearing inserts on the saddles and press into position.
NOTE: *Ensure that the oil holes align.* Press the corresponding bearing inserts into the main bearing caps. Lubricate the upper main bearings, and lay the crankshaft in position. Place a strip of Plastigage on each of the crankshaft journals, install the main caps, and torque to specifications. Remove the main caps, and compare the Plastigage to the scale on the Plastigage envelope. If clearances are within tolerances, remove the Plastigage, turn the crankshaft 90°, wipe off all oil and retest. If all clearances are correct, re-

Cylinder Block Reconditioning

Procedure	Method

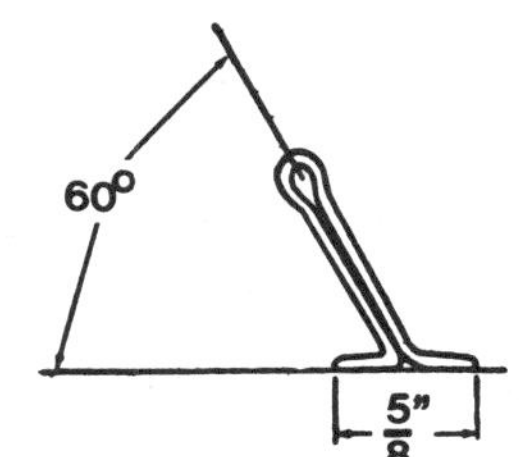

Home-made bearing roll-out pin

move all Plastigage, thoroughly lubricate the main caps and bearing journals, and install the main caps. If clearances are not within tolerance, the upper bearing inserts may be removed, without removing the crankshaft, using a bearing roll out pin (see illustration). Roll in a bearing that will provide proper clearance, and retest. Torque all main caps, excluding the thrust bearing cap, to specifications. Tighten the thrust bearing cap finger tight. To properly align the thrust bearing, pry the crankshaft the extent of its axial travel several times, the last movement held toward the front of the engine, and torque the thrust bearing cap to specifications. Determine the crankshaft end-play (see below), and bring within tolerance with thrust washers.

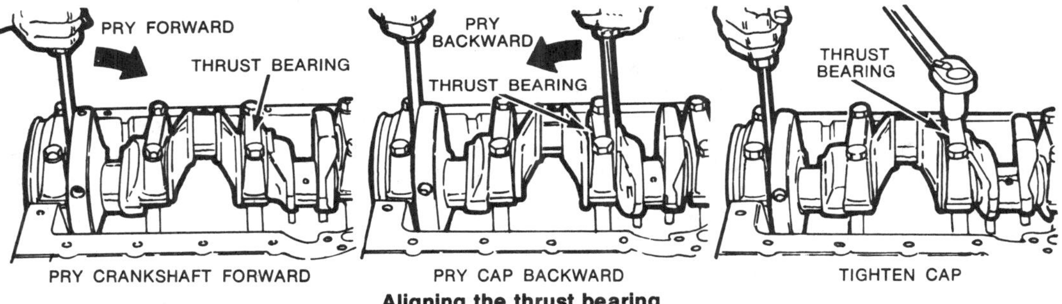

Aligning the thrust bearing

Measure crankshaft end-play:

Mount a dial indicator stand on the front of the block, with the dial indicator stem resting on the nose of the crankshaft, parallel to the crankshaft axis. Pry the crankshaft the extent of its travel rearward, and zero the indicator. Pry the crankshaft forward and record crankshaft end-play.
NOTE: *Crankshaft end-play also may be measured at the thrust bearing, using feeler gauges* (see illustration).

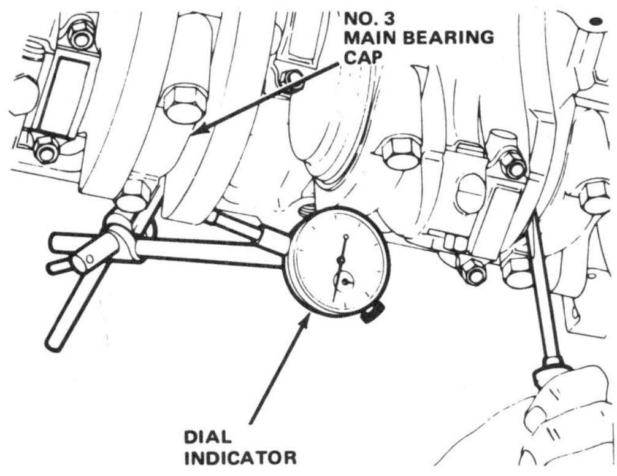

Check the crankshaft end-play with a dial indicator

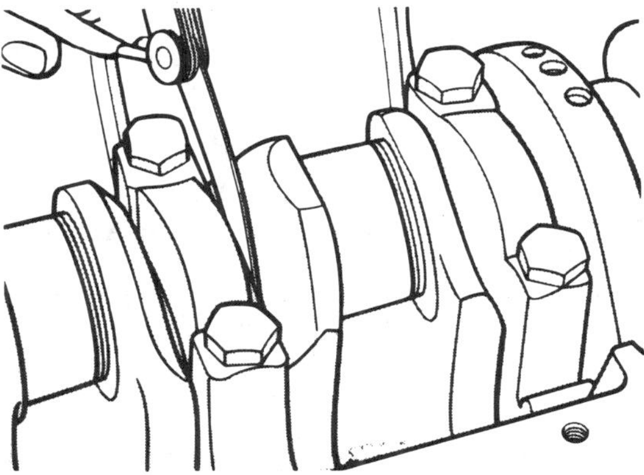

Check the crankshaft end-play with a feeler gauge

Cylinder Block Reconditioning

Procedure	Method

Install the pistons:

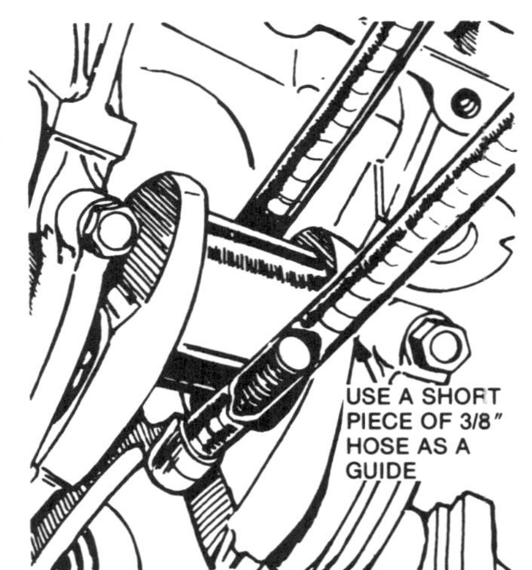

Use lengths of vacuum hose or rubber tubing to protect the crankshaft journals and cylinder walls during piston installation

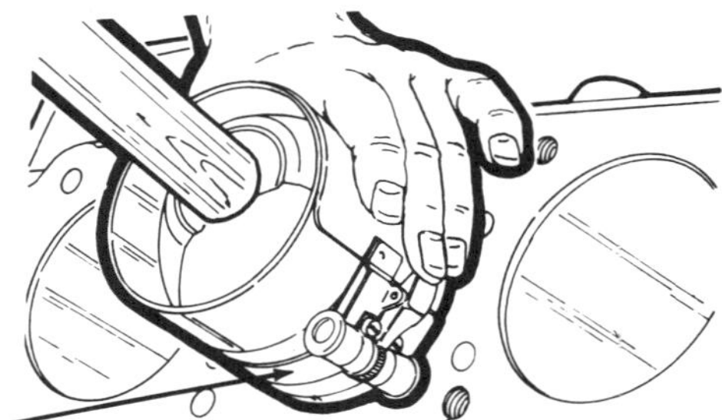

Install the piston using a ring compressor

Press the upper connecting rod bearing halves into the connecting rods, and the lower halves into the connecting rod caps. Position the piston ring gaps according to specifications (see car section), and lubricate the pistons. Install a ring compresser on a piston, and press two long (8″) pieces of plastic tubing over the rod bolts. Using the tubes as a guide, press the pistons into the bores and onto the crankshaft with a wooden hammer handle. After seating the rod on the crankshaft journal, remove the tubes and install the cap finger tight. Install the remaining pistons in the same manner. Invert the engine and check the bearing clearance at two points (90° apart) on each journal with Plastigage.

NOTE: *Do not turn the crankshaft with Plastigage installed.* If clearance is within tolerances, remove *all* Plastigage, thoroughly lubricate the journals, and torque the rod caps to specifications. If clearance is not within specifications, install different thickness bearing inserts and recheck.

CAUTION: *Never shim or file the connecting rods or caps.* Always install plastic tube sleeves over the rod bolts when the caps are not installed, to protect the crankshaft journals.

Check connecting rod side clearance:

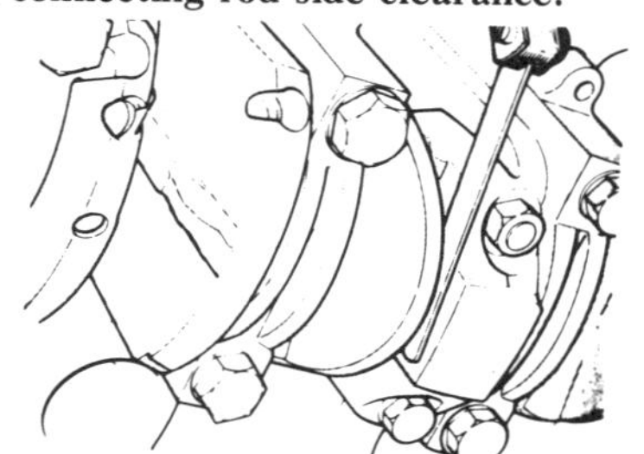

Check the connecting rod side clearance with a feeler gauge

Determine the clearance between the sides of the connecting rods and the crankshaft, using feeler gauges. If clearance is below the minimum tolerance, the rod may be machined to provide adequate clearance. If clearance is excessive, substitute an unworn rod, and recheck. If clearance is still outside specifications, the crankshaft must be welded and reground, or replaced.

Inspect the timing chain (or belt):

Visually inspect the timing chain for broken or loose links, and replace the chain if any are found. If the chain will flex sideways, it must be replaced. Install the timing chain as specified. Be sure the timing belt is not stretched, frayed or broken.

NOTE: *If the original timing chain is to be reused, install it in its original position.*

Cylinder Block Reconditioning

Procedure	Method
Check timing gear backlash and runout (OHV engines):	Mount a dial indicator with its stem resting on a tooth of the camshaft gear (as illustrated). Rotate the gear until all slack is removed, and zero the indicator. Rotate the gear in the opposite direction until slack is removed, and record gear backlash. Mount the indicator with its stem resting on the edge of the camshaft gear, parallel to the axis of the camshaft. Zero the indicator, and turn the camshaft gear one full turn, recording the runout. If either backlash or runout exceed specifications, replace the worn gear(s).

Check the camshaft gear backlash

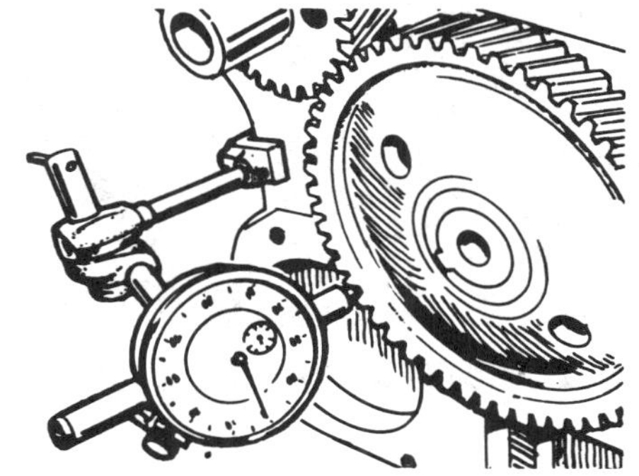

Check the camshaft gear run-out

Completing the Rebuilding Process

Following the above procedures, complete the rebuilding process as follows:

Fill the oil pump with oil, to prevent cavitating (sucking air) on initial engine start up. Install the oil pump and the pickup tube on the engine. Coat the oil pan gasket as necessary, and install the gasket and the oil pan. Mount the flywheel and the crankshaft vibration damper or pulley on the crankshaft. NOTE: *Always use new bolts when installing the flywheel.* Inspect the clutch shaft pilot bushing in the crankshaft. If the bushing is excessively worn, remove it with an expanding puller and a slide hammer, and tap a new bushing into place.

Position the engine, cylinder head side up. Lubricate the lifters, and install them into their bores. Install the cylinder head, and torque it as specified. Insert the pushrods (where applicable), and install the rocker shaft(s) (if so equipped) or position the rocker arms on the pushrods. Adjust the valves.

Install the intake and exhaust manifolds, the carburetor(s), the distributor and spark plugs. Adjust the point gap and the static ignition timing. Mount all accessories and install the engine in the car. Fill the radiator with coolant, and the crankcase with high quality engine oil.

Break-in Procedure

Start the engine, and allow it to run at low speed for a few minutes, while checking for leaks. Stop the engine, check the oil level, and fill as necessary. Restart the engine, and fill the cooling system to capacity. Check the point dwell angle and adjust the ignition timing and the valves. Run the engine at low to medium speed (800–2500 rpm) for approximately ½ hour, and retorque the cylinder head bolts. Road test the car, and check again for leaks.

Follow the manufacturer's recommended engine break-in procedure and maintenance schedule for new engines.

Emission Controls and Fuel System

EMISSION CONTROLS

There are three sources of automotive pollutants: crankcase fumes, exhaust gases and gasoline evaporation. The pollutants formed from these substances fall into three categories: unburnt hydrocarbons (HC), carbon monoxide (CO), and oxides of nitrogen (NO_x). The equipment that is used to limit these pollutants is commonly called emission control equipment.

Crankcase Emission Controls

The crankcase emission control equipment consists of a positive crankcase ventilation valve (PCV), a closed or open oil filler cap and hoses to connect this equipment.

When the engine is running, a small portion of the gases which are formed in the combustion chamber during combustion leak by the piston rings and enter the crankcase. Since these gases are under pressure they tend to escape from the crankcase and enter into the atmosphere. If these gases were allowed to remain in the crankcase for any length of time, they would contaminate the engine oil and cause sludge to build up. If the gases were allowed to escape into the atmosphere, they would pollute the air, as they contain unburned hydrocarbons. The crank-

case emission control equipment recycles these gases back into the engine combustion chamber where they are burned.

Crankcase gases are recycled in the following manner: while the engine is running, clean filtered air is drawn into the crankcase through the carburetor air filter and then through a hose leading to the rocker cover. As the air passes through the crankcase it picks up the combustion gases and carries them out of the crankcase, up through the PCV valve and into the intake manifold. After they enter the intake manifold they are drawn into the combustion chamber and burned.

The most critical component in the system is the PCV valve. This vacuum controlled valve regulates the amount of gases which are recycled into the combustion chamber. At low engine speeds the valve is partially closed, limiting the flow of gases into the intake manifold. As engine speed increases, the valve opens to admit greater quantities of the gases into the intake manifold. If the valve should become blocked or plugged, the gases will be prevented from escaping from the crankcases by the normal route. Since these gases are under pressure, they will find their own way out of the crankcase. This alternate route is usually a weak oil seal or gasket in the engine. As the gas escapes by the

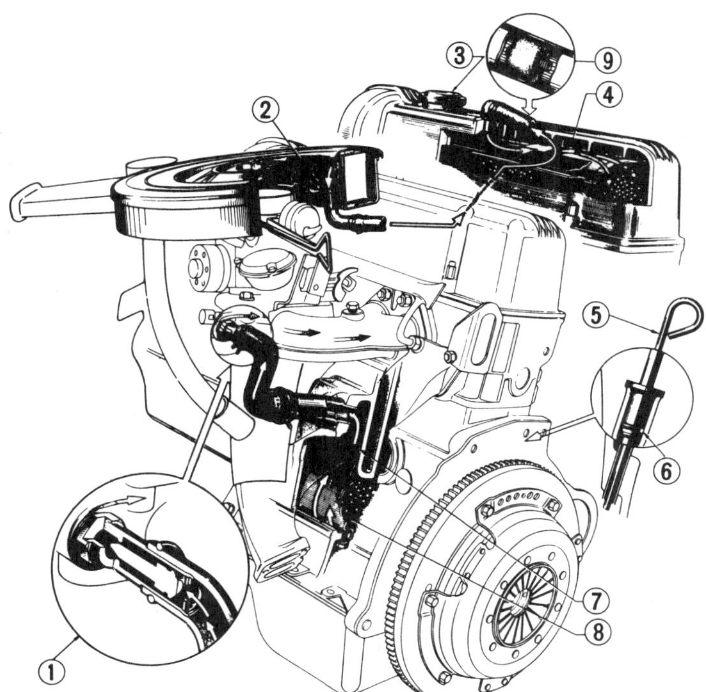

1. Crankcase ventilation
   control valve
2. Flame arrester
3. Sealed filler cap
4. Baffle plate
5. Oil level gauge
6. O-ring
7. Oil separator
8. Baffle plate
9. Flame arrester

Positive crankcase ventilation system

gasket, it also creates an oil leak. Besides causing oil leaks, a clogged PCV valve also allows these gases to remain in the crankcase for an extended period of time, promoting the formation of sludge in the engine.

The above explanation and the trouble-shooting procedure which follows applies to all engines with PCV systems.

TESTING

Check the PCV system hoses and connections, to see that there are no leaks; then replace or tighten, as necessary.

To check the valve, remove it and blow through both of its ends. When blowing from the side which goes toward the intake manifold, very little air should pass through it. When blowing from the crankcase side, air should pass through freely.

NOTE: *Do not attempt to clean or adjust the valve; replace it with a new one.*

REMOVAL AND INSTALLATION

To remove the PCV valve, simply loosen the hose clamp and remove the valve from the manifold-to-crankcase hose and intake manifold. Install the PCV valve in the reverse order of removal.

Evaporative Emission Control System

When raw fuel evaporates, the vapors contain hydrocarbons. To prevent these nasties from escaping into the atmosphere, the fuel evaporative emission control system was developed.

There are two different evaporative emission control systems used on Datsun pickups.

The system, until 1974, consists of a sealed fuel tank, a vapor-liquid separator, a flow guide (check) valve, and all of the hoses connecting these components, in the above order, leading from the fuel tank to the PCV hose, which connects the crankcase to the PCV valve.

In operation, the vapor formed in the fuel tank passes through the vapor separator, into the flow guide valve and the crankcase. When the engine is not running, if the fuel vapor pressure in the vapor separator goes above 0.4 in. Hg, the flow guide valve opens and allows the vapor to enter the engine crankcase. Otherwise the flow guide valve is closed to the vapor separator while the engine is not running. When the engine is running, and a vacuum is developed in the fuel tank or in the engine crankcase and the difference of pressure between the relief side

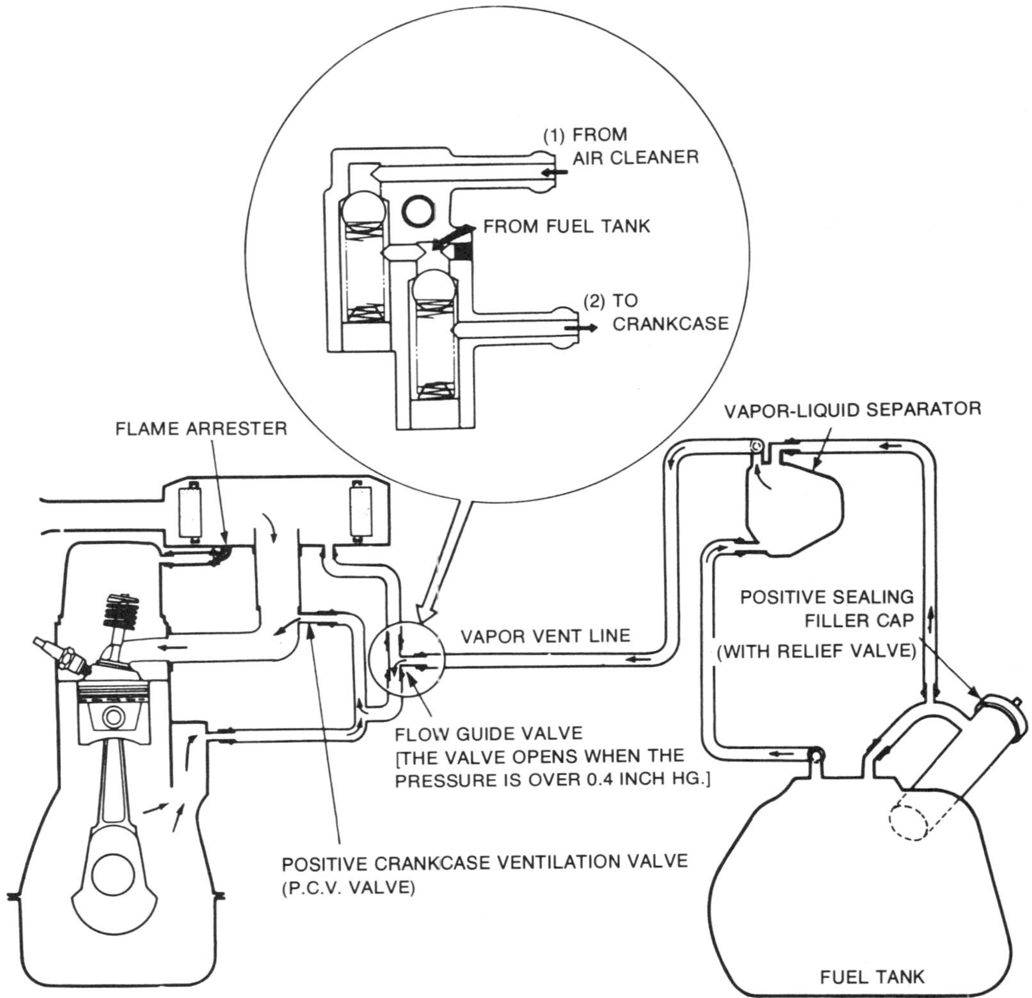

Evaporative emission control system

and the fuel tank or crankcase becomes 2 in. Hg, the relief valve opens and allows ambient air from the air cleaner into the fuel tank or the engine crankcase. This ambient air replaces the vapor within the fuel tank or crankcase, bringing the fuel tank or crankcase back into a neutral or positive pressure range.

The system used on 1975 and later models consists of a sealed fuel tank, vapor-liquid separator, vapor bent line, carbon canister, vacuum signal line, and a canister purge line.

In operation, fuel vapors and/or liquid are routed to the liquid/vapor separator where liquid fuel is directed back into the fuel tank as fuel vapors flow into the charcoal-filled canister. The charcoal absorbs and stores the fuel vapors when the engine is not running or is at idle. When the throttle valves in the carburetor are opened, vacuum from above the throttle valves is routed through a vacuum signal line to the purge control valve on the canister. The control valve opens and allows the fuel vapors to be drawn from the canister through a purge line and into the intake manifold and combustion chambers.

INSPECTION AND SERVICE

Check the hoses for proper connections and damage. Replace as necessary. Check the vapor separator tank for fuel leaks, distortion and dents, and replace as necessary.

Flow Guide Valve Pre-1975

Remove the flow guide valve and inspect it for leakage by blowing air into the ports in the valve. When air is applied from the fuel tank side, the flow guide valve is normal if air

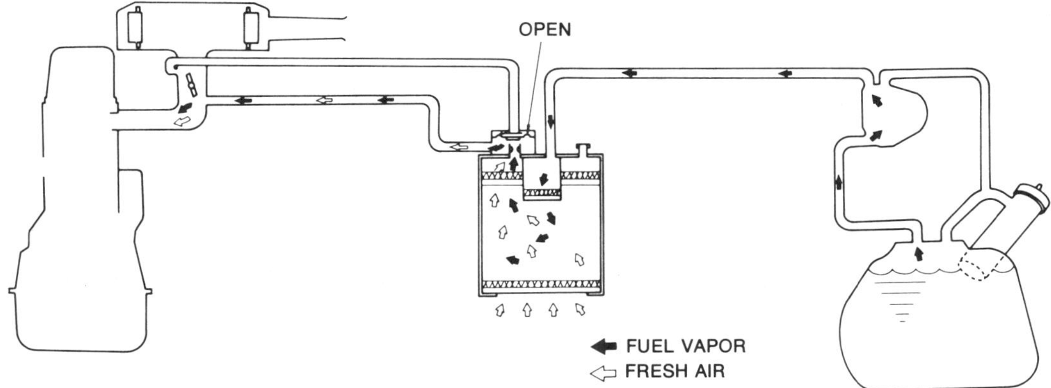

The evaporative emission control system on 1975 and later models—engine running above idle

passes into the check side (crankcase side), but not leaking into the relief side (air cleaner side). When air is applied from the check side, the valve is normal if the passage of air is restricted. When air is applied from the relief side (air cleaner side), the valve is normal if air passes into the fuel tank side or into the check side.

Carbon Canister and Purge Control Valve—1975 and Later

To check the operation of the carbon canister purge control valve, disconnect the rubber hose between the canister control valve and the T-fitting, at the T-fitting. Apply vacuum to the hose leading to the control valve. The vacuum condition should be maintained indefinitely. If the control valve leaks, remove the top cover of the valve and check for a dislocated or cracked diaphragm. If the diaphragm is damaged, a repair kit containing a new diaphragm, retainer, and spring is available and should be installed.

The carbon canister has an air filter in the bottom of the canister. The filter element should be checked once a year or every 12,000 miles; more frequently if the truck is operated in dusty areas. Replace the filter by pulling it out of the bottom of the canister and installing a new one.

REMOVAL AND INSTALLATION

Removal and installation of the various evaporative emission control system components consists of disconnecting the hoses, loosening retaining screws, and removing the part which is to be replaced or checked. Install in the reverse order. When replacing hose, make sure that it is fuel and vapor resistant.

Exhaust Emission Control

SPARK TIMING CONTROL SYSTEM DUAL POINT DISTRIBUTOR 1970–73

The dual point distributor has two sets of breaker points which operate independently of each other and are positioned with a relative phase angle of 5° (1970), 10° (1971), 7° (1972–73) apart. This makes one set the advanced points and the other set the retarded points.

The two sets of points, which mechanically operate continuously, are connected in parallel to the primary side of the ignition circuit. One set of points controls the firing of the spark plugs and hence, the ignition timing, depending on whether or not the retarded set of points is energized.

When both sets of points are electrically energized, the first set to open (the advanced set, 4° or 1° sooner) has no control over breaking the ignition coil primary circuit because the retarded set is still closed and maintaining a complete circuit to ground. When the retarded set of points opens, the advanced set is still open, and the primary circuit is broken causing the electromagnetic field in the coil to collapse and the ignition spark is produced.

When the retarded set of points is removed from the primary ignition circuit through the operation of a distributor relay inserted into the retarded points circuit, the advanced set of points controls the primary circuit. The retarded set of points is activated as follows:

On 1970 and 1971 models, the retarded set of points is activated only while cruising or accelerating with the throttle partially open and the transmission in Third gear. Under all

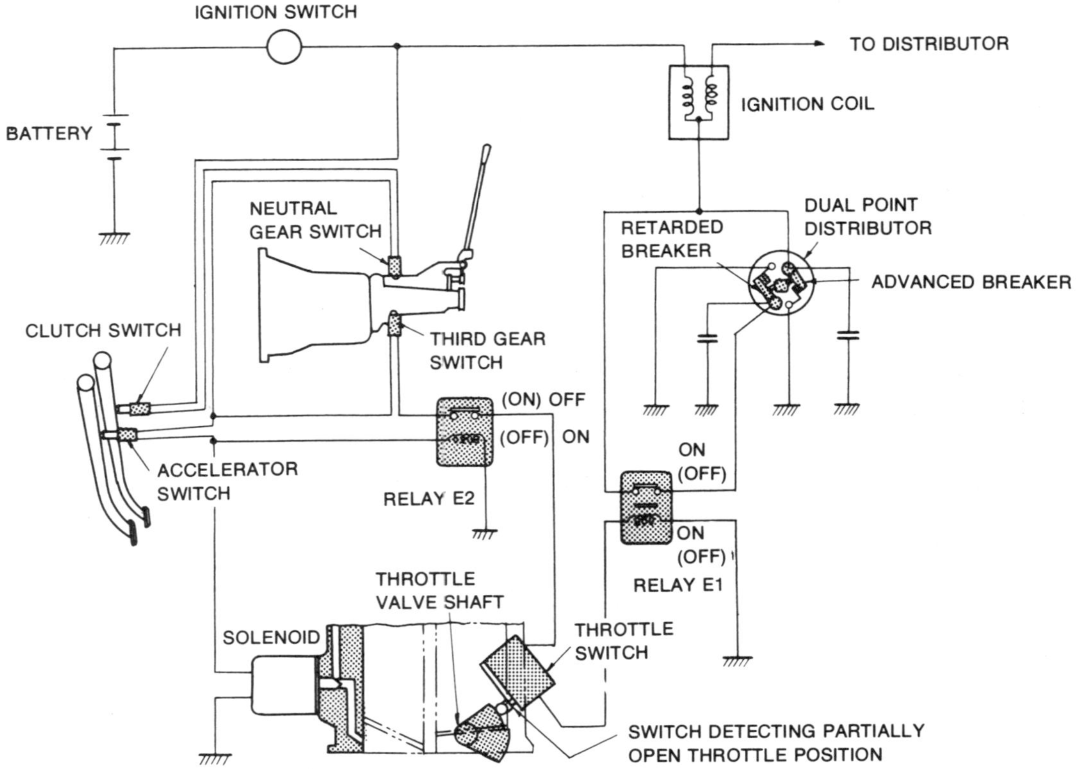

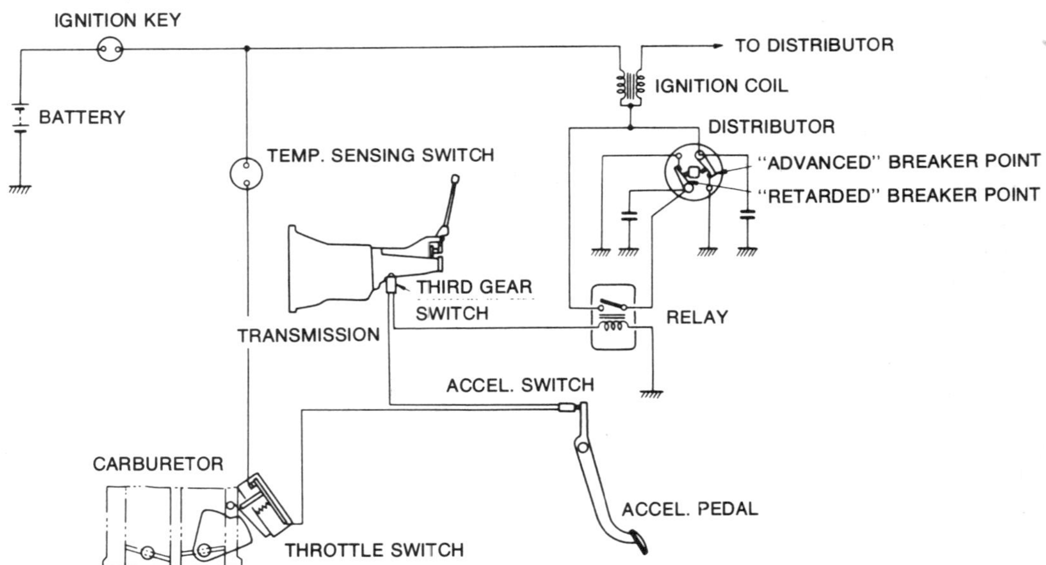

Spark timing control system (dual point distributor) for 1970–71 models

Spark timing control system (dual point distributor) for 1972 models with manual transmission

other conditions the retarded set of points is removed from the ignition circuit.

On 1972 models, the retarded set of points is activated only while cruising or accelerating with the throttle partially open, the transmission in Third gear, and with the ambient temperature above 50°F.

On 1973 models, the retarded set of points is activated only while the throttle is partially open, the temperature is above 50°F, and the

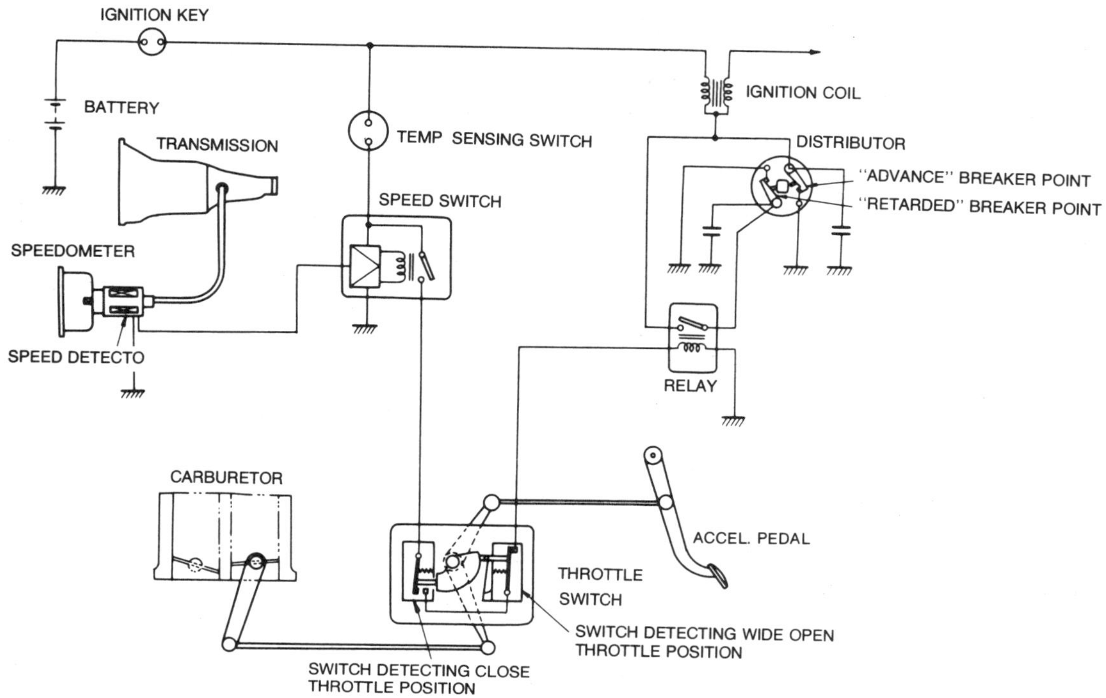

Spark timing control system (dual point distributor) for 1972 models with automatic transmission

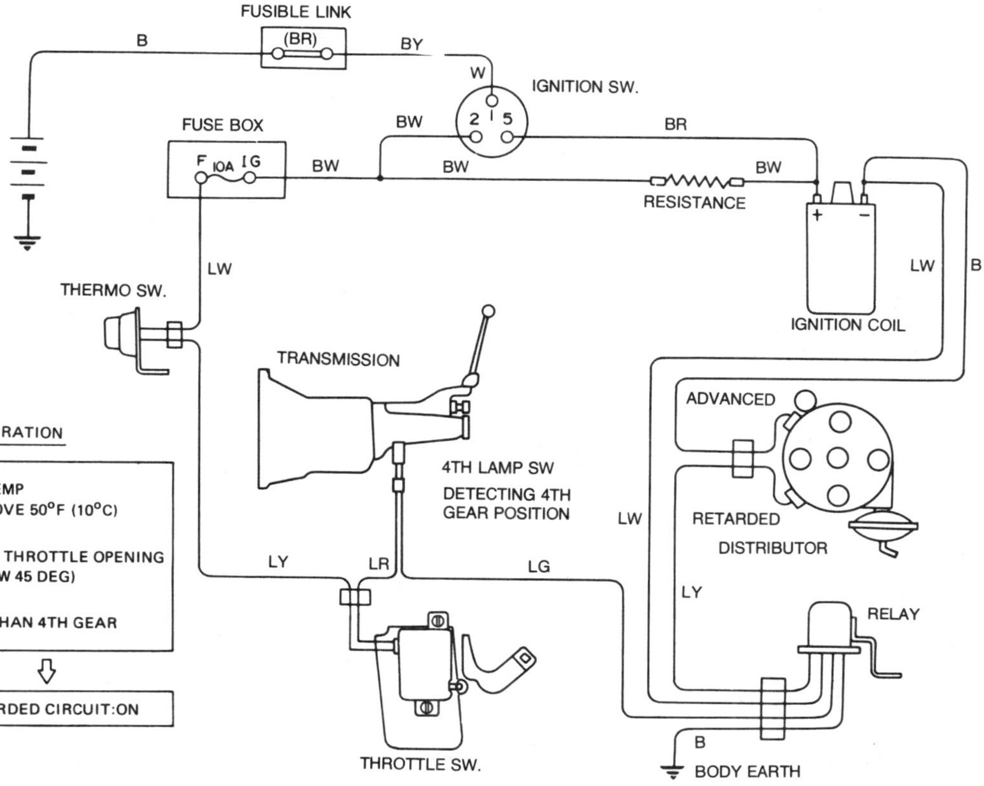

Spark timing control system (dual point distributor) for 1973 models with manual transmission

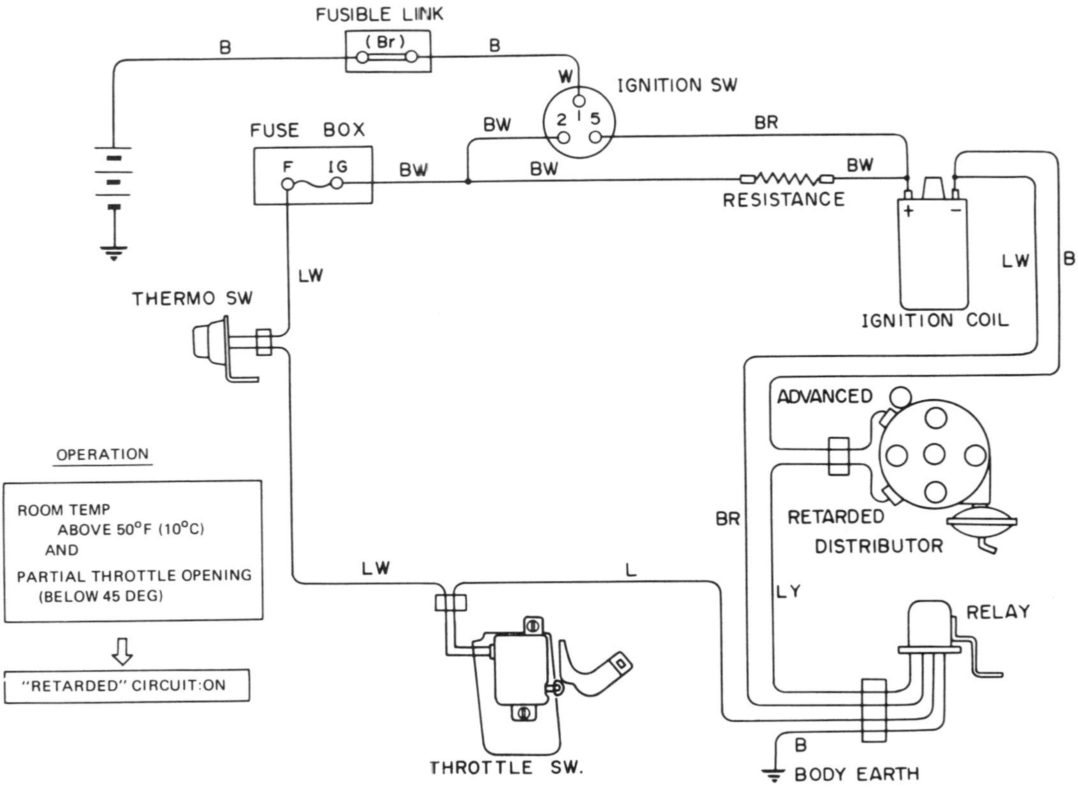

Spark timing control system (dual point distributor) for 1973 models with automatic transmission

transmission is in any gear but Fourth gear.

NOTE: *When the ambient temperature is below 34°F, the retarded set of points is removed from the ignition circuit no matter what switch is on.*

In the case of an automatic transmission, the retarded set of points is activated at all times except under heavy acceleration and high-speed cruising (wide open throttle) with the ambient temperature above 50°F.

There are three switches which control the operation of the distributor relay on 1972–73 models; five switches on 1970–71 models. All of the switches must be ON in order to energize the distributor relay thus energizing the retarded set of points.

The switches and their operation are as follows:

A transmission switch located in the transmission closes an electrical circuit when the transmission is in Third gear on 1970–71 models and in all gears except Fourth gear on 1973 models. On 1970–71 models only, there is a transmission neutral switch which is ON when the transmission is in all gears except Neutral.

A clutch switch mounted against the clutch pedal on 1970–71 models only is ON when the clutch pedal is released (clutch engaged).

A throttle switch located on the throttle linkage at the carburetor is ON when the throttle valve is moved within a predetermined angle: up to 35° on 1970–71 models, 40° on 1972 models, and 45° on 1973 models.

An accelerator switch mounted to the accelerator pedal linkage on all models except 1973 models is ON when the accelerator pedal is nearly completely released and OFF when the pedal is opened farther.

The temperature sensing switch on 1972–73 models is located near the hood release lever inside the passenger compartment. The

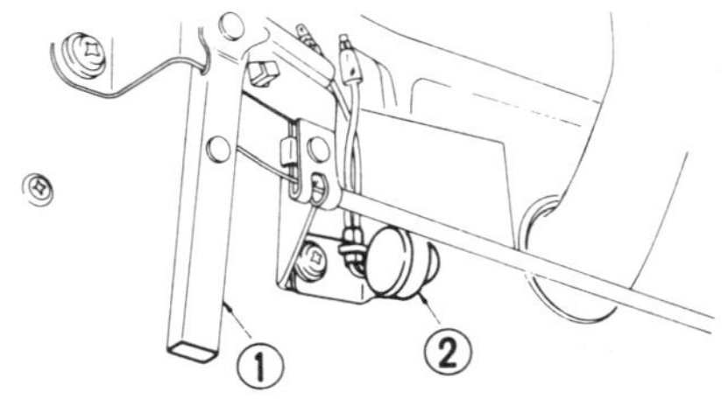

1. Hood release lever
2. Temperature sensing switch

Temperature sensing switch location

temperature sensing switch comes ON between 41°F and 55°F when the temperature is rising and goes OFF at about 34°F when the temperature falls.

The distributor vacuum advance mechanism produces a spark advance based on the amount of vacuum in the intake manifold. With a high vacuum, less air/fuel mixture enters the engine cylinders and the mixture is therefore less highly compressed. Consequently, this mixture burns more slowly and the advance mechanism gives it more time to burn. This longer burning time results in higher combustion temperatures at peak pressure and hence, more time for nitrogen to react with oxygen and form nitrogen oxides (NO_x). At the same time, this advance timing results in less complete combustion due to the greater area of cylinder wall (quench area) exposed at the instant of ignition. This "cooled" fuel will not burn as readily and hence, results in higher unburned hydrocarbons (HC). The production of NO_x and HC resulting from vacuum advance is highest during idle and moderate acceleration in lower gears.

Retardation of the ignition timing is necessary to reduce NO_x and HC emissions. Various ways of retarding the ignition spark have been used in automobiles, all of which remove vacuum to the distributor vacuum advance mechanism at different times under certain conditions. Another way of accomplishing the same goal is the dual point distributor system.

AIR INJECTION SYSTEM (AIS) 1975–80

In gasoline engines, it is difficult to burn the air/fuel mixture completely through normal combustion in the combustion chambers. Under certain operating conditions, unburned fuel is exhausted into the atmosphere.

The air injection reactor system is designed so that ambient air, pressurized by an air pump, is injected through the injection nozzles into exhaust ports near each exhaust valve. The exhaust gases are at high temperatures and ignite when brought into contact with the oxygen of the ambient air. Thus, the unburned fuel is burned in the exhaust ports and manifold.

A check valve is installed in the air pump discharge line to prevent the airflow from reversing due to a broken drive belt, relief valve spring failure, or backfire in the ex-

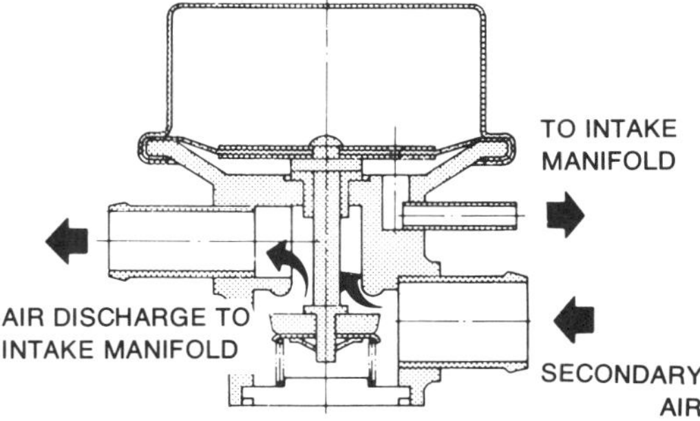

Anti-backfire valve operation—1975 and later

haust manifold. Reversed airflow could damage the air pump.

The air pump relief valve bleeds off excess air from the pump at high speeds. The valve is mounted on the carburetor air cleaner.

Trucks with a catalytic converter (1976 and later pick-ups sold in California) have protection devices to prevent converter overheating due to large quantities of injected air. 1976–77 models use an emergency air relief valve and an air control valve. The emergency valve has a diaphragm operated by engine vacuum. When intake manifold vacuum reaches a predetermined level, the valve opens, diverting air from the pump into the atmosphere. When vacuum drops, the valve closes allowing normal AIS operation.

The air control valve is also controlled by engine vacuum. High vacuum and high pressure from the air pump open the control valve, venting air from the pump into the air cleaner.

1978–79 models have a combined air control valve instead of the relief valve, emergency valve, and air control valve. The combined air control valve regulates the amount of injected air according to intake manifold and air pump discharge pressure, to prevent the converter from overheating.

An anti-backfire valve is installed in an air delivery hose. The purpose of the valve is to prevent backfiring in the exhaust manifold during deceleration. When the throttle closes suddenly, an overly rich air/fuel mixture exists in the intake manifold due to the lack of air getting past the throttle valves. This rich mixture will not completely burn in the combustion chamber. If the unburned gases were to come in contact with the oxygen pumped into the exhaust ports by the air pump, they would ignite and cause backfiring and possible damage.

The anti-backfire valve is connected to the intake manifold by a vacuum line and when the vacuum rises, the valve opens a port in the intake manifold, allowing extra filtered air from the air cleaner to be admitted into the combustion chambers, leaning out the overly rich mixture.

1979 cab and chassis models have a transistorized programmed control unit, a vacuum switching valve, and an air control switch which govern air flow through the AIS. The air control switch, located between the intake manifold and the control unit, turns off when manifold vacuum is high, and on when vacuum is low. This provides a signal to the control unit, which determines when to turn the vacuum switching valve on or off accordingly. The vacuum switching valve controls the upper chamber of the CAC valve diaphragm, opening or closing the CAC valve according to signals received from the control unit. Thus, the amount of air injected into the AIS is monitored and adjusted as conditions warrant.

AIR INDUCTION SYSTEM 1981

The air induction system is designed to send secondary air to the exhaust manifold, utilizing a vacuum caused by exhaust pulsation in the exhaust manifold.

The exhaust pressure in the exhaust manifold usually pulsates in response to the opening and closing of the exhaust valve and it decreases below atmospheric pressure periodically.

If a secondary air intake pipe is opened to the atmosphere under vacuum conditions, secondary air can be drawn into the exhaust manifold in proportion to the vacuum.

Therefore, the air induction system reduces CO and HC emissions in exhaust gases. The system consists of two air induction valves, a filter, hoses and E.A.I. tube(s).

Air Induction Valve Case

The air induction valve case consists of two reed valves, a rubber seal and a filter and is attached to the air cleaner. There are two types of air induction valve cases. Type-A is equipped with two hose connectors and is installed on California models, while Type-B is equipped with one connector and is installed on non-California models.

Air Induction Valve

Two reed valve type check valves are installed in the air cleaner. When the exhaust

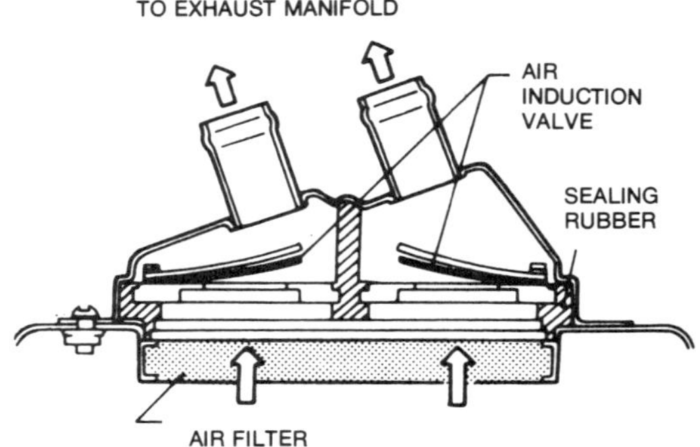

California type air induction case

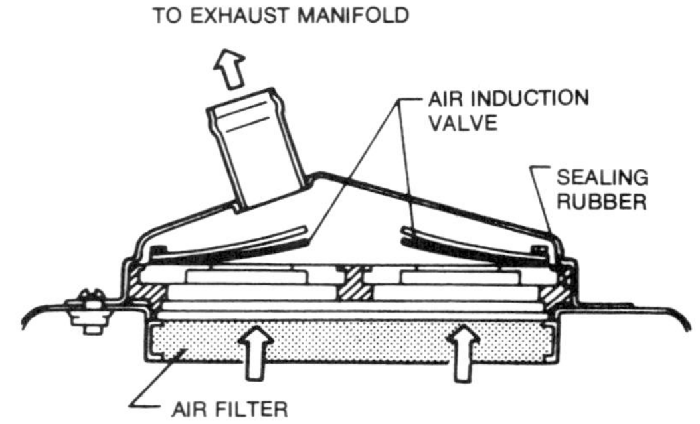

49 states air induction case

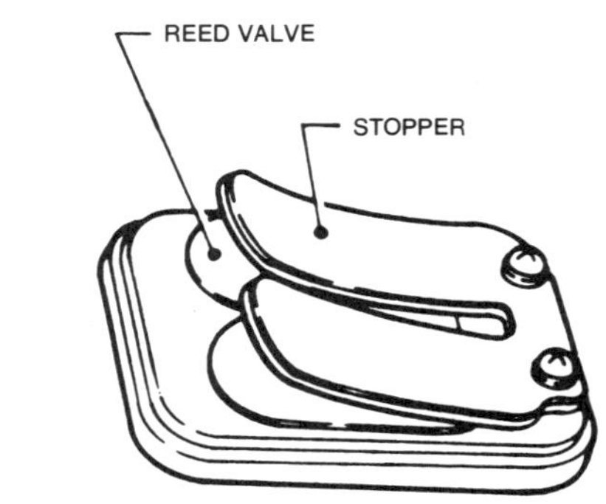

Air induction valve

pressure is below atmospheric pressure (negative pressure), secondary air is sent to the exhaust manifold.

When the exhaust pressure is above atmospheric pressure, the reed valves prevent secondary air from being sent back to the air cleaner.

Air Induction Valve Filter

The air induction valve filter is installed at the dust side of the air cleaner. It purifies secondary air to be sent to the exhaust manifold.

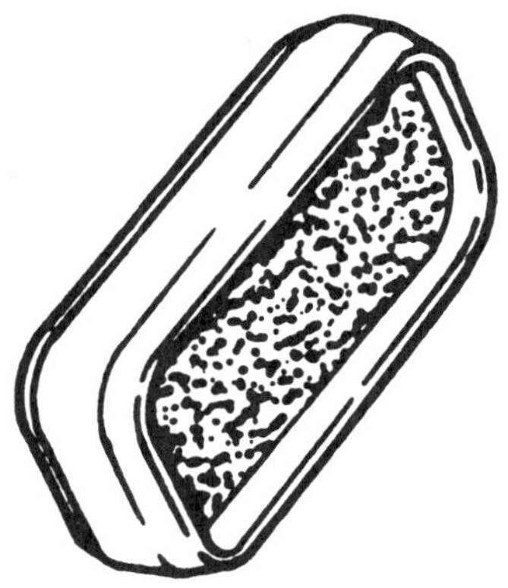

Air induction filter

Air Induction Pipe

The secondary air fed from the air induction valve goes through the E.A.I. pipe to the exhaust manifold.

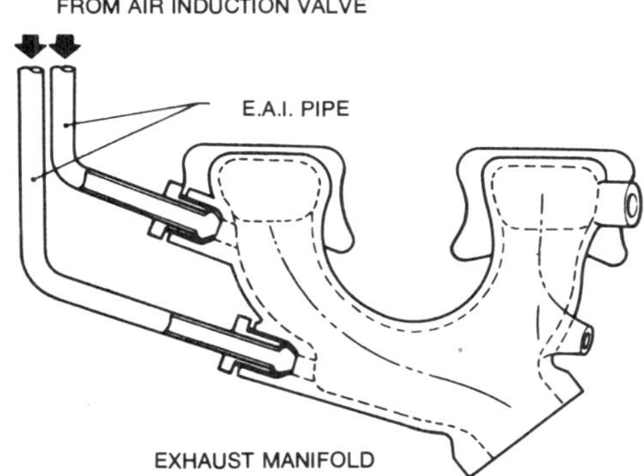

Air induction pipes on California trucks

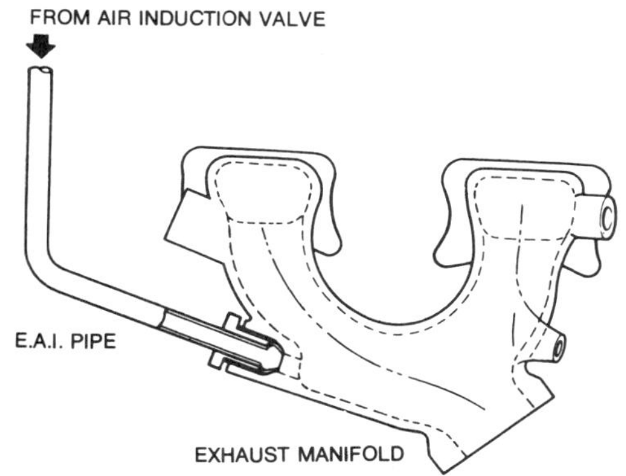

Air induction pipe on non-California trucks

Anti-backfire (A.B.) Valve

This valve is actuated by intake manifold vacuum to prevent backfire in the exhaust system at the initial period of deceleration.

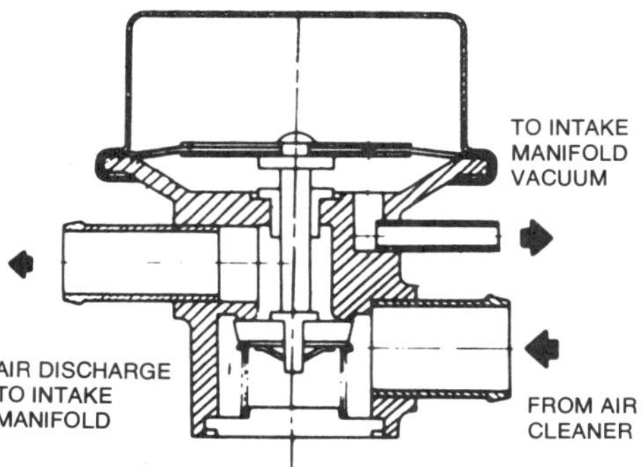

Anti-backfire valve

At this period, the mixture in the intake manifold becomes too rich to ignite and burn in the combustion chamber and burns easily in the exhaust system with injected air in the exhaust manifold.

The A.B. valve provides air to the intake manifold to make the air-fuel mixture leaner and prevents backfire.

The correct function of this valve reduces hydrocarbon emission during deceleration.

BOOST CONTROL DECELERATION DEVICE (BCCD)

The BCDD reduces hydrocarbon emissions during coasting conditions.

High manifold vacuum during coasting prevents the complete combustion of the air/fuel mixture because of the reduced amount of air. This condition will result in large HC emissions. Enriching the air/fuel mixture for a short time (during the high vacuum condition) will reduce the emission of HC in conjunction with the AIR system.

However, enriching the air/fuel mixture with only the mixture adjusting screw will cause poor engine idle, or invite an increase in the carbon monoxide (CO) content of the exhaust gases.

The BCDD consists of an independently operated auxiliary fuel system. This system functions when the engine is coasting to enrich the air-fuel mixture which minimizes the hydrocarbon content of the exhaust gases through more efficient combustion. This is accomplished without adversely affecting engine idle and the carbon monoxide content of the exhaust gases.

When intake manifold vacuum exceeds a predetermined value, a vacuum-actuated diaphragm opens an air passage allowing additional air to enter the intake manifold. When the additional air passage is opened, vacuum is brought to bear on another diaphragm which opens a fuel passage allowing additional fuel to enter the intake manifold.

When the engine changes from a coasting condition to that of idling, the transmission speed sensor closes an electrical circuit, energizing the vacuum control solenoid valve. When energized, the vacuum control solenoid valve vents the intake manifold vacuum to the atmosphere, thus causing the two diaphragms to return to their normal positions, closing off the additional air and fuel mixture. The transmission switch is not used on 1978–79 models.

AUTOMATIC TEMPERATURE CONTROLLED (ATC) AIR CLEANER

The rate of fuel atomization varies with the temperature of the air with which the fuel is being mixed. The air/fuel ratio cannot be held constant for efficient fuel combustion with a wide range of air temperatures. Cold air being drawn into the engine causes a denser and richer air/fuel mixture, inefficient fuel atomization, and thus, more hydrocarbons in the exhaust gas. Hot air being drawn into the engine causes a leaner air/fuel mixture and more efficient atomization and combustion for less hydrocarbons in the exhaust gases.

The automatic temperature controlled air cleaner is designed so that the temperature of the ambient air being drawn into the engine is automatically controlled, to hold the temperature of the air and, consequently, the fuel/air ratio at a constant rate for efficient fuel combustion.

A temperature sensing vacuum switch controls vacuum applied to a vacuum motor op-

erating a valve in the intake snorkle of the air cleaner. When the engine is cold or the air being drawn into the engine is cold, the vacuum motor opens the valve, allowing air heated by the exhaust manifold to be drawn into the engine. As the engine warms up, the temperature sensing unit shuts off the vacuum applied to the vacuum motor which allows the valve to close, shutting off the heated air and allowing cooler, outside (underhood) air to be drawn into the engine.

EXHAUST GAS RECIRCULATION SYSTEM (EGR)

Exhaust gas recirculation is used to reduce combustion temperatures in the engine, thereby reducing the oxides of nitrogen emissions.

An EGR valve is mounted on the center of the intake manifold. The recycled exhaust gas is drawn into the bottom of the intake manifold riser portion through the exhaust manifold heat stove and EGR valve. A vacuum diaphragm is connected to a timed signal port at the carburetor flange.

As the throttle valve is opened, vacuum is applied to the EGR valve vacuum diaphragm. When the vacuum reaches about 2 in. Hg, the diaphragm moves against spring pressure and is in a fully up position at 8 in. Hg of vacuum. As the diaphragm moves up, it opens the exhaust gas metering valve which allows exhaust gas to be pulled into the engine intake manifold. The system does not operate when the engine is idling because the exhaust gas recirculation would cause a rough idle.

On pre-1975 models, an electrically-operated solenoid is located in the vacuum line between the EGR valve and the carburetor. The operation of the solenoid is controlled by a temperature sensing switch mounted in the coolant outlet housing. When the temperature of the coolant is below normal operating temperature, the solenoid is electrically activated and blocks the vacuum line leading to the EGR valve, thus preventing exhaust gas recirculation. When the temperature of the engine coolant reaches operating temperature, the solenoid is deactivated and the vacuum is allowed to act upon the EGR valve diaphragm and exhaust gas recirculation takes place.

On 1975 and later models, a thermal vacuum valve inserted in the engine thermostat housing controls the application of vacuum to the EGR valve. When the engine coolant

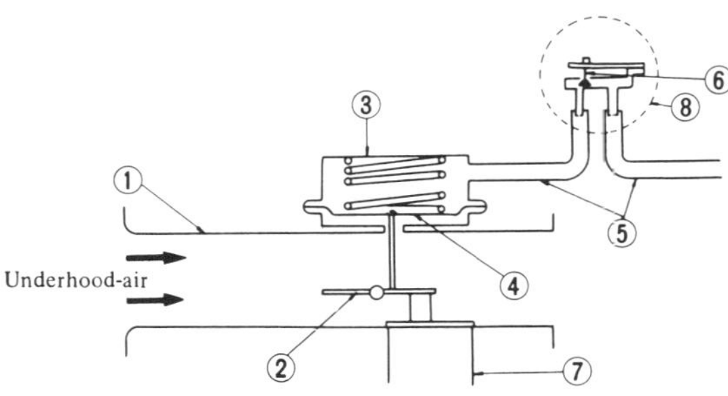

1. Air inlet pipe
2. Air control valve
3. Diaphragm spring
4. Diaphragm
5. Vacuum hoses
6. Air bleed valve (fully open)
7. Hot air pipe
8. Temperature sensor assembly

Automatic temperature control air cleaner system open to underhood air—warm engine operation

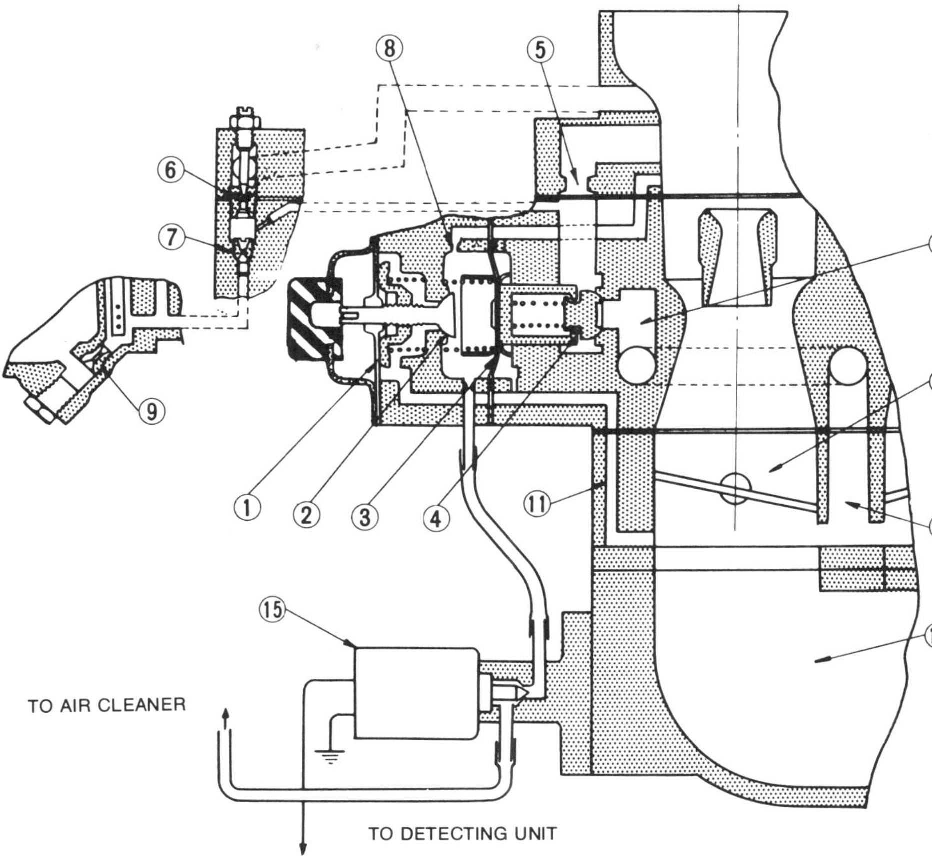

1. Diaphragm I
2. Vacuum control valve
3. Diaphragm II
4. Mixture control valve
5. Coasting air bleed II
6. Coasting air bleed I
7. Coasting jet
8. Air jet
9. Secondary main jet
10. Mixture air passage
11. Boost passage
12. Secondary barrel
13. Mixture outlet
14. Intake manifold
15. Vacuum control
    solenoid valve

Boost Controlled Deceleration Device (BCDD). 1973 shown; later years similar

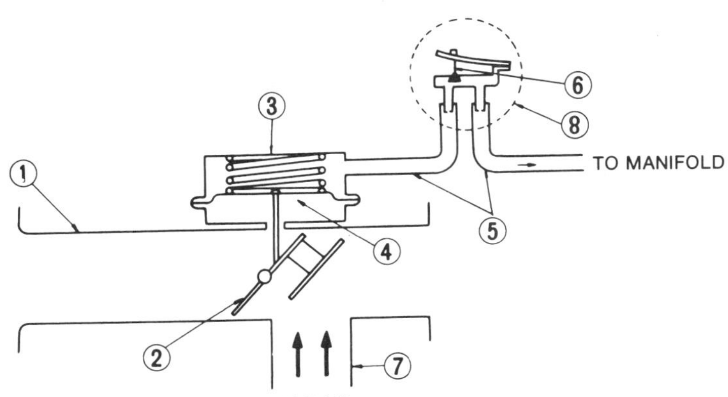

1. Air inlet pipe
2. Air control valve
3. Diaphragm spring
4. Diaphragm
5. Vacuum hose
6. Air bleed valve
    (partially open)
7. Hot air pipe
8. Temperature sensor
    assembly

Automatic temperature control air cleaner system closed to underhood air—cold engine operation

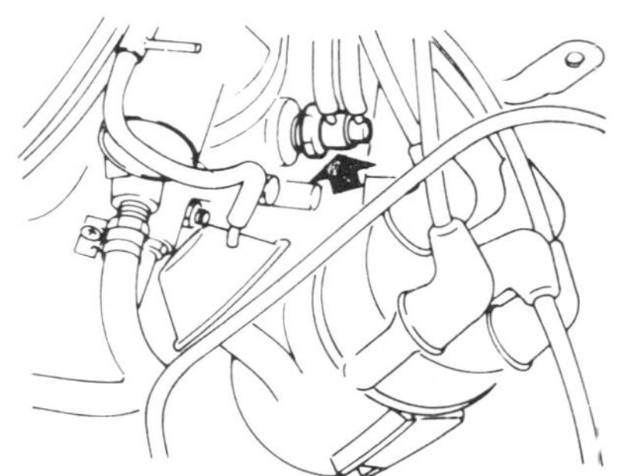

Installation of the EGR thermal vacuum valve—1975

reaches a predetermined temperature, the thermal vacuum valve opens and allows vacuum to be routed to the EGR valve. Below the predetermined temperature, the thermal vacuum valve closes and blocks vacuum to the EGR valve.

1978–79 models have a B.P.T. valve installed between the EGR valve and the thermal vacuum valve. The B.P.T. valve has a diaphragm raised or lowered by exhaust back pressure. The diaphragm opens or closes an air bleed, which is connected into the EGR vacuum line. High pressure results in higher levels of EGR, because the diaphragm is raised, closing off the air bleed, which allows more vacuum to reach and open the EGR valve. Thus, the amount of recirculated exhaust gas varies with exhaust pressure.

1978–79 California models have a vacuum delay valve installed in the line between the thermal vacuum valve and the EGR valve. This valve delays rapid drops in vacuum in the EGR line, thus effecting a longer EGR time.

On all 1975 model trucks (except Canadian models) and all 1976–77 49 States model trucks, the EGR system is equipped with a warning system which monitors the distance the pick-up has traveled and activates a warning light when the EGR system must be

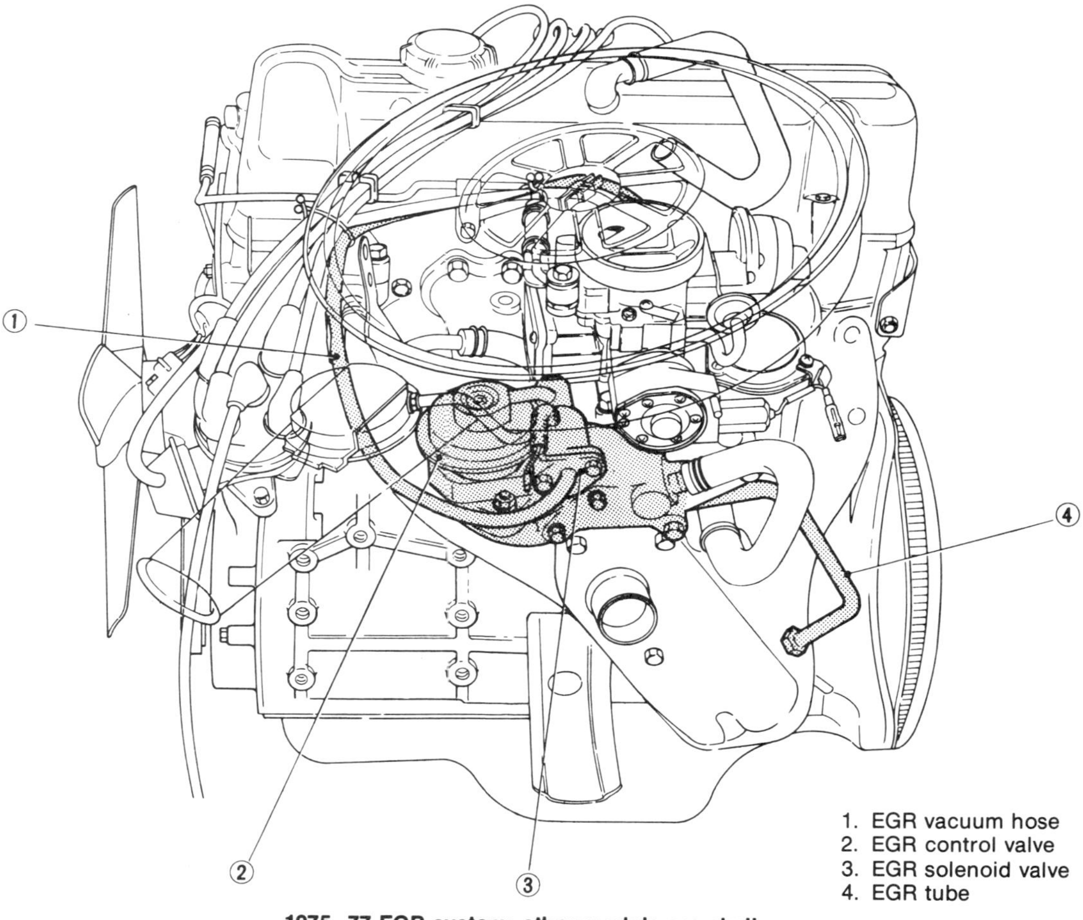

1. EGR vacuum hose
2. EGR control valve
3. EGR solenoid valve
4. EGR tube

1975–77 EGR system; other models are similar

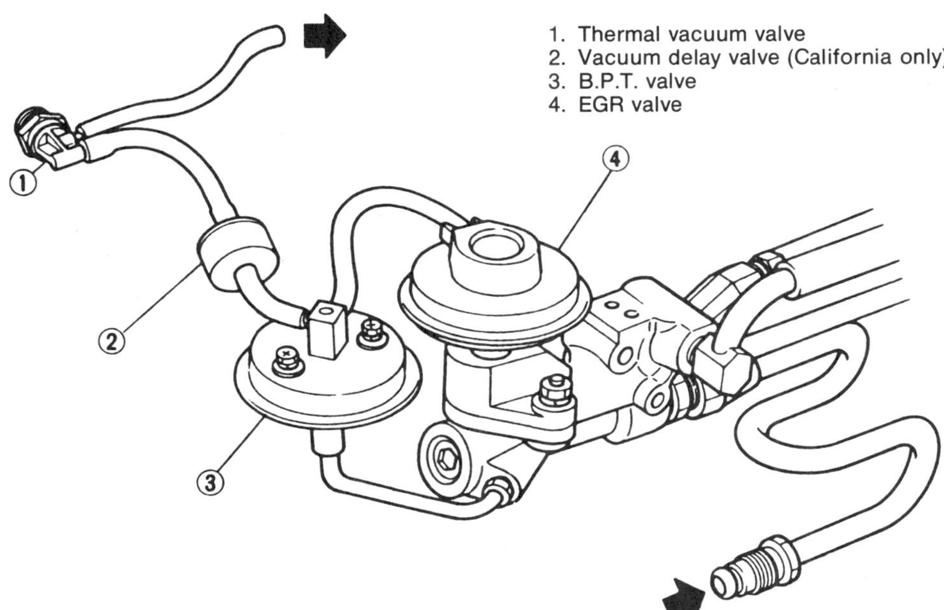

1978–79 EGR components; the hose at the top runs to the carburetor, and the tube at the bottom connects to the exhaust manifold

checked and possibly serviced. The EGR warning light, mounted on top of the dash, comes on when a predetermined number of miles has been traveled and every time the starter is engaged as a check for a burned-out bulb.

To reset the counter, which is mounted on the right fender apron under the hood, remove the grommet installed in the side of the counter and insert the tip of a small screwdriver into the hole. Press down on the knob inside the hole. Reinstall the grommet.

ELECTRIC CHOKE

The purpose of the electric choke installed on the Datsun pick-up since 1972 is to shorten the time that the choke is in operation after the engine is started, thus shortening the time of high HC output.

An electric heater warms the bimetal spring which controls the opening and closing of the choke valve. The heater starts to heat as soon as the engine starts.

CATALYTIC CONVERTER

1976 and later trucks sold in California have a catalytic converter, which is a muffler-shaped device installed into the exhaust system. The converter is filled with a monolithic substrate coated with small amounts of platinum and palladium. Through catalytic action, a chemical change converts carbon monoxide and hydrocarbons into carbon dioxide and water. The catalytic process is aided by the injection of air from the air pump system, which oxidizes the HC and CO before they reach the converter.

1976–78 catalyst-equipped trucks have a floor temperature warning system, consisting of a temperature sensor, installed onto the floor of the cab above the converter; a relay, located with the other relays on the right fender of the engine compartment; and a light, installed on the instrument panel. The lamp turns on when floor temperatures become abnormally high, due to converter or engine malfunction. The light also comes on when the ignition switch is turned to Start, to check its operation. 1979 models do not have the warning system.

Trucks with the catalytic converter also have a combined air control valve in 1978 and 1979, which controls the amount of secondary air injected into the exhaust manifold. It is regulated by engine vacuum and air pump pressure, and works to keep the converter temperatures within proper limits. The combined air control valve replaces the air pump relief valve, found in the air pump system of trucks not equipped with a catalytic converter. 1976–77 models have an emergency air relief valve for catalyst protection. See the AIS section for a description.

SPARK TIMING CONTROL SYSTEM

A spark timing control system is added to manual transmission models sold in the U.S. in 1979. The system controls distributor vacuum advance, giving full vacuum advance when the transmission is in 4th or 5th, and partial advance in the first three gears. This provides better control of the combustion process, lowering emissions of HC and NO_x.

The system components include a top detecting switch, installed into the transmission, and a vacuum switching valve spliced into the distributor vacuum advance hose by means of a three way connector. When the transmission is shifted into either of the two top gears, the transmission switch goes on, thus activating the vacuum switching valve which closes its air bleed, giving full advance. Shifting into any gear but 4th or 5th turns the transmission switch off, deactivating the vacuum switching valve. The valve opens a vacuum leak, providing only partial vacuum advance to the distributor.

INSPECTION AND ADJUSTMENTS

Spark Timing Control System Dual Point Distributor Phase Difference

1. Disconnect the wiring harness of the distributor from the engine harness.

2. Connect the black wire of the engine harness with the black wire of the distributor harness with a jumper wire. This connects the advanced set of points.

3. With the engine idling, adjust the ignition timing by rotating the distributor.

4. Disconnect the jumper wire from the black wire of the distributor harness and connect it to the yellow wire of the distributor harness. The retarded set of points is now activated.

5. With the engine idling, check the igni-

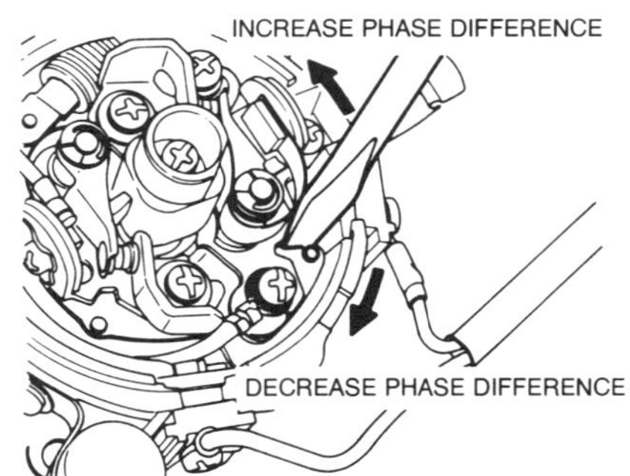

Adjusting the phase angle on a dual point distributor

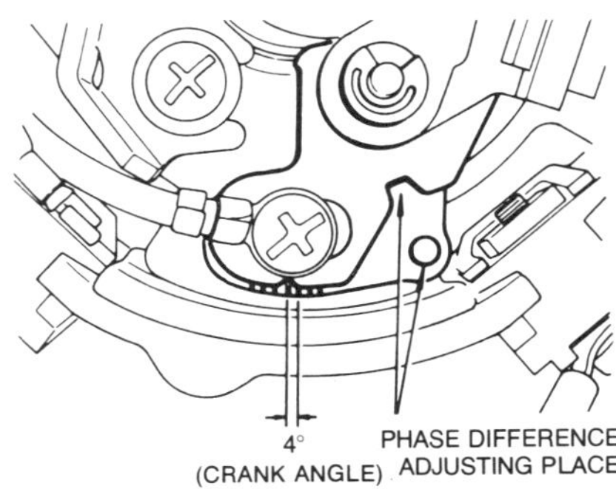

Phase angle adjusting scale

tion timing. The timing should be retarded from the advanced setting as follows: 5°—1970; 10°—1971; 7°—1972–73.

6. To adjust the out-of-phase angle of the ignition timing, loosen the adjuster plate set screws on the same side as the retarded set of points.

7. Place the blade of a screwdriver in the adjusting notch of the adjuster plate and turn the adjuster plate as required to obtain the correct retarded ignition timing specification. The ignition timing is retarded when the adjuster plate is turned counterclockwise. There are graduations on the adjuster plate to make the adjustment easier; one graduation is equal to 4° of crankshaft rotation.

8. Replace the distributor cap, start the engine, and check the ignition timing with the retarded set of points activated (yellow wire of the distributor wiring harness connected to the black wire of the engine wiring harness).

9. Repeat the steps above as necessary to properly set the retarded ignition timing.

Transmission Switch

Disconnect the electrical leads at the switch and connect a self-powered test light to the

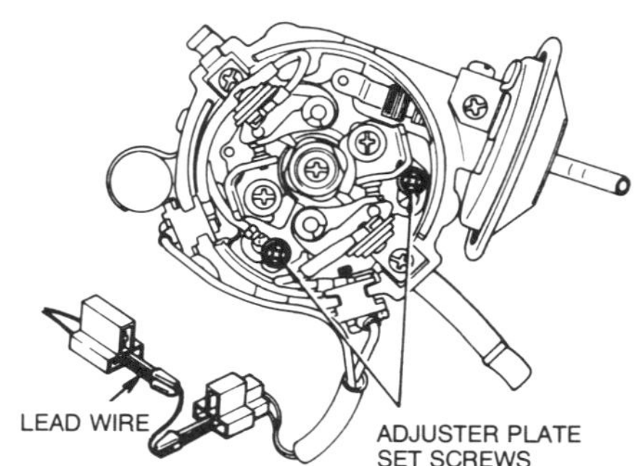

Connect a jumper wire between the two wiring harnesses to activate just one set of points

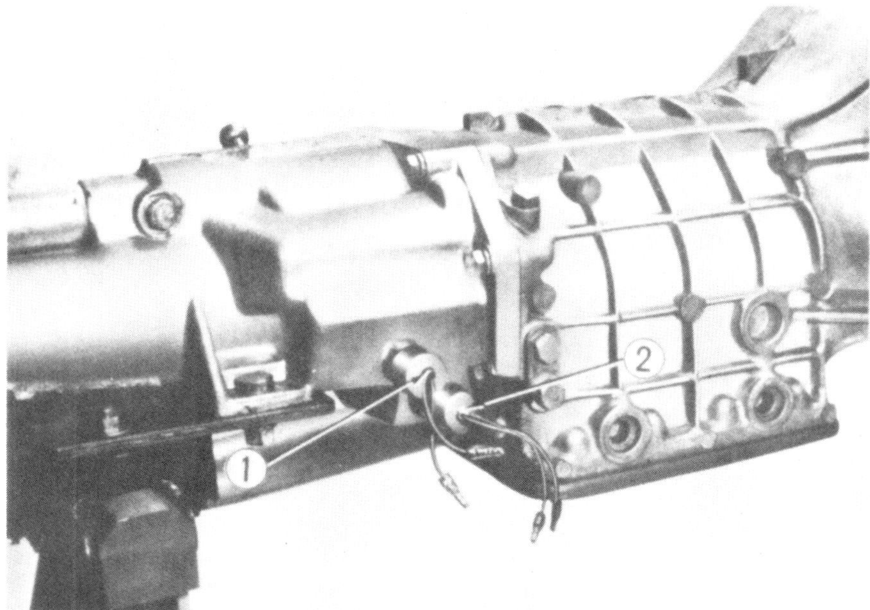

1. Third gear switch　　2. Reverse gear switch

Transmission switch location

electrical leads. The switch should conduct electricity only when the gearshift is moved to the corresponding gear for that particular model year vehicle: Third gear on 1970–72 models and Fourth gear on 1973 models. The neutral switch on 1970–71 models should conduct current when the transmission is shifted into Neutral.

If the switch fails to perform in the above manner, replace it with a new one.

Clutch Switch

Test the clutch switch on 1970–71 models in the same manner as the transmission switch (self-powered test light). The switch should conduct current when the clutch pedal is released (clutch engaged).

Accelerator Switch

The accelerator switch mounted on the accelerator pedal linkage on all models except 1973 vehicles is checked with a self-powered test light in the same manner as outlined for the transmission switch. The switch should conduct current when the accelerator pedal is nearly completely released.

Throttle Switch

The throttle switch located on the throttle linkage at the carburetor is checked with a self-powered test light. Disconnect the elec-

trical leads of the switch and connect the test light. The switch should not conduct current when the throttle valve is closed or opened as follows: the throttle valve opened up to 35° on 1970–71 models; 40° on 1972 models; and 45° on 1973 models. When the throttle is fully opened, the switch should conduct current.

Temperature Sensing Switch

The temperature sensing switch mounted in the passenger compartment (near the hood release lever) should not conduct current when the temperature is above 55°F when connected to a self-powered test light as previously outlined for the throttle switch.

Air Injection System

AIR PUMP

If the air pump makes an abnormal noise and cannot be corrected without removing the pump from the vehicle, check the following in sequence:

1. Turn the pulley ¾ of a turn in the clockwise direction and ¼ of a turn in the counterclockwise direction. If the pulley is binding and if rotation is not smooth, a defective bearing is indicated.

2. Check the inner wall of the pump body, vanes, and rotor for wear. If the rotor has abnormal wear, replace the air pump.

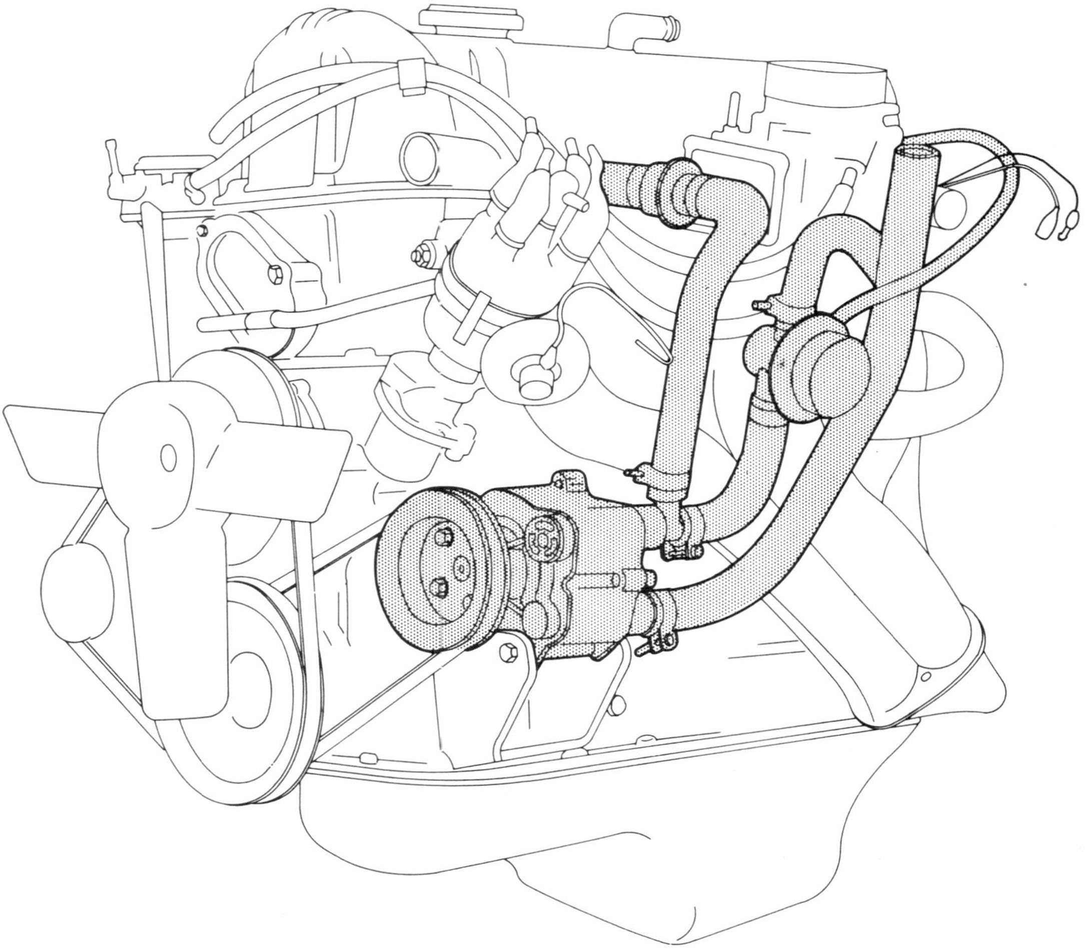

The air injection system

3. Check the needle roller bearing for wear and damage. If the bearings are defective, the air pump should be replaced.

4. Check and replace the rear side seal if abnormal wear or damage is noticed.

5. Check and replace the carbon shoes holding the vanes if they are found to be worn or damaged.

6. A deposit of carbon particles on the inner wall of the pump body and vanes is normal, but should be removed with compressed air before reassembling the air pump.

CHECK VALVE

Remove the check valve from the air pump discharge line. Test it for leakage by blowing air into the valve from the air pump side and from the air manifold side. Air should only pass through the valve from the air pump side if the valve is functioning normally. A

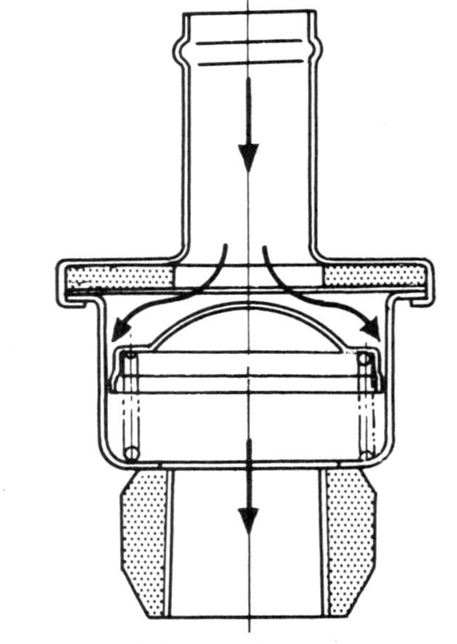

The check valve of the air injection system

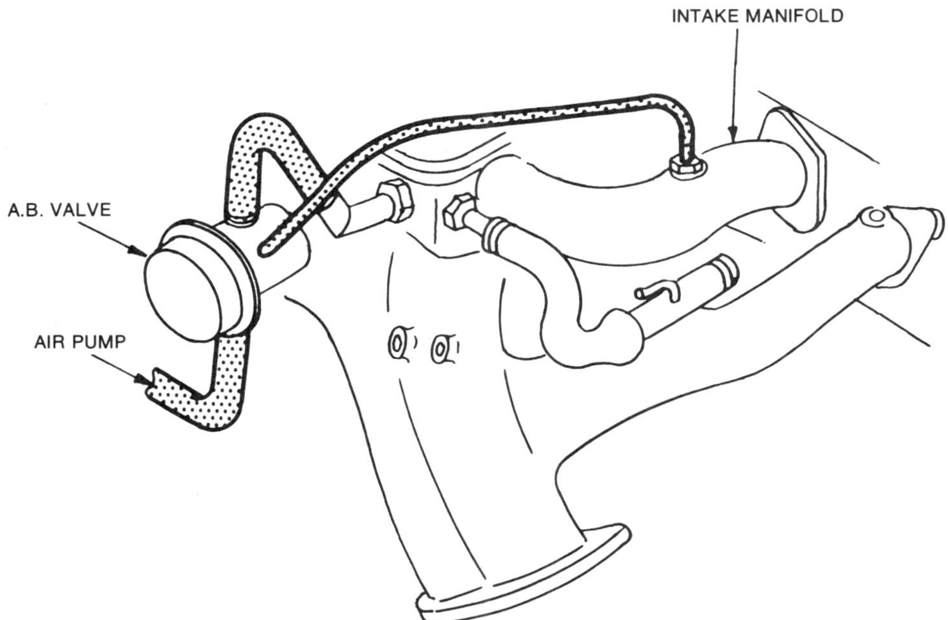

The anti-backfire valve of the air injection system

small amount of air leakage from the manifold side can be overlooked. Replace the check valve if it is found to be defective.

ANTI-BACKFIRE VALVE

To check the valve, disconnect the hose from the air cleaner and place a finger on the end. Run the engine up to about 3,000 rpm, then quickly release the throttle. If the valve is performing correctly, suction should be felt at the end of the hose. If no suction is felt, replace the anti-backfire valve.

AIR PUMP RELIEF VALVE

1. Disconnect the hoses leading to the check valve (on the air injection manifold) and the air control valve from the air hose connector. Plug the connector.

2. Start the engine and increase the engine speed to about 3000 rpm. Place your finger on the outlet of the relief valve (inside the air cleaner housing) and check for air discharge. If you do not feel any air coming out, the relief valve is faulty, and must be replaced.

AIR INJECTION NOZZLES

Check around the air manifold for air leakage with the engine running at 2,000 rpm. If air is leaking from the eye joint bolt, retighten or replace the gasket. Check the air nozzles for restrictions by blowing air into the nozzles.

HOSES

Check and replace hoses if they are found to be weakened or cracked. Check all hose connections and clips. Be sure that the hoses are not in contact with other parts of the engine.

EMERGENCY AIR RELIEF VALVE

1. Warm up the engine.

2. Check all hoses for leaks, kinks, improper connections, etc.

3. Run the engine up to 2000 rpm under no load. No air should be discharged from the valve.

4. Disconnect the vacuum hose from the valve. This is the hose which runs to the intake manifold. Run the engine up to 2000 rpm. Air should be discharged from the valve. If not, replace it.

COMBINED AIR CONTROL VALVE

1. Check all hoses for leaks, kinks, and improper connections.

2. Thoroughly warm up the engine.

3. With the engine idling, check for air discharge from the relief opening in the air cleaner case.

4. Disconnect and plug the vacuum hose from the valve. Air should be discharged from the valve with the engine idling. If the disconnected vacuum hose is not plugged, the engine will stumble.

5. Connect a hand-operated vacuum

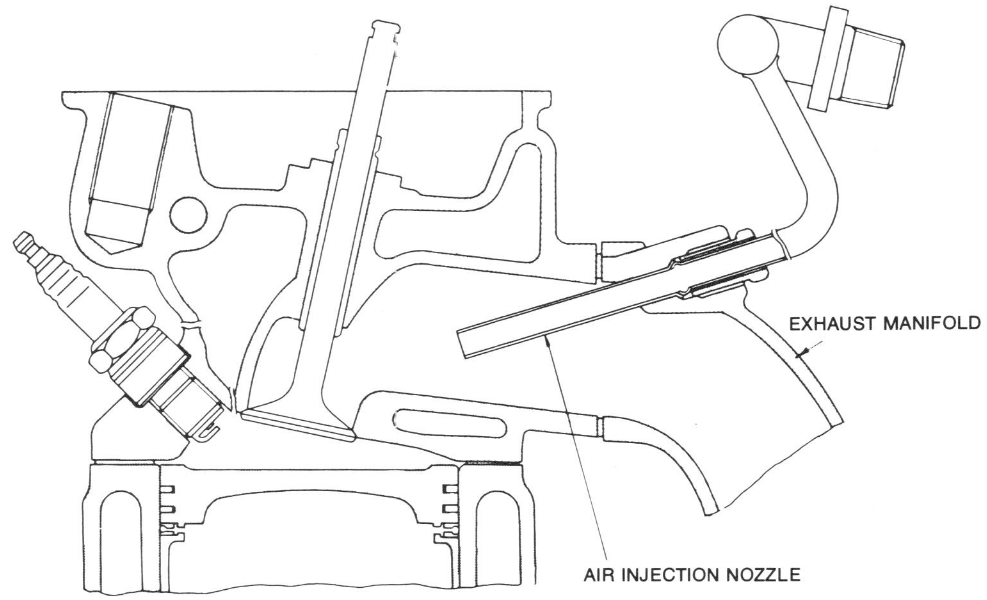

The air injection nozzles

pump to the vacuum fitting on the valve and apply 7.8–9.8 in. Hg. of vacuum. Run the engine speed up to 3000 rpm. No air should be discharged from the valve.

6. Disconnect and plug the air hose at the check valve, with the conditions as in the preceding step. This should cause the valve to discharge air. If not, or if any of the conditions in this procedure are not met, replace the valve.

Air Induction System

AIR INDUCTION VALVE AND FILTER

Remove the valve and filter on the air cleaner. The air induction valve and valve filter can then be taken out easily. Installation is in the reverse sequence of removal.

AIR INDUCTION PIPE

Remove nut securing the pipe to the exhaust manifold. At the same time, remove the screws securing the bracket and rubber hose clamp.

The air induction pipe can then be taken out. Installation is in the reverse sequence of removal.

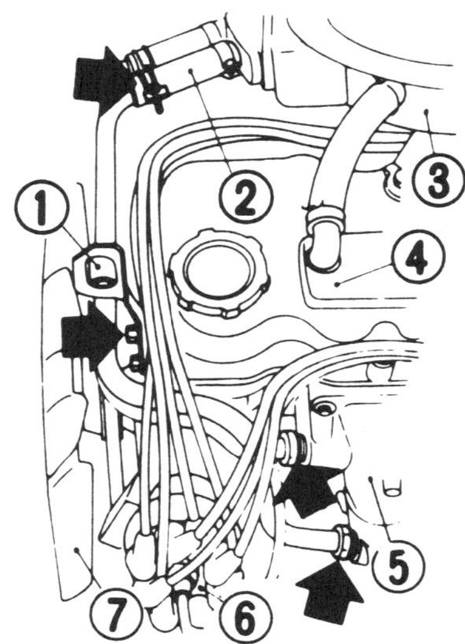

1. Bracket
2. Rubber hose
3. Air cleaner
4. Rocker cover
5. Exhaust manifold
6. Distributor
7. Fan

California air induction pipe layout

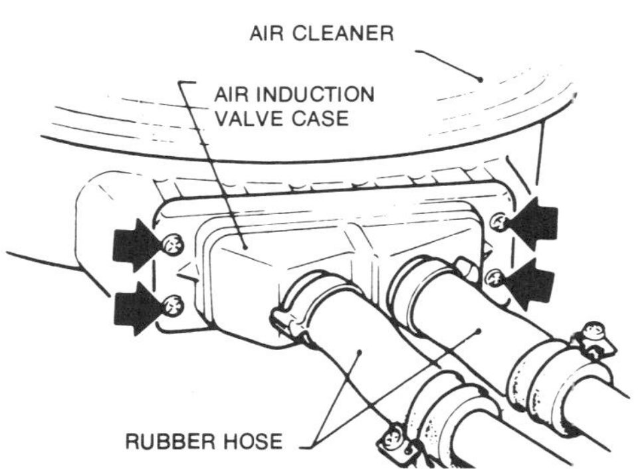

Air induction valve and filter location

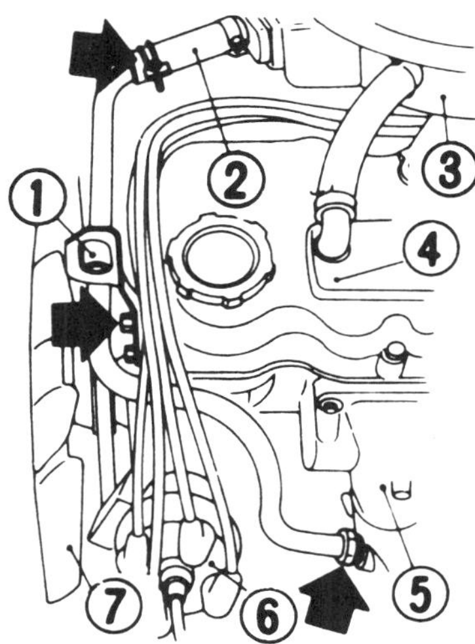

1. Bracket
2. Rubber hose
3. Air cleaner
4. Rocker cover
5. Exhaust manifold
6. Distributor
7. Fan

Non-California air induction pipe layout

A.B. VALVE

1. Remove air cleaner.
2. Remove air hoses and vacuum tube. Then the A.B. valve can be taken out.

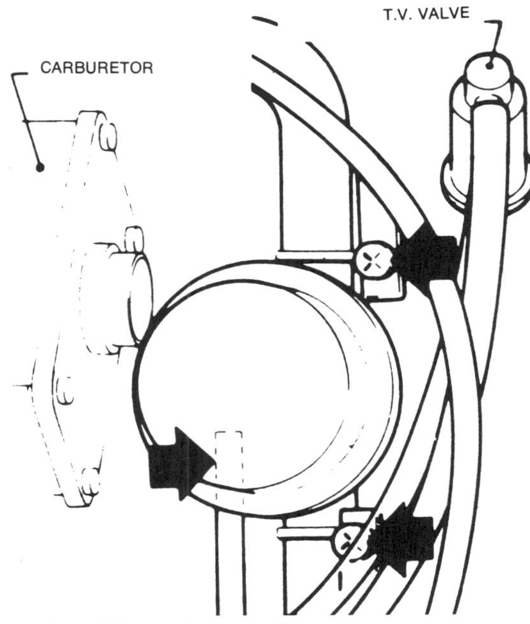

Anti-backfire valve mounting

Boost Controlled Deceleration Device (BCDD)

Normally, the BCDD never needs adjustment. However, if the need should arise be-

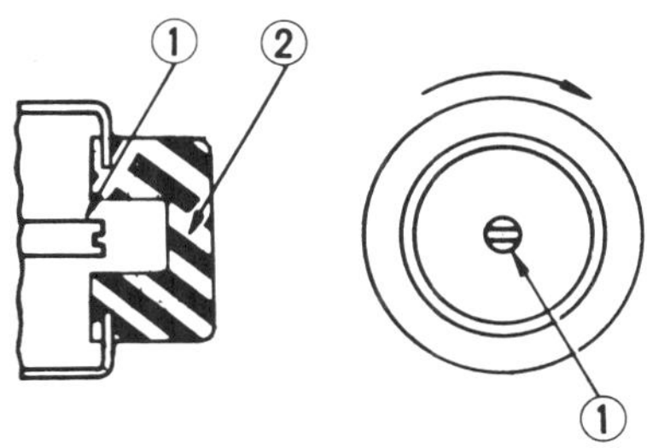

1. Adjusting screw "S"   2. Cover "C"

The operating pressure adjusting screw of the BCDD

cause of suspected malfunction of the system, proceed as follows:

1. Connect a tachometer to the engine.

2. Connect a quick-response vacuum gauge to the intake manifold.

3. Disconnect the BCDD solenoid valve electrical leads.

4. Start and warm up the engine until it reaches normal operating temperature.

5. Adjust the idle speed to the proper specification.

6. Raise the engine speed to 3,000–3,500 rpm under no-load (transmission in Neutral or Park), then allow the throttle to close quickly. Take notice as to whether or not the engine rpm returns to idle speed and if it does, how long the fall in rpm is interrupted before it reaches idle speed.

At the moment the throttle is snapped closed at high engine rpm, the vacuum in the intake manifold reaches −27.7 in. Hg on pre-1975 vehicles and −23.6 in. Hg on 1975 models, and then gradually falls to about −16.5 in. Hg at idle speed. The process of the fall of intake manifold vacuum and engine rpm will take one of the following three forms:

a. When the operating pressure of the BCDD is too high, the system remains inoperative, and the vacuum in the intake manifold decreases without interruption just like that of an engine without a BCDD;

b. When the operating pressure is lower than that of the case given, but still higher than the properly set pressure, the fall of vacuum in the intake manifold is interrupted and kept constant at a certain level (operating pressure) for about one second and then gradually falls down to the normal vacuum at idle speed;

c. When the set operating pressure of the BCDD is lower than the intake manifold vacuum when the throttle is suddenly released, the engine speed will not lower to idle speed.

To adjust the set operating pressure of the BCDD, remove the adjusting screw cover from the BCDD mechanism mounted on the side of the carburetor.

The adjusting screw is a left-hand threaded screw. Turning the screw ⅛ of a turn in either direction will change the operation pressure about 0.79 in. Hg. Turning the screw counterclockwise will increase the amount of vacuum needed to operate the mechanism and turning the screw clockwise will decrease the amount of vacuum needed to operate the mechanism.

The operating pressure for the BCDD on a vehicle with a manual transmission is −19.7 ±0.79 in. Hg and for a vehicle with an automatic transmission −18.9 ±0.79 in. Hg. through 1974. The decrease in intake manifold vacuum should be interrupted at these levels for about one second when the BCDD is operating correctly. The figures for later years are:
- 1975–76:
  −20.7 to −21.1, manual transmission;
  −19.9 to −20.3, automatic transmission;
- 1977:
  −20.1 to −21.7, manual transmission;
  −19.3 to −20.9, automatic transmission;
- 1978:
  −22.05±0.79, all models;
- 1979:
  −21.65±0.75, all models.

Don't forget to install the adjusting screw cover when the system is adjusted.

Automatic Temperature Controlled Air Cleaner

When the air around the temperature sensor of the unit mounted inside the air cleaner housing reaches 100°F, the sensor should block the flow of vacuum to the air control valve vacuum motor. When the temperature around the temperature sensor is below 100°F, the sensor should allow vacuum to pass onto the air valve vacuum motor, thus blocking off the air cleaner snorkle to underhood (unheated) air.

When the temperature around the sensor is about 118°F, the air control valve should be completely open to underhood air.

When the engine is operating under a heavy load (wide open throttle acceleration), the air control valve fully opens to underhood air to obtain full power no matter what the temperature is around the temperature sensor.

Exhaust Gas Recirculation

PRE-1975

Check the operation of the EGR system as follows:

1. Visually inspect the entire EGR control system. Clean the mechanism free of oil and dirt. Replace any rubber hoses found to be cracked or broken.

2. Make sure that the EGR solenoid valve is properly wired.

3. Increase the engine speed from idling to 3,000−3,500 rpm. The plate of the EGR control valve diaphragm and the valve shaft should move upward as the engine speed is increased.

4. Disconnect the EGR solenoid valve electrical leads and connect them directly to the vehicle's 12 volt electrical supply (battery). Race the engine again with the EGR solenoid valve connected to a 12 volt power source. The EGR control valve should remain stationary.

5. With the engine running at idle, push up the EGR control valve diaphragm by pressing it up with your finger. When this is done, the engine idle should become rough and uneven.

Inspect the two components of the EGR system as necessary in the following manner:

1. Remove the EGR control valve from the intake manifold.

2. Apply 4.7−5.1 in. Hg of vacuum to the EGR control valve by sucking on a tube attached to the outlet on top of the valve. The valve should move to the full up position. The valve should remain open for more than 30 seconds after the application of vacuum is discontinued and the vacuum hose is blocked.

3. Inspect the EGR valve for any signs of warpage or damage.

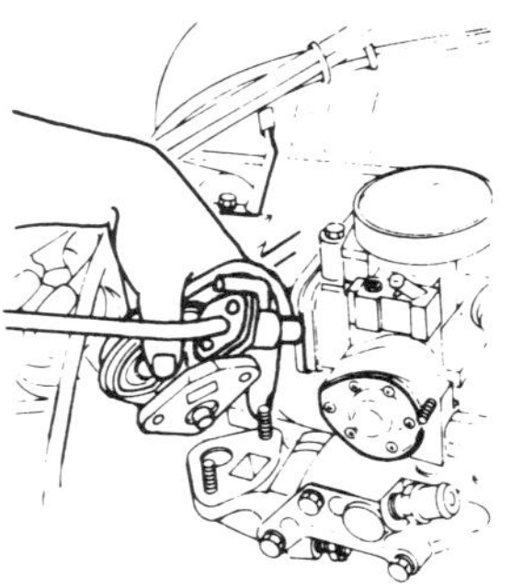

Removing the EGR valve

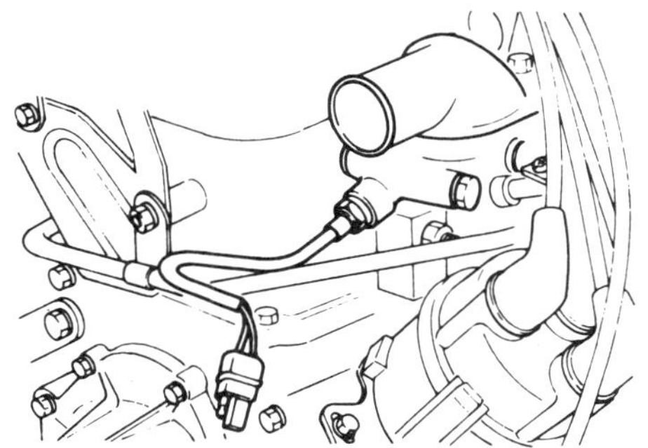

The EGR water temperature sensing switch

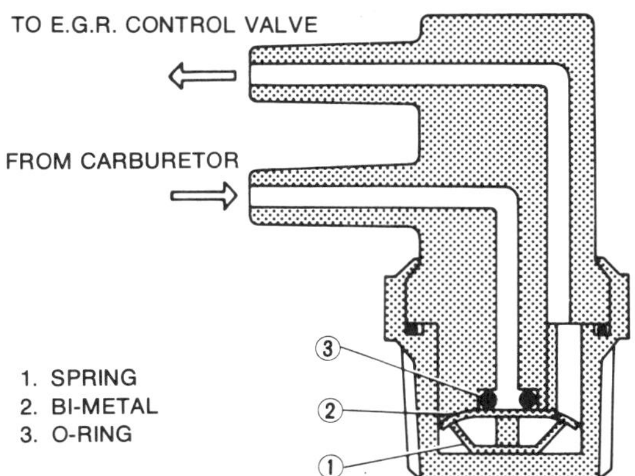

Cutaway view of the operation of the EGR thermal vacuum valve—1975

4. Clean the EGR valve seat with a brush and compressed air to prevent clogging.

5. Connect the EGR solenoid valve to a 12 volt DC power source and notice if the valve clicks when intermittently electrified. If the valve clicks, it is considered to be working properly.

6. Check the EGR temperature sensing switch by removing it from the engine and placing it in a container of water together with a thermometer. Connect a self-powered test light to the two electrical leads of the switch.

7. Heat the container of water.

8. The switch should conduct current when the water temperature is below 77°F and stop conducting current when the water reaches a temperature somewhere between 88–106°F. Replace the switch if it behaves otherwise.

1975 AND LATER

1. Remove the EGR valve and apply enough vacuum to the diaphragm to open the valve.

2. The valve should remain open for over 30 seconds after the vacuum is removed.

3. Check the valve for damage, such as warpage, cracks, and excessive wear around the valve and seat.

4. Clean the seat with a brush and compressed air and remove any deposits from around the valve and port (seat).

5. To check the operation of the thermal vacuum valve, remove the valve from the engine and apply vacuum to the ports of the valve. The valve should not allow vacuum to pass.

6. Place the valve in a container of water with a thermometer and heat the water. When the temperature of the water reaches 134°–145°F, remove the valve and apply vacuum to the ports; the valve should allow vacuum to pass through it.

7. To test the B.P.T. valve installed on 1978 and later models, disconnect the two vacuum hoses from the valve. Plug one of the ports. While applying pressure to the bottom of the valve, apply vacuum to the unplugged port and check for leakage. If any exists, replace the valve.

8. To test the check valve installed in some 1978 and later models, remove the valve and blow into the side which connects to the EGR valve. Air should flow. When air is applied to the other side, air flow resistance should be greater. If not, replace the valve.

Spark Timing Control System

1. Check all hoses and electrical wires for proper connections, leaks or corrosion, and so on.

2. Check the distributor vacuum advance unit for proper operation. This can be checked by hooking up a timing light, starting the engine, then increasing engine speed and observing whether or not the timing marks advance. If not, the advance unit must be checked for binding or leaks.

3. With the timing light installed, increase the engine speed to 2000 rpm. Have an assistant disengage the clutch, then shift between 3rd, 4th, and 5th, then back down and into neutral. Spark timing should vary when the transmission is in 4th or 5th (advance should be greater). If this is not the case, check the vacuum switching valve.

VACUUM SWITCHING VALVE

1. Disconnect the valve's electrical connectors. With the timing light installed, run

the engine up to about 2000 rpm and keep it there. Check the timing.

2. Connect the valve's electrical connectors directly to the battery with a pair of jumper wires. Be sure to observe correct polarity. If spark timing varies, the valve is ok. If not, replace it.

TRANSMISSION SWITCH

1. The switch can be checked easily with an ohmmeter. Connect the ohmmeter leads to the switch leads on the transmission. Shift back and forth between either 4th or 5th and one of the other gears. If the resistance does not change, replace the switch.

GASOLINE ENGINE FUEL SYSTEM

If the fuel pump is suspected as being faulty, tests for both pressure and volume should be performed. Never replace the pump without performing these simple tests first. Always check all hoses for leaks or clogs before testing the pump.

Mechanical Fuel Pump

The fuel pump is a mechanically-operated, diaphragm-type driven by the fuel pump eccentric cam on the front of the camshaft.

Design of the fuel pump permits disassembly, cleaning, and repair or replacement of defective parts.

TESTING

1. Disconnect the line between the carburetor and the pump at the carburetor.

2. Connect a fuel pump pressure gauge into the line.

3. Start the engine. The pressure should be between 3.0 and 3.9 psi. There is usually enough gas in the float bowl to perform this test.

4. If the pressure is ok, perform a capacity test. Remove the gauge from the line. Use a graduated container to catch the gas from the fuel line. Fill the carburetor float bowl with gas. Run the engine for one minute at about 1000 rpm. The pump should deliver 1000cc in one minute or less.

REMOVAL AND INSTALLATION

1. Disconnect the two fuel lines from the fuel pump. Be sure to keep the line leading from the fuel tank up high to prevent the excessive loss of fuel.

2. Remove the two fuel pump mounting nuts and remove the fuel pump assembly from the side of the engine.

3. Install the fuel pump in the reverse order of removal, using a new gasket and sealer on the mating surface.

Electric Fuel Pump

An electric fuel pump is used on 1977 and later models with factory-installed air conditioning, and also on 1979 and later cab and chassis models. The pump is mounted on a bracket located on the right frame rail next to the fuel tank. There is a filter mounted in the body of the pump, which does not normally require service. The pump can be disassembled, if necessary, but all electronic parts within the body (one transistor, two diodes, and three resistors) must be replaced as an assembly.

TESTING

1. Disconnect the hose from the pump outlet at the pump.

2. Connect a length of hose to the outlet. The hose should have an inside diameter of ¼ in. (6mm). The diameter of the hose is important for accurate measurements.

3. Raise the end of the hose above the level of the pump. Turn the ignition switch on and catch the gasoline in a graduated container. Pump output should be 1400cc in one minute or less.

4. Fuel pump pressure should be 4.6 psi

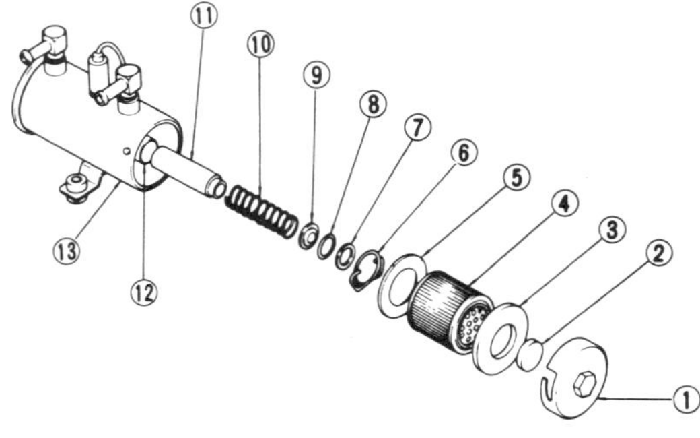

1. End cover	8. O-ring
2. Magnet	9. Inlet valve
3. Gasket	10. Return spring
4. Filter	11. Plunger
5. Gasket	12. Plunger cylinder
6. Retainer	13. Body
7. Washer	

Exploded view of the electric fuel pump

through 1978, and between 3.1 and 3.8 psi in 1979 and later.

REMOVAL AND INSTALLATION

1. Remove the inlet and outlet hoses, catching the fuel that drains in a metal container.

2. Disconnect the wiring at the connector.

3. Remove the two bolts securing the pump to the bracket and remove the pump.

4. Installation is the reverse. Replace the hose clamps if their condition warrants.

Carburetor

The carburetor used on the Datsun pick-up is a two-barrel, downdraft-type with a low-speed (primary) side and a high-speed (secondary) side.

The carburetor installed on 1973–81 models has an electrically-operated anti-dieseling solenoid. As the ignition switch is turned off, the valve is energized and shuts off the supply of fuel to the idle circuit of the carburetor.

REMOVAL AND INSTALLATION

1. Remove the air cleaner.

2. Disconnect and label the fuel and vacuum lines from the carburetor.

3. Remove the throttle linkage.

4. Remove the four nuts and washers retaining the carburetor to the manifold.

5. Lift the carburetor from the manifold.

6. Remove and discard the gasket used between the carburetor and the manifold.

7. Install the carburetor in the reverse order of removal using a new carburetor base gasket.

AUTOMATIC CHOKE ADJUSTMENT

1. With the engine cold, make sure the choke is fully closed (press the gas pedal all the way to the floor and release, or pull the

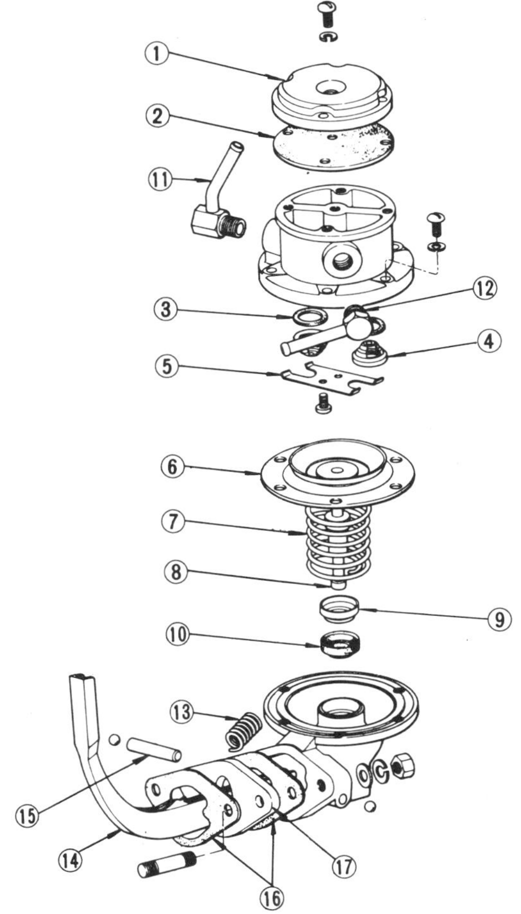

1. Fuel pump cap
2. Cap gasket
3. Valve packing
4. Fuel pump valve
5. Valve retainer
6. Diaphragm
7. Diaphragm spring
8. Pull rod
9. Lower body seal washer
10. Lower body seal
11. Inlet connector
12. Outlet connector
13. Rocker arm spring
14. Rocker arm
15. Rocker arm side pin
16. Fuel pump packing
17. Spacer—fuel pump to
     cylinder block

Exploded view of the mechanical fuel pump

choke knob out on early models with that system).

2. Check the choke linkage for binding. The choke plate should be easily opened and closed with your finger. If the choke sticks or binds, it can usually be freed with a liberal application of a carburetor cleaner made for

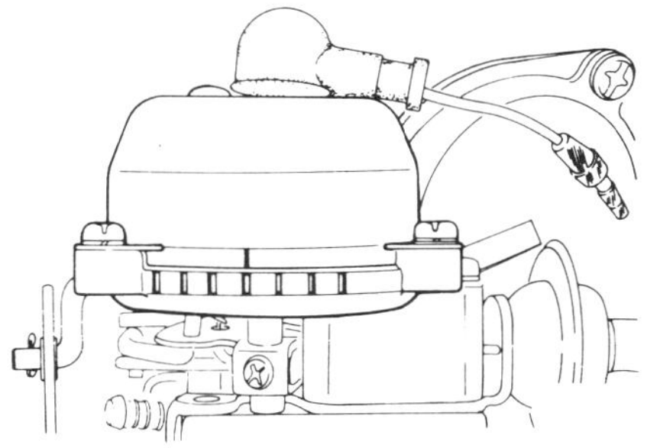

Align the choke cover mark with the center notch on the carburetor

the purpose. A couple of quick shots from a spray can of this stuff normally does the trick. If not, the carburetor will have to be disassembled for repairs.

3. The choke is correctly adjusted when the index mark on the choke housing (notch) aligns with the center mark on the carburetor body. If the setting is incorrect, loosen the three screws clamping the choke body in place and rotate the choke cover left or right until the marks align. Tighten the screws carefully to avoid cracking the housing.

THROTTLE LINKAGE ADJUSTMENT

When the primary throttle valve is opened to an angle of 50° from its closed position, the adjust plate which is integral with the primary throttle valve, is brought into contact with portion A (see illustration) of the return plate. When the primary throttle valve is opened farther, the return plate is pulled apart from the stopper (B in the illustration), allowing the secondary throttle valve to open.

To adjust the linkage:

1. Measure the clearance between the primary throttle valve and the wall of the throttle chamber at the center of the throttle valve when the adjust plate is brought into contact with portion A of the return plate. Standard clearance is 0.26–0.32 in.

2. If necessary, make the adjustment by bending the portion A of the return plate.

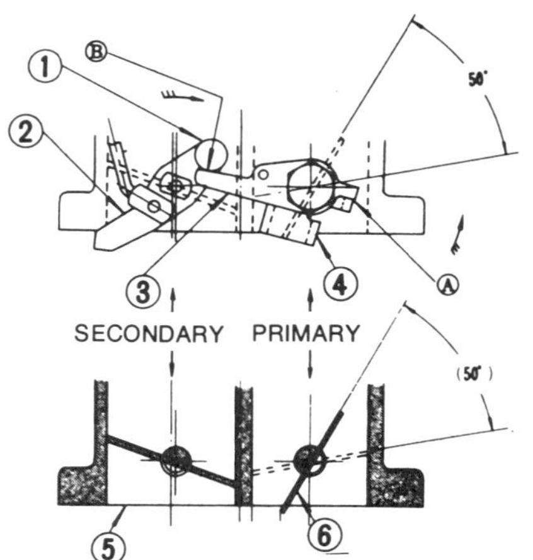

1. Roller
2. Connecting lever
3. Return plate
4. Adjust plate
5. Throttle chamber
6. Throttle valve

Throttle linkage adjustment

FLOAT LEVEL ADJUSTMENT

The fuel level is normal if it is within the lines on the window glass of the float chamber when the vehicle is resting on level ground and the engine is off.

If the fuel level is outside the lines, remove the float housing cover. Have an absorbent cloth under the cover to catch the fuel from the fuel bowl. Adjust the float level by bending the needle seat on the float.

The needle valve should have an effective stroke of about 0.0591 in. When necessary, the needle valve stroke can be adjusted by bending the float stopper.

NOTE: *Be careful not to bend the needle valve rod when installing the float and baffle plate, if removed.*

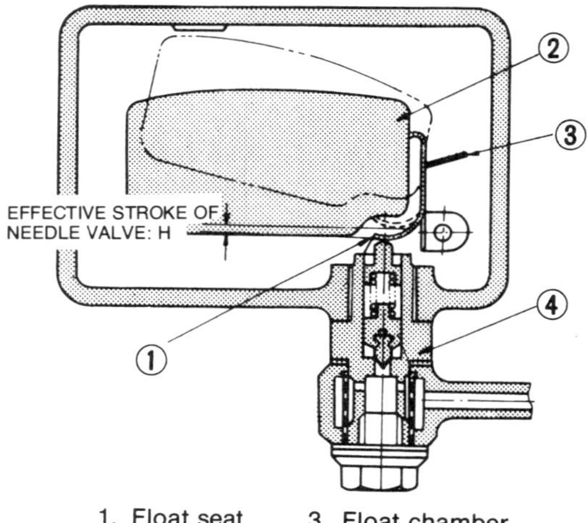

1. Float seat
2. Float
3. Float chamber
4. Needle valve

Float level adjustment

FAST IDLE ADJUSTMENT

1. With the carburetor removed from the vehicle, place the upper side of the fast idle screw on the second step of the fast idle cam and measure the clearance between the throttle valve and the wall of the throttle valve chamber at the center of the throttle valve (A in the illustration). The clearance should be 0.035–0.039 in. with manual transmission, or 0.044–0.048 in. with automatic transmission through 1974. For 1975–76 the figures are 0.040–0.048 in., manual, 0.048–0.052 in. automatic.

For 1977–81 models, the fast idle screw should be placed on the first step of the fast idle cam. This is the highest of the four. Clearance A should be 0.052–0.058 in.,

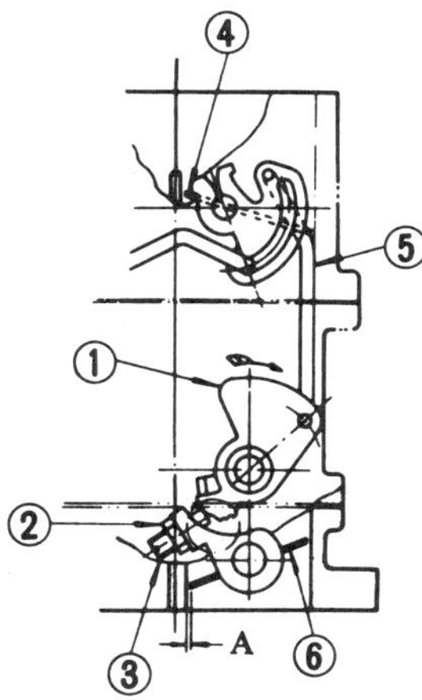

1. Fast idle cam
2. Nut
3. Fast idle screw
4. Choke valve
5. Choke connecting rod
6. Throttle valve

Fast idle adjustment

manual transmission, or 0.062–0.068 in. automatic.

2. Install the carburetor on the engine. Place the fast idle screw on the second step of the cam for all years, including 1977–81 models.

3. Start the engine and measure the fast idle rpm with the engine at operating temperature. The rpm should be about 2,000 rpm with a manual transmission and 2,400 rpm with an automatic transmission through 1976. For 1977–81, the idle should be between 1900 and 2800 for manual transmission, 2200–3200 for automatic. For 1977–81, there is a "reference" value given for distance A, with the carburetor mounted on the engine and the fast idle screw on the *second* step of the cam: 0.037–0.046 in., manual transmission, 0.046–0.055 in., automatic transmission.

4. To adjust the fast idle speed, loosen the locknut and turn the fast idle adjusting screw.

NOTE: *The first step of the fast idle adjustment procedure is not absolutely necessary; it should be used as a guide to correct adjustment at overhaul.*

CHOKE UNLOADER ADJUSTMENT

1. Close the choke valve completely.

2. Hold the choke valve closed by stretching a rubber band between the choke piston lever and a stationary part of the carburetor.

3. Open the throttle lever fully.

4. With the throttle lever fully open, ad-

just the clearance between the choke valve and the carburetor body to 0.173 in. through 1974, or 0.096 in., 1975 and later by bending the unloader tongue.

NOTE: *Make sure that the throttle valve opens completely when the carburetor is mounted on the engine.*

DASHPOT ADJUSTMENT

Only trucks with an automatic transmission have a dashpot through 1974 and 1981. All 1975–81 models have a dashpot. The purpose of this device is to prevent the throttle from suddenly snapping shut. The dashpot has a plunger which extends when the throttle is closed suddenly. The plunger contacts a tab on the throttle lever and holds the throttle open slightly for a second, then closes the throttle slowly over the period of another second or so.

1. Adjust the idle speed and mixture before making adjustments to the dashpot. Warm the engine to operating temperature, and connect a tachometer to the engine.

2. Move the throttle lever by hand, and note the engine speed when the dashpot plunger just touches the throttle lever.

3. The engine speed should be 1600–1800

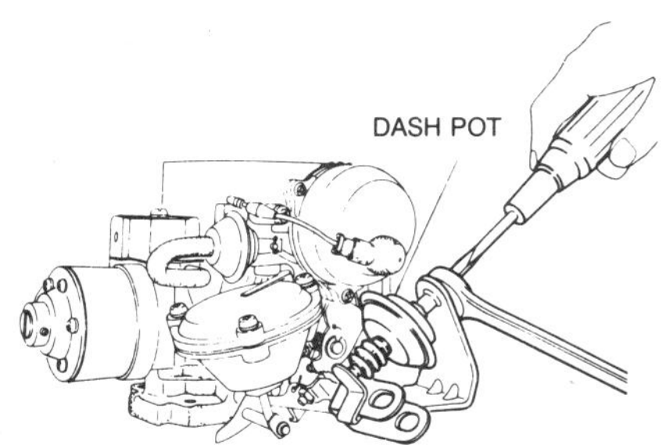

Dashpot adjustment on all models except those with air conditioning

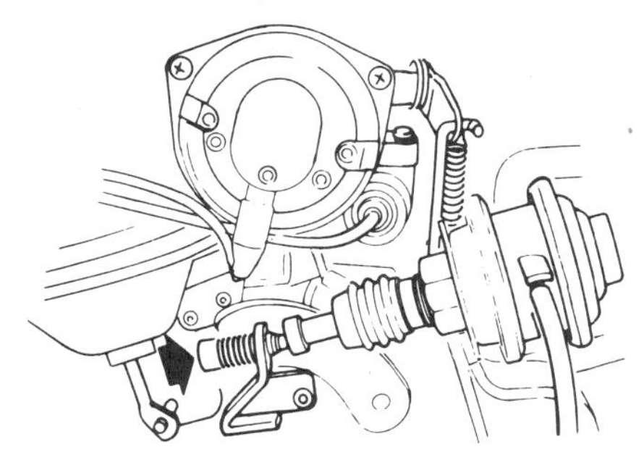

Arrow indicates the screw used for dashpot adjustment on all air conditioned models

rpm through 1974. For 1975 and later trucks, the speed should be 1650–1850 rpm with automatic transmission, or 1900–2100 rpm with manual transmission. On 1981 models the speed is 1600 rpm.

4. If not, loosen the locknut and turn the adjusting screw until the engine speed is in the proper range. Tighten the locknut. On 1978–79 models with air conditioning, a different dashpot is used. Adjustment is made by turning the screw on the throttle lever which contacts the plunger.

1. Choke chamber
2. Float chamber
3. Throttle chamber
4. Throttle adjusting screw
5. Idle adjusting screw
6. Adjusting spring screw
7. Adjusting screw spring
8. Chamber diaphragm
9. Diaphragm
10. Diaphragm spring
12–23. Screw
24–26. Washer
28. Gasket
29. Washer
30. Pin
31. Gasket
32. Washer
33. Gasket
34. Screw
35. Nut
36–37. Washer
38. Inlet valve
40–48. Gasket
49. Choke wire bracket
50. Spring hanger
51. Choke connecting rod
52. Choke connecting rod
53. Accelerator switch holder
54. Plate
55. Throttle lever
56. Accelerator switch lever
57. Return spring
58. Accelerator switch bracket
59. Accelerator pump piston
60. Pump return spring
61. Injector weight
62. Primary small venturi
63. Outer emulsion tube
64. Pump lever
65. Pump connecting lever
66. Return spring
67. Piston plate
68. Secondary small venturi
69. Float
70. Dust cover
71. Choke connecting lever
72. Sleeve (B)
73. Spring
74. Rubber seal
75. Level gauge
76. Level gauge cover
77. Adjust lever
78. Return plate
79. Sleeve (A)
80. Setscrew
81. Joint nipple
82. Spring

83. Float collar
84. Secondary emulsion tube
85. Injector weight plug
86. Main jet plug
87. Strainer
88. Primary slow jet plug
89. Float needle valve assy.
90. Collar
91. Primary main air bleed
92. Accelerator air bleed
93. Secondary slow air bleed
94. Secondary main air bleed

95. Primary slow air bleed
96. Coasting jet
97. Vacuum jet
98. Coasting air bleed
99. Primary main jet
100. Secondary main jet
101. Primary slow jet
102. Secondary slow jet
103. Power valve
104. Anti dieseling solenoid
105. Coasting valve solenoid
106. Accelerator switch

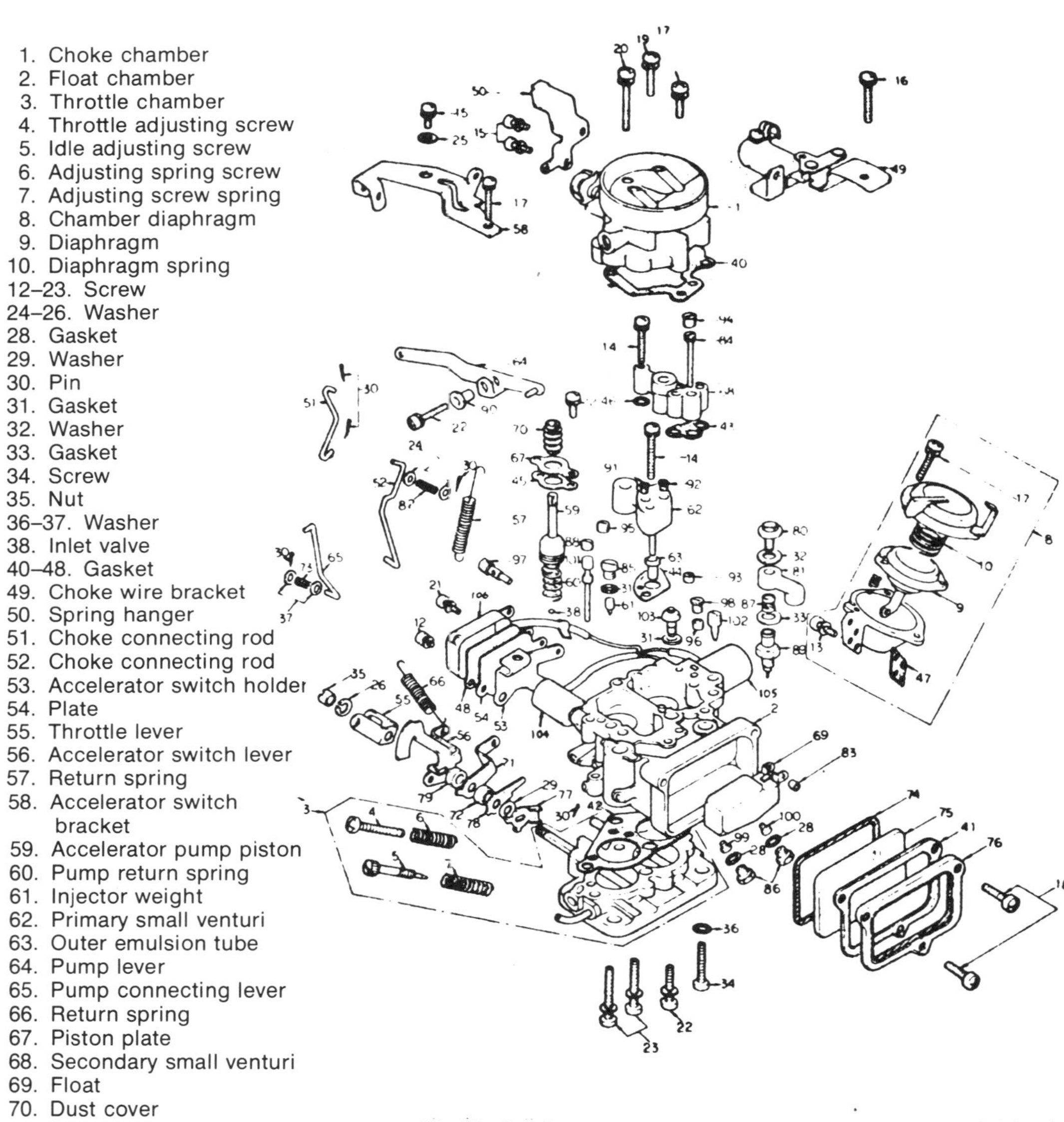

Exploded view of a typical carburetor—1974 shown

Carburetor Specifications

Year	Engine Displacement cc (cu in.)	Model Number	Main Jet		Main Air Bleed		Slow Jet		Slow Jet Air Bleed		Float Level (in.)	Power Jet
			Primary	Secondary	Primary	Secondary	Primary	Secondary	Primary	Secondary		
1970–71	1595(97.3)	DAF328	#115	#155	#240	#120	#48	#180	#180	#100	0.9055 ± 0.0394	—
1972	1595(97.3)	DAH328-3B	#112	#155	#240	#120	#48	#180	#150	#100	0.906	#50
1973	1595(97.3) and 1770 (108)	①	#97.5	#170	#65	#60	#48	#90	#145	#100	0.906	#53
1974	1770 (108)	②	#100	#170	#60	#60	#45	#90	#145	#100	0.910	#41
1975	1952(119)	③	#99	#160	#70	#60	#48	#80	NA	NA	0.906	#43
1976	1952(119)	④	#99	#160	#70	#60	#48	#100	NA	NA	0.906	#43
		⑤	#101	#160	#70	#60	#48	#80	NA	NA	0.906	#40
1977	1952(119)	⑥	#99	#160	#70	#60	#48	#100	NA	NA	0.906	#43
		⑦	#101	#160	#70	#60	#48	#100	NA	NA	0.906	#40
		⑧	#101	#160	#70	#60	#48	#80	NA	NA	0.906	#40
1978	1952(119)	⑨	#101	#158	#70	#60	#48	#70	NA	NA	0.910	#40
		⑩	#103	#160	#60	#60	#48	#70	NA	NA	0.910	#43
1979	1952(119)	⑪	#103 ⑫	#158	#70	#60	#48	#70 ⑬	NA	NA	0.910	#35
		⑭	#103 ⑮	#160	#60	#60	#48	#70	NA	NA	0.910	#43

Carburetor Specifications (cont.)

Year	Engine Displacement cc (cu in.)	Model Number	Main Jet		Main Air Bleed		Slow Jet		Slow Jet Air Bleed		Float Level (in.)	Power Jet
			Primary	Secondary	Primary	Secondary	Primary	Secondary	Primary	Secondary		
		⑯	#103	#150	#60	#60	#48	#50	NA	NA	0.910	#35
1980	1952(119)	DCH340–111, 112	#104	#158	#70	#60	#48	#70	—	—	0.910	#35
		DCH340–115	#103	#140	#70	#60	#48	#60	—	—	0.910	#35
		DCH340–113	#105	#158	#60	#60	#48	#70	—	—	0.910	#35
		DCH340–116	#103	#140	#60	#60	#48	#50	—	—	0.910	#35
		DCH340–114	#105	#158	#60	#60	#48	#70	—	—	0.910	#35
		DCH340–117, 118	#106	#160	#60	#60	#48	#70	—	—	0.910	#43
1981	2187(133.5)	DCR342–11, DCR342–12, DCR342–13	#112	#155	#90	#60	#47	#100	—	—	0.910	#35
		DCR342–21 DCR342–22	#110	#145	#90	#60	#47	#80	—	—	0.910	#35
		DCR342–17, DCR342–18	#110	#155	#90	#60	#47	#100	—	—	0.910	#35

| DCR342−14,
DCR342−15 | #105 | #155 | #80 | #60 | #47 | #100 | — | — | 0.910 | #40 |
| DCR342−23,
DCR342−24 | #105 | #155 | #80 | #60 | #47 | #100 | — | — | 0.910 | #40 |

① DCH340−8 on L16 in a 620 with automatic transmission
  DCH340−9 on L16 in a 620 with manual transmission
② DCH340−12 with manual transmission
  DCH340−13 with automatic transmission
③ DCH340−47 with manual transmission
  DCH340−48 with automatic transmission
④ DCH340−47—49 States and Canada, manual transmission
  DCH340−48—49 States and Canada, automatic transmission
  DCH340−45A—California, manual transmission
⑤ DCH340−46—California, automatic transmission
⑥ DCH340−47A—49 States and Canada, manual transmission
  DCH340−48A—49 States and Canada, automatic transmission
⑦ DCH340−45B—California, manual transmission
⑧ DCH340−46A—California, automatic transmission
⑨ DCH340−95—California, manual transmission
  DCH340−96—California, automatic transmission
⑩ DCH340−97—49 States and Canada, manual transmission
  DCH340−98—49 States and Canada, automatic transmission
⑪ DCH340−95C—California, manual transmission except cab and chassis
  DCH340−65A—California, manual transmission, cab and chassis
  DCH340−96C—California, automatic transmission
⑫ #101, DCH340−65A
⑬ #60, DCH340−65A
⑭ DCH340−97C—49 States, manual transmission
  DCH340−57A—Canada, manual transmission
  DCH340−98B—49States, automatic transmission
  DCH340−58A—Canada, automatic transmission
⑮ #106, DCH340−57A and B
⑯ DCH340−67A—49 States, manual transmission, cab and chassis
NA Not available
—Not used

5. Open the throttle and allow it to close by itself. The dashpot should smoothly reduce the idling speed from 2000 to 1000 rpm in about three seconds.

OVERHAUL

Efficient carburetion depends greatly on careful cleaning and inspection during overhaul, since dirt, gum, water, or varnish in or on the carburetor parts are often responsible for poor performance.

Overhaul your carburetor in a clean, dust-free area. Carefully disassemble the carburetor, referring often to the exploded views. Keep all similar and look-alike parts segregated during disassembly and cleaning to avoid accidental interchange during assembly. Make a note of all jet sizes.

When the carburetor is disassembled, wash all parts (except diaphragms, electric choke units, pump plunger, and any other plastic, leather, fiber, or rubber parts) in clean carburetor solvent. Do not leave parts in the solvent any longer than is necessary to sufficiently loosen the deposits. Excessive cleaning may remove the special finish from the float bowl and choke valve bodies, leaving these parts unfit for service. Rinse all parts in clean solvent and blow them dry with compressed air or allow them to air dry. Wipe clean all cork, plastic, leather, and fiber parts with a clean, lint-free cloth.

Blow out all passages and jets with compressed air and be sure that there are no restrictions or blockages. Never use wire or similar tools to clean jets, fuel passages, or air bleeds. Clean all jets and valves separately to avoid accidental interchange.

Check all parts for wear or damage. If wear or damage is found, replace the defective parts. Especially check the following:

1. Check the float needle and seat for wear. If wear is found, replace the complete assembly.

2. Check the float hinge pin for wear and the float(s) for dents or distortion. Replace the float if fuel has leaked into it.

3. Check the throttle and choke shaft bores for wear or an out-of-round condition. Damage or wear to the throttle arm, shaft, or shaft bore will often require replacements of the throttle body. These parts require a close tolerance of fit; wear may allow air leakage, which could affect starting and idling.

NOTE: *Throttle shafts and bushings are not included in overhaul kits. They can be purchased separately.*

4. Inspect the idle mixture adjusting needles for burrs or grooves. Any such condition requires replacement of the needle, since you will not be able to obtain a satisfactory idle.

5. Test the accelerator pump check valves. They should pass air one way but not the other. Test for proper seating by blowing and sucking on the valve. Replace the valve if necessary. If the valve is satisfactory, wash the valve again to remove breath moisture.

6. Check the bowl cover for warped surfaces with a straightedge.

7. Closely inspect the valves and seats for wear and damage, replacing as necessary.

8. After the carburetor is assembled, check the choke valve for freedom of operation.

Carburetor overhaul kits are recommended for each overhaul. These kits contain all gaskets and new parts to replace those that deteriorate most rapidly. Failure to replace all parts supplied with the kit (especially gaskets) can result in poor performance later.

Some carburetor manufacturers supply overhaul kits of three basic types: minor repair; major repair; and gasket kits.

After cleaning and checking all components, reassemble the carburetor, using new parts and referring to the exploded view. When reassembling, make sure that all screws and jets are tight in their seats, but do not overtighten as the tips will be distorted. Tighten all screws gradually, in rotation. Do not tighten needle valves into their seats; uneven jetting will result. Always use new gaskets. Be sure to adjust the float level when reassembling.

DIESEL ENGINE FUEL SYSTEM

Injection Pump

REMOVAL AND INSTALLATION

NOTE: *In some applications, this procedure is best done with the engine removed from the vehicle.*

1. Remove the inlet and outlet lines from the oil cooler.

2. Remove the bolts (4) and separate the oil filter and lines from the cooler.

3. Remove the coolant hose between the oil cooler and the head.

4. Remove the bolts (10) and separate the cooler from the block.

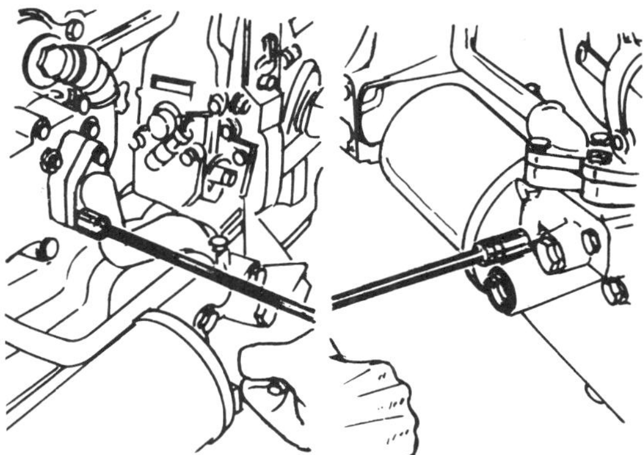

Water passage removal

Fuel filter removal

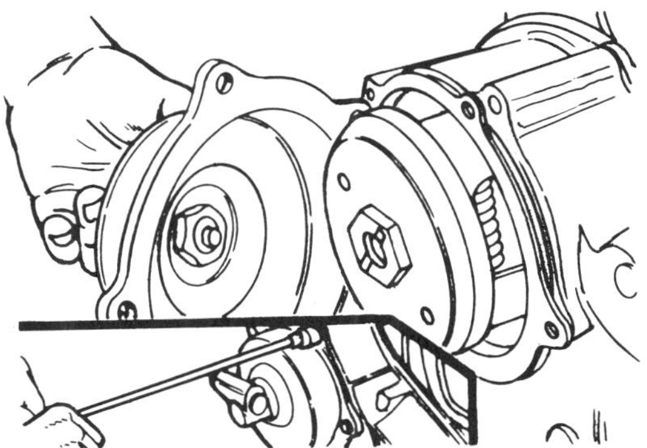

Timing cover removal

Injection line removal

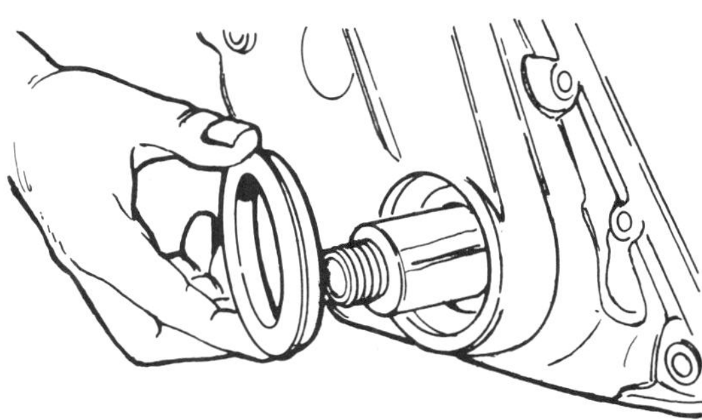

Oil seal removal

5. Disconnect the fuel lines and remove the fuel filter from the bracket.

6. Remove the injection lines from the nozzles and pump. Cover all openings immediately.

7. Remove the fan, spacer and pulley from the water pump.

8. Remove the bypass hose from the pump and thermostat housing.

9. Remove the three bolts and lift off the water pump and gasket.

10. Remove the inspection cover and pointer from the flywheel housing and lock the flywheel in place with a locking tool.

11. Flatten the lockwasher and remove the crankshaft pulley nut.

12. Tap evenly around the edge of the pulley using a brass drift, until the cone protrudes from the pulley. Remove the cone.

13. Drive the pulley and damper from the crankshaft with a soft mallet.

14. Remove the inner cover from the timing gear case.

15. Pry out the oil seal.

16. Remove the mounting bolts and tap the case loose with a soft mallet.

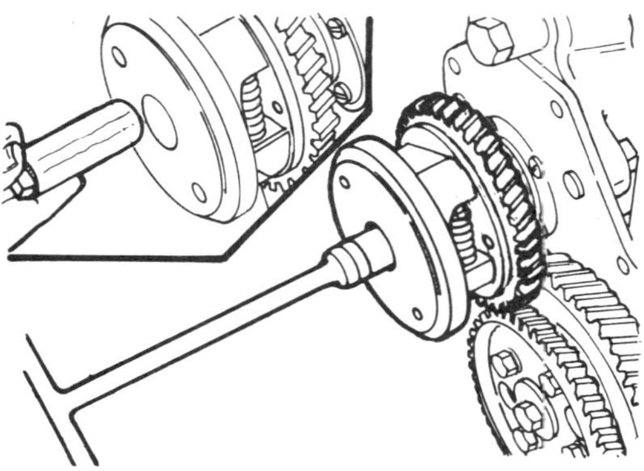

Removing timer round nut

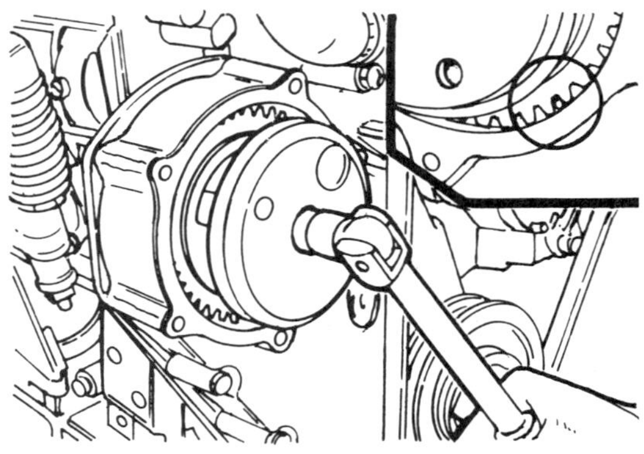

Timer installation

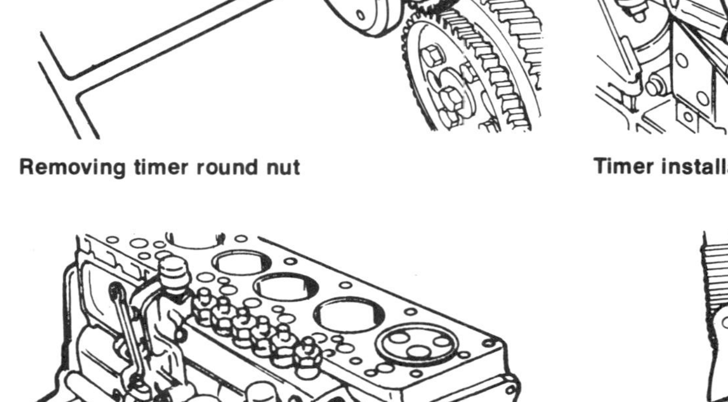

Injection pump and endplate assembly

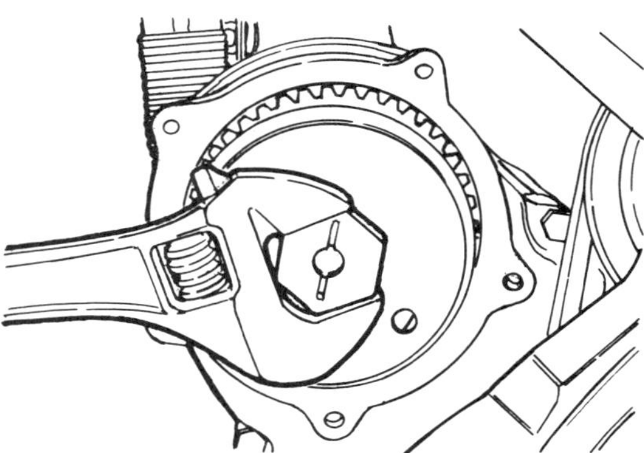

Tachometer drive coupling installation

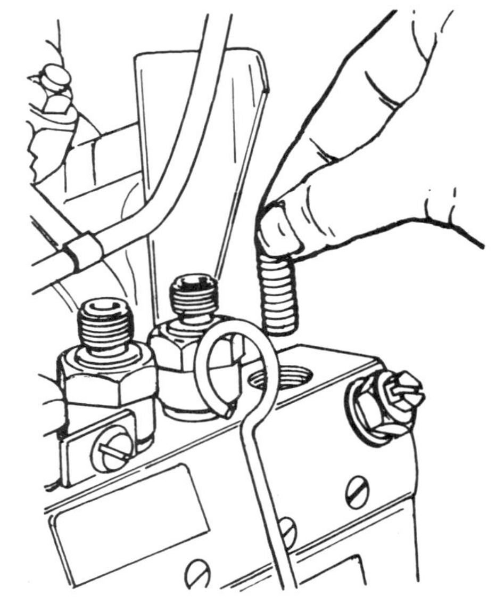

Delivery valve spring removal

17. Remove the tachometer drive support nuts.

18. Remove the timer round nut.

19. Thread the timer extractor, special tool # 57926-581 into the timer weight holder. Remove the timer assembly by tightening the extractor bolt.

20. Unbolt and separate the injection pump from the front end plate.

21. Temporarily install the injection pump and gasket on the front plate.

22. Check the timing marks and bring the #1 piston to TDC.

23. Mesh the injection pump drive gear and idler gear at the timing marks.

24. After aligning the injection pump keyway, install the lockwasher and round nut and torque to 50–58 ft. lbs.

25. Install the tachometer drive coupling.

26. Check the backlash between the pump drive gear and the idler gear. Backlash should be .0028–0079″. Adjust if necessary.

27. Remove the #1 cylinder holder clamp, loosen the delivery valve and pull out

the delivery spring. Tighten the valve holder to 22–25 ft. lb.

28. Connect the fuel supply lines.

29. Bring the #1 piston to 20° BTDC.

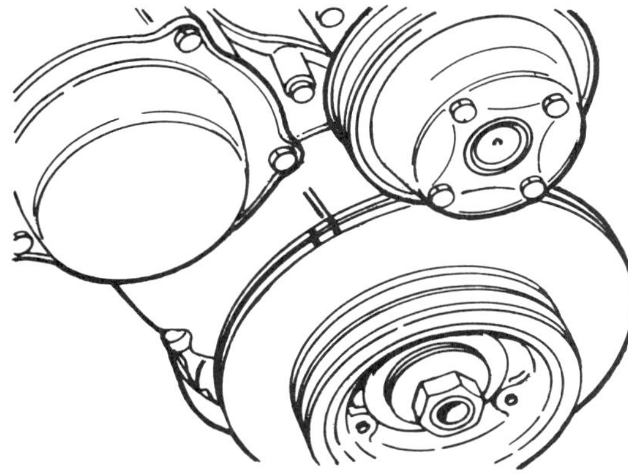

#1 piston at 20°BTDC

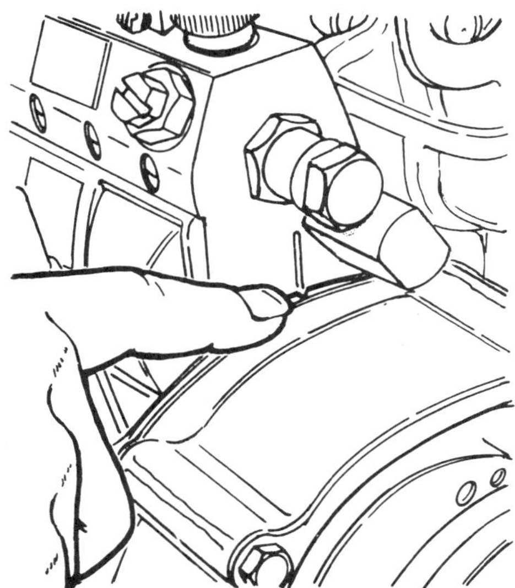

Lock the pump at this point, which is the beginning of injection

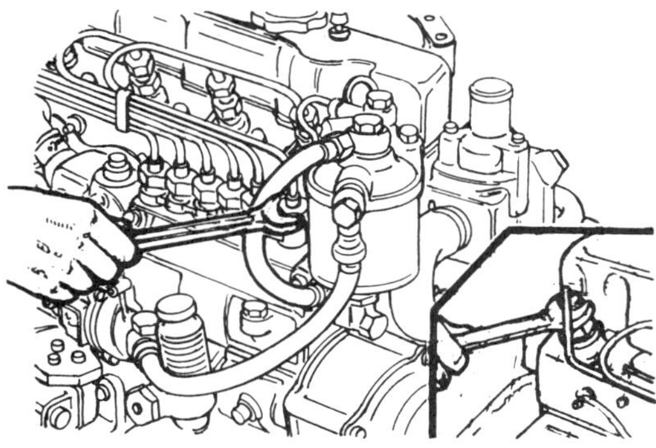

Connecting injection lines

This can be done by aligning the first mark, in normal rotation, on the crankshaft pulley with the raised line on the gear case.

30. Hand prime the pump. Push the pump in all the way toward the block. Move the pump slowly away from the block until the fuel just stops flowing from the valve holder. Lock the pump in place.

31. Remove the delivery holder and assemble the spring. Torque the holder to 22–25 ft. lb.

32. Install remaining parts in reverse order of removal.

NOTE: *Oil filter bolt torque is 15–18 ft. lb.*

Injection Nozzle

REMOVAL, OVERHAUL AND INSTALLATION

1. Loosen the injection lines at the pump and nozzles and remove the lines. Cap the openings immediately.

2. Unscrew the injector and holder from the head.

3. Secure the nozzle holder in a vise and remove the lock nut.

4. Remove the nipple.

5. Remove the nozzle holder body from the nozzle nut.

6. Remove the spacer collar and push-rod.

7. Remove the nozzle holder body from the vise and remove the nozzle spring and adjusting shims.

NOTE: *The adjusting shims may be removed with a piece of wire, but great care must be taken to avoid damage to the nozzle tip.*

8. Clean fuel oil may be used to clean all parts. Inspect all parts for damage and good fit.

9. Assemble the nozzle in reverse order of disassembly.

10. Install the nozzle in a tester.

11. Operate the tester lever at 1 stroke per second and read the pressure at injection. The pointer will oscillate slightly during injection.

12. Increase or decrease the thickness of the nozzle spring adjusting shims until opening pressure is 1,422.3 psi. A total of 31 different shims are available. A shim thickness of .05 mm equals a difference of 85.338 psi.

13. Install the nozzles and lines. Torque the nozzles to 50–65 ft. lb.

Governor

RSV AND RAD MECHANICAL TYPE REMOVAL

1. Remove the injection pump and place it in a holding fixture.

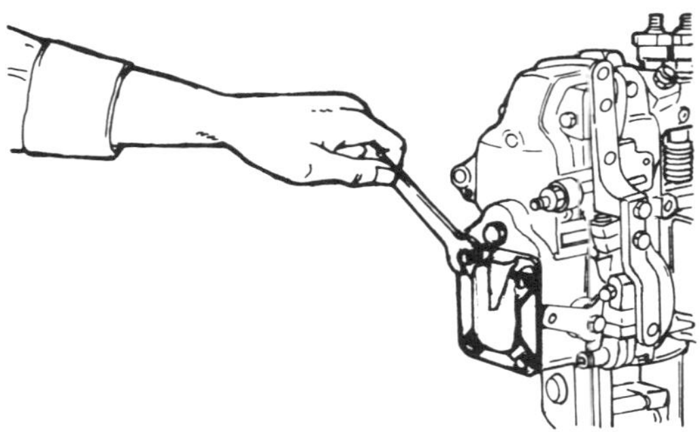

Removing the cam cover

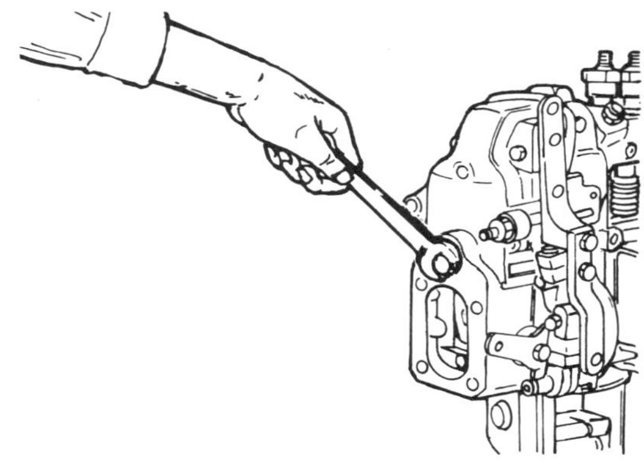

Removing the idling sub spring

Feed pump removal

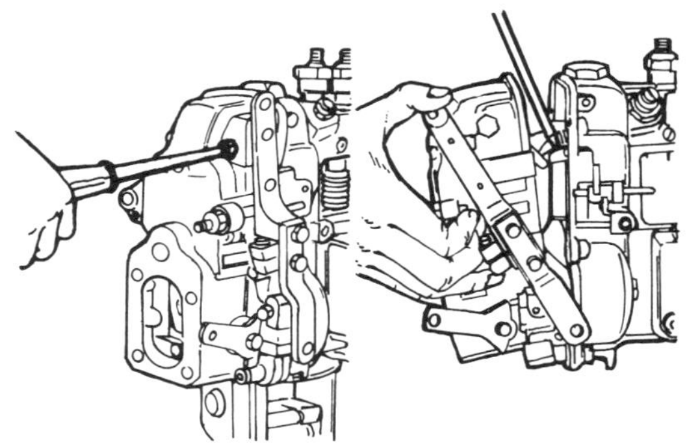

Removing the governor cover and link

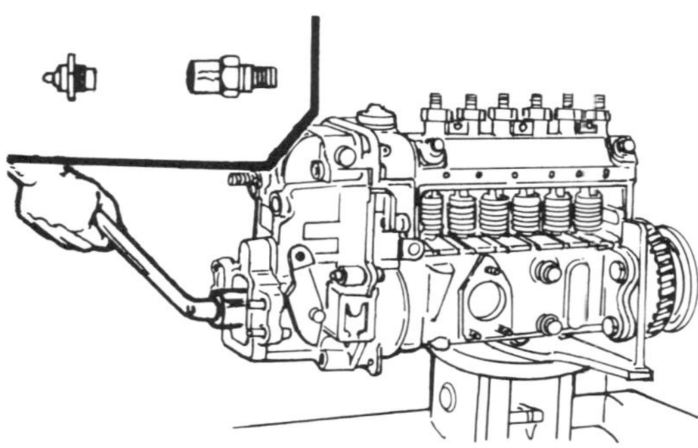

Removing the idling spring, type RAD

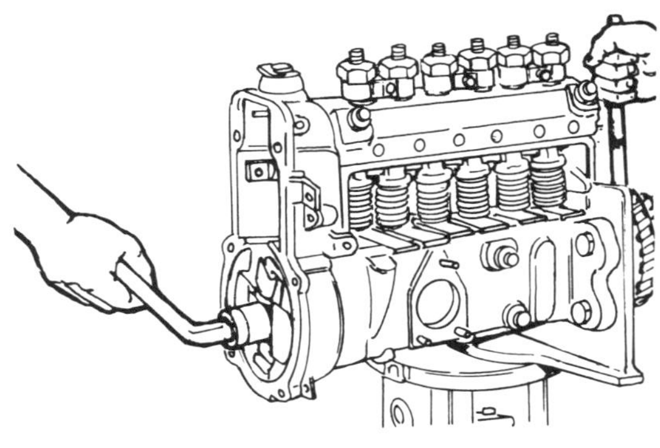

Removing the idling spring, type RSV

2. Install the timer.
3. Drain the cam and governor chambers.
4. Remove the supply pump.

Removing counterweights

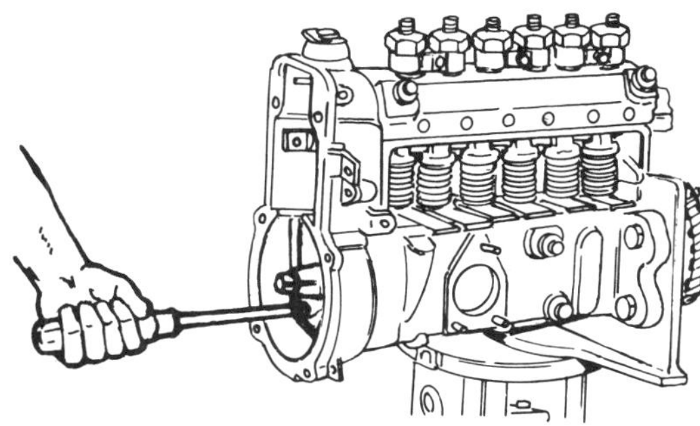

Removing governor body

5. Remove the cam cover.

6. Using a special wrench, ST-57916-432, on the timer, turn the camshaft until all the tappets are raised to TDC. Place a tappet holder, 57931-210, between the tappet adjusting bolt and nut for each cylinder.

7. Remove the rear cover and dipstick.

8. Loosen the balance idler spring and the auxiliary idler spring lock nut.

9. Loosen the governor cover lock screw.

10. Unbolt and remove the governor cover from the governor body. Remove the link from the control rack.

11. Remove the start spring from the spring eye.

12. Remove the counterweights from the camshaft.

13. Hold the timer and remove the slotted nuts and lockwashers.

14. Using a puller, ST57926-512, remove the flyweight assembly.

15. Remove the timer.

16. Unbolt and remove the governor body.

MZ PNEUMATIC TYPE REMOVAL

1. Follow steps 1 through 6 of RSV Type Removal.

2. Unbolt and remove the diaphragm housing and main spring.

3. Remove the diaphragm ring with a screwdriver.

Removing the timer

4. Remove the cotter pin from the connecting rod bolt with a needle-nosed pliers.

5. Remove the diaphragm assembly from the control rack.

6. Remove the five set screws and remove the governor body by applying force with a screwdriver blade in the slit between the governor and pump housings.

7. Remove the timer.

RSV TYPE INSTALLATION

1. Apply RTV silicone gasket material to the governor body and install the governor on the injection pump.

2. Tighten the upper spring eye screw holding the starting spring.

3. Install the timer.

4. Install the flyweight assembly.

5. Apply RTV silicone gasket material to the governor cover and install the starting spring on the housing side of the spring eye.

6. Install the link leaf spring in the hole in the end of the control rack.

7. Install the cover and set screws.

MZ TYPE INSTALLATION

1. Apply RTV silicone gasket material to the governor body, position it on the pump body and tap it into position with a plastic mallet. Install the set service.

2. Install the diaphragm and balance spring on the control rack connecting bolt and lock with a new cotter pin. Apply chassis lube to the diaphragm ring.

3. Insert the main spring and install diaphragm housing with the four bolts.

Injection Pump Service
REPLACING THE DELIVERY VALVE

1. Thoroughly clean the area around the nozzle tube and delivery valve.

2. Remove the nozzle tube.

3. Remove the delivery valve holder lock plate.

4. Remove the delivery valve holder and spring.

5. Using ST57930-032, remove the delivery valve.

6. Position the delivery valve in the pump housing making sure no dirt gets between the top of the plunger barrel and the delivery valve.

7. Install a new delivery valve gasket. The gasket is installed with the larger face downward and may be tapped into place through the extractor.

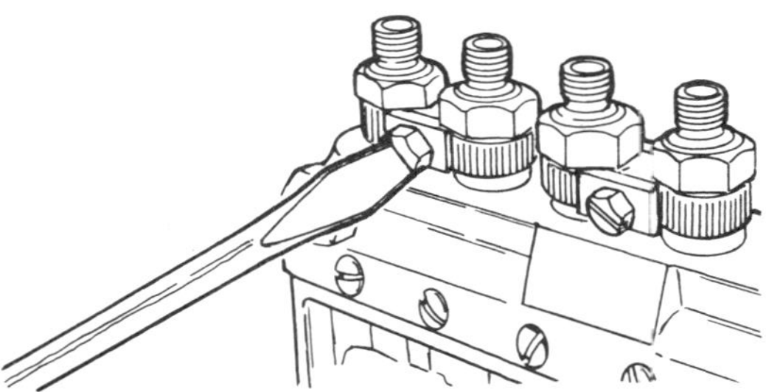

Removing delivery valve holder lock plate

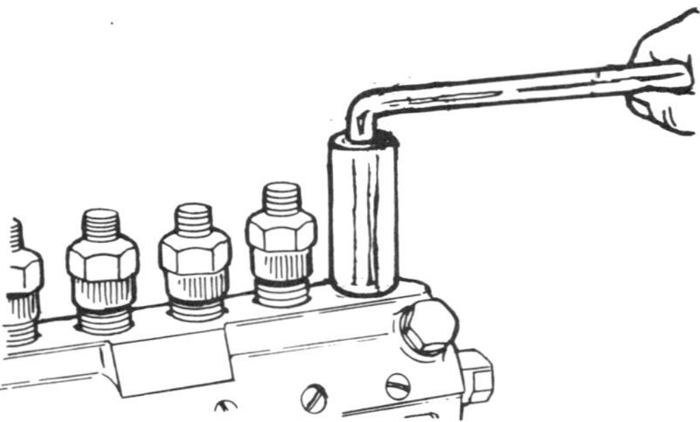

Loosening the delivery valve holder

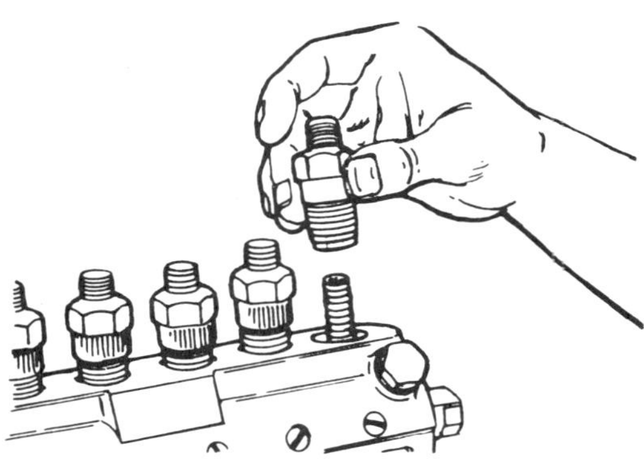

Removing the delivery valve holder and spring

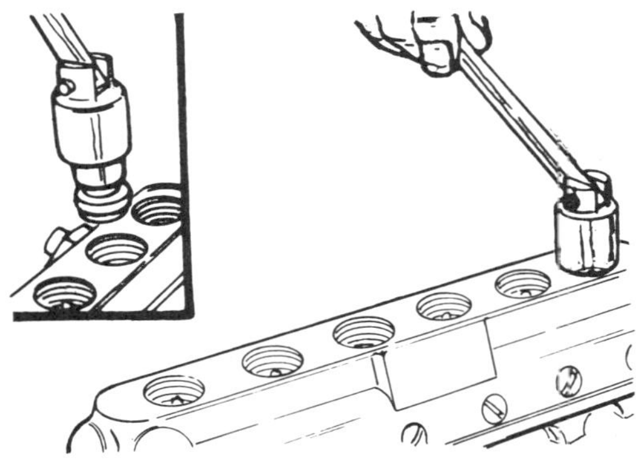

Removing the delivery valve with tool 57930-032

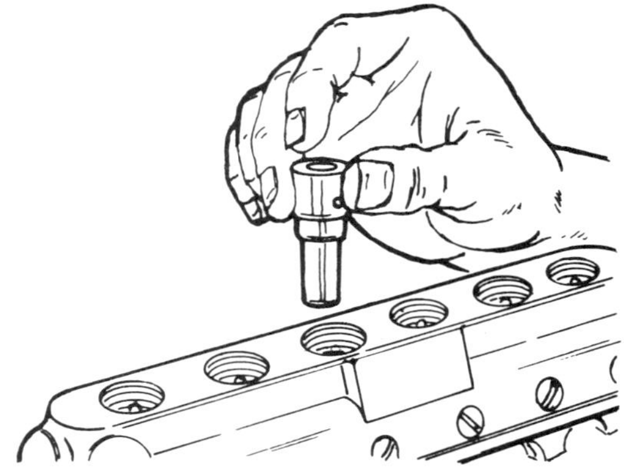

Plunger barrel removal, with delivery valves removed

8. Install the delivery valve spring.

9. Install the delivery valve holder and torque it to 22–25 ft. lb.

10. Loosen the holder and retorque it.

11. Install the lock plate, nozzle tube and nozzle clamp.

REPLACING THE PLUNGER

1. Remove the delivery valve.

2. Push the plunger spring up with two screwdrivers and remove the lower spring seat from the plunger.

3. Insert a hooked wire through the top of the pump housing, down through the plunger opening and hook it on the lead unit of the plunger. Pull up to remove the plunger and barrel.

4. Immerse a new plunger in clean fuel oil and thoroughly wash off the rust preventive.

NOTE: *The plunger was lapped at the factory. Do not hold it by the lapped section.*

5. Operate the plunger in clean fuel oil to check its operation.

6. Slowly insert the plunger and barrel

Inserting the plunger spring holder

into the pump with the barrel groove and plunger notch facing forward. Make certain the plunger piston pin is properly seated in the groove in the control sleeve.

7. Push the plunger spring up with two screwdrivers and insert the lower spring seat.

8. Install the remaining parts in reverse of disassembly.

CAMSHAFT REMOVAL AND INSTALLATION

1. Remove the key from the camshaft and remove the four setscrews on the drive side of the pump housing. Remove the bearing cover.

2. Lay the pump on its side and remove the screw plugs from the bottom of the housing with tool 57910-112 or a ratchet handle.

3. Remove the set screws from the center bearing.

4. Remove the camshaft from the housing.

5. Install the bearing cover on the timer side of the housing.

6. Install the camshaft so that alignment mark "A" is toward the timer.

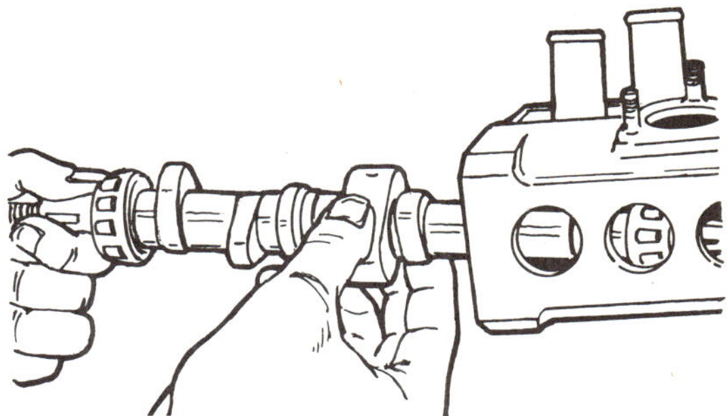

Camshaft removal

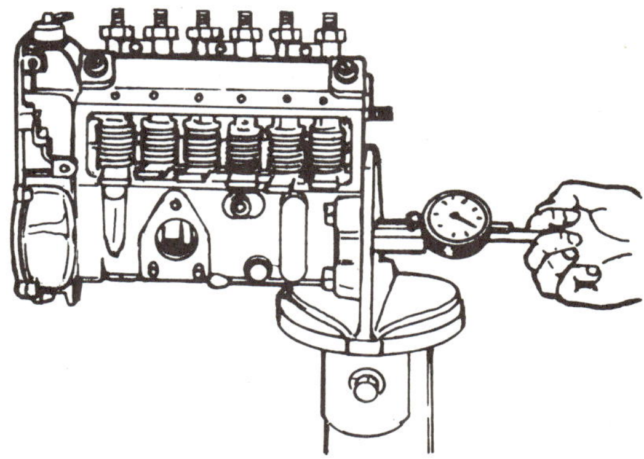

Measuring end play

7. Check end play as illustrated. End play is 0-.0012 in.

TAPPET REMOVAL AND INSTALLATION

1. Remove the camshaft.

2. Install a tappet insert tool 57912-012 and remove the tappet holder, at the same time, removing the tappet from the holder. Be careful to avoid damaging the housing plug hole threads.

3. Loosen the tappet insert tool, withdraw the tappet assembly into the camshaft chamber and remove the tappet from the housing with a tappet clamp.

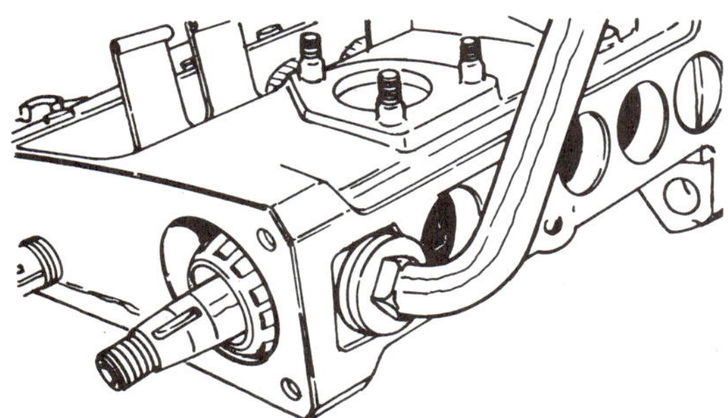

Screw plug removal and installation

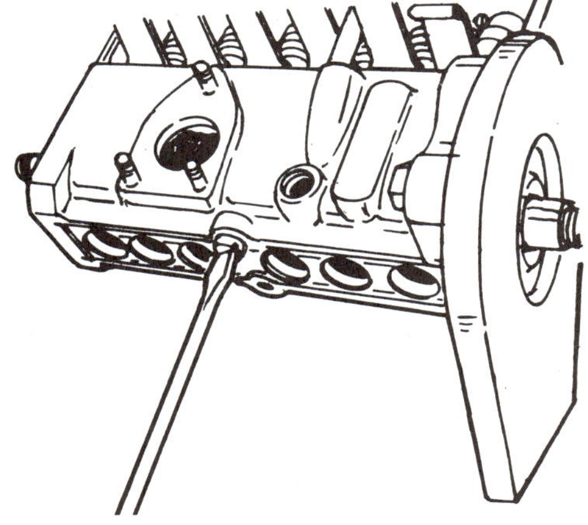

Set screw removal and installation

Inserting tappet holder and insert tool

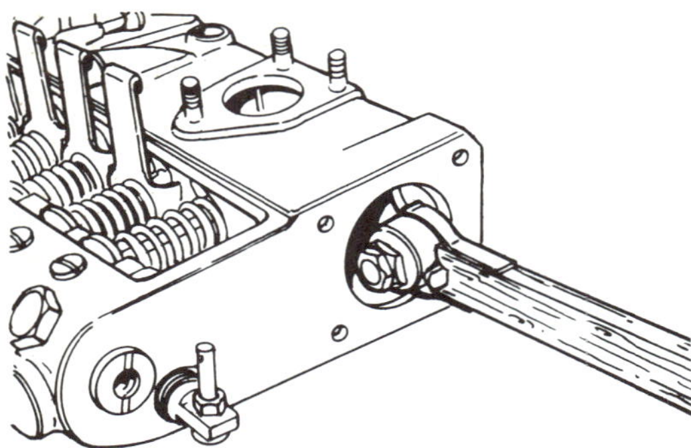

Removing tappet with tappet clamp

4. Installation is the reverse of removal. Make sure that the mechanism works smoothly.

PLUNGER REMOVAL AND INSTALLATION

1. Remove the tappets.
2. Remove the plungers using plunger pinchers 57921-412 or equivalent, together with the lower spring seat.

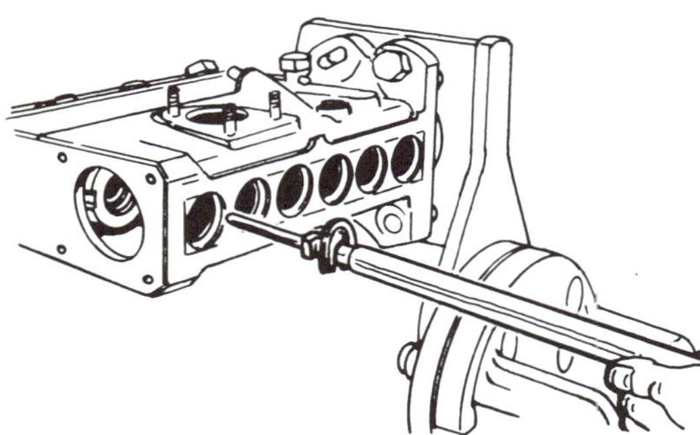

Removing plungers with tool 57921-412

3. Place any reusable plungers in the corresponding plunger barrels and immerse them in kerosene or safe solvent.
4. Remove the plunger springs, upper spring seats and control sleeves.
5. Lay the housing on its side and position the control rack so that the punch marks on both sides are the same distance from each end of the housing.
6. Install the control sleeve with the gap in the sleeve facing straight upward. At the same time install the spring seats.
7. Install the plunger spring in the housing.
8. Install the plunger in the barrel using tool 57921-412, along with the lower spring seat.

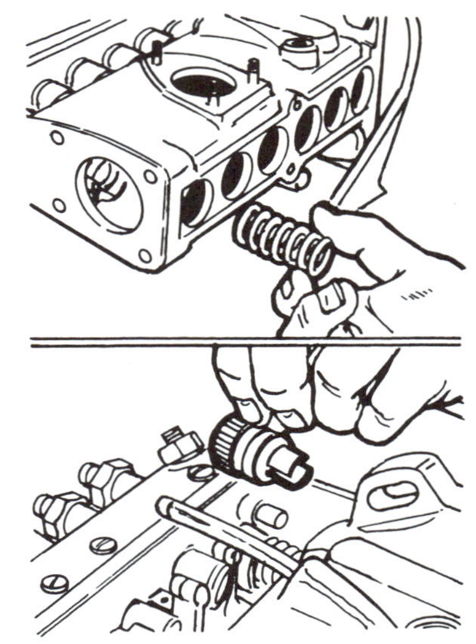

Removing plunger springs, control sleeves and upper spring seats

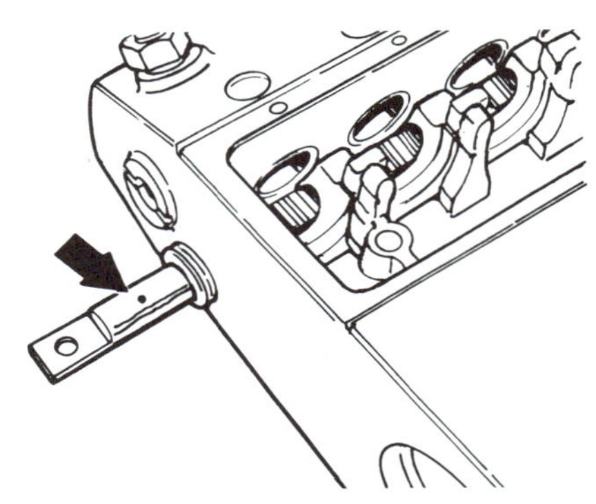

Control rack; punch mark is indicated by arrow

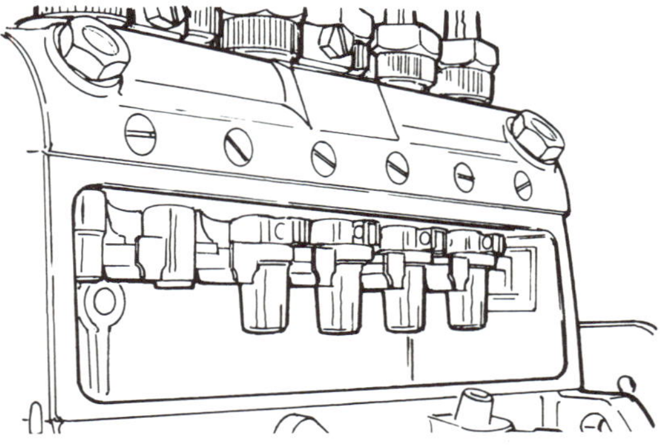

Control sleeves installed

9. Make sure the mechanism functions smoothly.

TESTING AND ADJUSTING THE FUEL INJECTION PUMP

NOTE: *It is necessary to inspect and adjust the pump, using a pump tester, when-*

CHILTON'S
FUEL ECONOMY & TUNE-UP TIPS

Tune-Up • Spark Plug Diagnosis • Emission Controls

Fuel System • Cooling System • Tires and Wheels

General Maintenance

CHILTON'S FUEL ECONOMY & TUNE-UP TIPS

Fuel economy is important to everyone, no matter what kind of vehicle you drive. The maintenance-minded motorist can save both money and fuel using these tips and the periodic maintenance and tune-up procedures in this Repair and Tune-Up Guide.

There are more than 130,000,000 cars and trucks registered for private use in the United States. Each travels an average of 10-12,000 miles per year, and, in total they consume close to 70 billion gallons of fuel each year. This represents nearly $\frac{2}{3}$ of the oil imported by the United States each year. The Federal government's goal is to reduce consumption 10% by 1985. A variety of methods are either already in use or under serious consideration, and they all affect your driving and the cars you will drive. In addition to "down-sizing", the auto industry is using or investigating the use of electronic fuel delivery, electronic engine controls and alternative engines for use in smaller and lighter vehicles, among other alternatives to meet the federally mandated Corporate Average Fuel Economy (CAFE) of 27.5 mpg by 1985. The government, for its part, is considering rationing, mandatory driving curtailments and tax increases on motor vehicle fuel in an effort to reduce consumption. The government's goal of a 10% reduction could be realized — and further government regulation avoided — if every private vehicle could use just 1 less gallon of fuel per week.

How Much Can You Save?

Tests have proven that almost anyone can make at least a 10% reduction in fuel consumption through regular maintenance and tune-ups. When a major manufacturer of spark plugs sur-

TUNE-UP

1. Check the cylinder compression to be sure the engine will really benefit from a tune-up and that it is capable of producing good fuel economy. A tune-up will be wasted on an engine in poor mechanical condition.

2. Replace spark plugs regularly. New spark plugs alone can increase fuel economy 3%.

3. Be sure the spark plugs are the correct type (heat range) for your vehicle. See the Tune-Up Specifications.

Heat range refers to the spark plug's ability to conduct heat away from the firing end. It must conduct the heat away in an even pattern to avoid becoming a source of pre-ignition, yet it must also operate hot enough to burn off conductive deposits that could cause misfiring.

The heat range is usually indicated by a number on the spark plug, part of the manufacturer's designation for each individual spark plug. The numbers in bold-face indicate the heat range in each manufacturer's identification system.

Manufacturer	Typical Designation
AC	R **45** TS
Bosch (old)	WA **145** T30
Bosch (new)	HR **8** Y
Champion	RBL **15** Y
Fram/Autolite	**415**
Mopar	P-**62** PR
Motorcraft	BR**F-42**
NGK	BP **5** ES-15
Nippondenso	W **16** EP
Prestolite	14GR **5** 2A

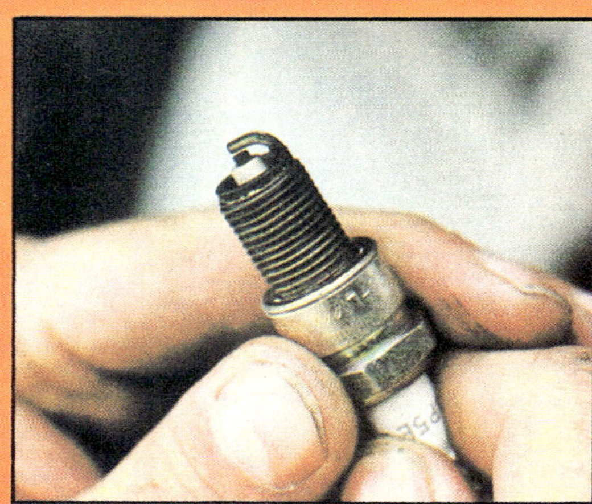

Periodically, check the spark plugs to be sure they are firing efficiently. They are excellent indicators of the internal condition of your engine.

On AC, Bosch (new), Champion, Fram/Autolite, Mopar, Motorcraft and Prestolite, a higher number indicates a hotter plug. On Bosch (old), NGK and Nippondenso, a higher number indicates a colder plug.

4. Make sure the spark plugs are properly gapped. See the Tune-Up Specifications in this book.

5. Be sure the spark plugs are firing efficiently. The illustrations on the next 2 pages show you how to "read" the firing end of the spark plug.

6. Check the ignition timing and set it to specifications. Tests show that almost all cars

veyed over 6,000 cars nationwide, they found that a tune-up, on cars that needed one, increased fuel economy over 11%. Replacing worn plugs alone, accounted for a 3% increase. The same test also revealed that 8 out of every 10 vehicles will have some maintenance deficiency that will directly affect fuel economy, emissions or performance. Most of this mileage-robbing neglect could be prevented with regular maintenance.

Modern engines require that all of the functioning systems operate properly for maximum efficiency. A malfunction anywhere wastes fuel. You can keep your vehicle running as efficiently and economically as possible, by being aware of your vehicles operating and performance characteristics. If your vehicle suddenly develops performance or fuel economy problems it could be due to one or more of the following:

PROBLEM	POSSIBLE CAUSE
Engine Idles Rough	Ignition timing, idle mixture, vacuum leak or something amiss in the emission control system.
Hesitates on Acceleration	Dirty carburetor or fuel filter, improper accelerator pump setting, ignition timing or fouled spark plugs.
Starts Hard or Fails to Start	Worn spark plugs, improperly set automatic choke, ice (or water) in fuel system.
Stalls Frequently	Automatic choke improperly adjusted and possible dirty air filter or fuel filter.
Performs Sluggishly	Worn spark plugs, dirty fuel or air filter, ignition timing or automatic choke out of adjustment.

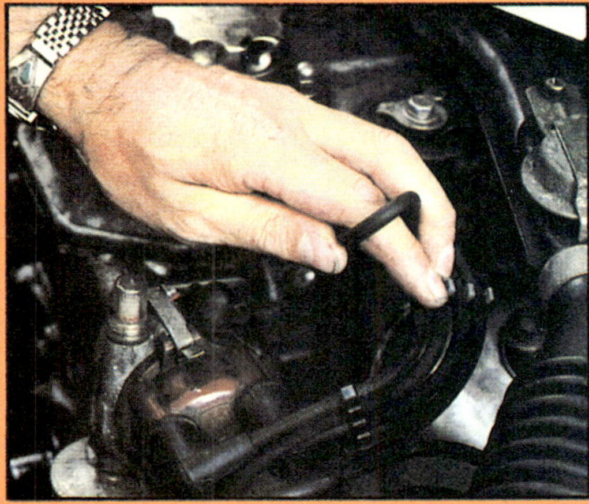

Check spark plug wires on conventional point type ignition for cracks by bending them in a loop around your finger.

Be sure that spark plug wires leading to adjacent cylinders do not run too close together. (Photo courtesy Champion Spark Plug Co.)

have incorrect ignition timing by more than 2°.

7. If your vehicle does not have electronic ignition, check the points, rotor and cap as specified.

8. Check the spark plug wires (used with conventional point-type ignitions) for cracks and burned or broken insulation by bending them in a loop around your finger. Cracked wires decrease fuel efficiency by failing to deliver full voltage to the spark plugs. One misfiring spark plug can cost you as much as 2 mpg.

9. Check the routing of the plug wires. Misfiring can be the result of spark plug leads to adjacent cylinders running parallel to each other and too close together. One wire tends to pick up voltage from the other causing it to fire "out of time".

10. Check all electrical and ignition circuits for voltage drop and resistance.

11. Check the distributor mechanical and/or vacuum advance mechanisms for proper functioning. The vacuum advance can be checked by twisting the distributor plate in the opposite direction of rotation. It should spring back when released.

12. Check and adjust the valve clearance on engines with mechanical lifters. The clearance should be slightly loose rather than too tight.

SPARK PLUG DIAGNOSIS

Normal

APPEARANCE: This plug is typical of one operating normally. The insulator nose varies from a light tan to grayish color with slight electrode wear. The presence of slight deposits is normal on used plugs and will have no adverse effect on engine performance. The spark plug heat range is correct for the engine and the engine is running normally.

CAUSE: Properly running engine.

RECOMMENDATION: Before reinstalling this plug, the electrodes should be cleaned and filed square. Set the gap to specifications. If the plug has been in service for more than 10-12,000 miles, the entire set should probably be replaced with a fresh set of the same heat range.

Oil Deposits

APPEARANCE: The firing end of the plug is covered with a wet, oily coating.

CAUSE: The problem is poor oil control. On high mileage engines, oil is leaking past the rings or valve guides into the combustion chamber. A common cause is also a plugged PCV valve, and a ruptured fuel pump diaphragm can also cause this condition. Oil fouled plugs such as these are often found in new or recently overhauled engines, before normal oil control is achieved, and can be cleaned and reinstalled.

RECOMMENDATION: A hotter spark plug may temporarily relieve the problem, but the engine is probably in need of work.

Incorrect Heat Range

APPEARANCE: The effects of high temperature on a spark plug are indicated by clean white, often blistered insulator. This can also be accompanied by excessive wear of the electrode, and the absence of deposits.

CAUSE: Check for the correct spark plug heat range. A plug which is too hot for the engine can result in overheating. A car operated mostly at high speeds can require a colder plug. Also check ignition timing, cooling system level, fuel mixture and leaking intake manifold.

RECOMMENDATION: If all ignition and engine adjustments are known to be correct, and no other malfunction exists, install spark plugs one heat range colder.

Carbon Deposits

APPEARANCE: Carbon fouling is easily identified by the presence of dry, soft, black, sooty deposits.

CAUSE: Changing the heat range can often lead to carbon fouling, as can prolonged slow, stop-and-start driving. If the heat range is correct, carbon fouling can be attributed to a rich fuel mixture, sticking choke, clogged air cleaner, worn breaker points, retarded timing or low compression. If only one or two plugs are carbon fouled, check for corroded or cracked wires on the affected plugs. Also look for cracks in the distributor cap between the towers of affected cylinders.

RECOMMENDATION: After the problem is corrected, these plugs can be cleaned and reinstalled if not worn severely.

MMT Fouled

APPEARANCE: Spark plugs fouled by MMT (Methycyclopentadienyl Maganese Tricarbonyl) have reddish, rusty appearance on the insulator and side electrode.

CAUSE: MMT is an anti-knock additive in gasoline used to replace lead. During the combustion process, the MMT leaves a reddish deposit on the insulator and side electrode.

RECOMMENDATION: No engine malfunction is indicated and the deposits will not affect plug performance any more than lead deposits (see Ash Deposits). MMT fouled plugs can be cleaned, regapped and reinstalled.

High Speed Glazing

APPEARANCE: Glazing appears as shiny coating on the plug, either yellow or tan in color.

CAUSE: During hard, fast acceleration, plug temperatures rise suddenly. Deposits from normal combustion have no chance to fluff-off; instead, they melt on the insulator forming an electrically conductive coating which causes misfiring.

RECOMMENDATION: Glazed plugs are not easily cleaned. They should be replaced with a fresh set of plugs of the correct heat range. If the condition recurs, using plugs with a heat range one step colder may cure the problem.

Ash (Lead) Deposits

APPEARANCE: Ash deposits are characterized by light brown or white colored deposits crusted on the side or center electrodes. In some cases it may give the plug a rusty appearance.

CAUSE: Ash deposits are normally derived from oil or fuel additives burned during normal combustion. Normally they are harmless, though excessive amounts can cause misfiring. If deposits are excessive in short mileage, the valve guides may be worn.

RECOMMENDATION: Ash-fouled plugs can be cleaned, gapped and reinstalled.

Detonation

APPEARANCE: Detonation is usually characterized by a broken plug insulator.

CAUSE: A portion of the fuel charge will begin to burn spontaneously, from the increased heat following ignition. The explosion that results applies extreme pressure to engine components, frequently damaging spark plugs and pistons.

Detonation can result by over-advanced ignition timing, inferior gasoline (low octane) lean air/fuel mixture, poor carburetion, engine lugging or an increase in compression ratio due to combustion chamber deposits or engine modification.

RECOMMENDATION: Replace the plugs after correcting the problem.

EMISSION CONTROLS

13. Be aware of the general condition of the emission control system. It contributes to reduced pollution and should be serviced regularly to maintain efficient engine operation.

14. Check all vacuum lines for dried, cracked or brittle conditions. Something as simple as a leaking vacuum hose can cause poor performance and loss of economy.

15. Avoid tampering with the emission control system. Attempting to improve fuel econ-

FUEL SYSTEM

Check the air filter with a light behind it. If you can see light through the filter it can be reused.

Extremely clogged filters should be discarded and replaced with a new one.

18. Replace the air filter regularly. A dirty air filter richens the air/fuel mixture and can increase fuel consumption as much as 10%. Tests show that ⅓ of all vehicles have air filters in need of replacement.

19. Replace the fuel filter at least as often as recommended.

20. Set the idle speed and carburetor mixture to specifications.

21. Check the automatic choke. A sticking or malfunctioning choke wastes gas.

22. During the summer months, adjust the automatic choke for a leaner mixture which will produce faster engine warm-ups.

COOLING SYSTEM

29. Be sure all accessory drive belts are in good condition. Check for cracks or wear.

30. Adjust all accessory drive belts to proper tension.

31. Check all hoses for swollen areas, worn spots, or loose clamps.

32. Check coolant level in the radiator or expansion tank.

33. Be sure the thermostat is operating properly. A stuck thermostat delays engine warm-up and a cold engine uses nearly twice as much fuel as a warm engine.

34. Drain and replace the engine coolant at least as often as recommended. Rust and scale

TIRES & WHEELS

38. Check the tire pressure often with a pencil type gauge. Tests by a major tire manufacturer show that 90% of all vehicles have at least 1 tire improperly inflated. Better mileage can be achieved by over-inflating tires, but never exceed the maximum inflation pressure on the side of the tire.

39. If possible, install radial tires. Radial tires deliver as much as ½ mpg more than bias belted tires.

40. Avoid installing super-wide tires. They only create extra rolling resistance and decrease fuel mileage. Stick to the manufacturer's recommendations.

41. Have the wheels properly balanced.

omy by tampering with emission controls is more likely to worsen fuel economy than improve it. Emission control changes on modern engines are not readily reversible.

16. Clean (or replace) the EGR valve and lines as recommended.

17. Be sure that all vacuum lines and hoses are reconnected properly after working under the hood. An unconnected or misrouted vacuum line can wreak havoc with engine performance.

23. Check for fuel leaks at the carburetor, fuel pump, fuel lines and fuel tank. Be sure all lines and connections are tight.

24. Periodically check the tightness of the carburetor and intake manifold attaching nuts and bolts. These are a common place for vacuum leaks to occur.

25. Clean the carburetor periodically and lubricate the linkage.

26. The condition of the tailpipe can be an excellent indicator of proper engine combustion. After a long drive at highway speeds, the inside of the tailpipe should be a light grey in color. Black or soot on the insides indicates an overly rich mixture.

27. Check the fuel pump pressure. The fuel pump may be supplying more fuel than the engine needs.

28. Use the proper grade of gasoline for your engine. Don't try to compensate for knocking or "pinging" by advancing the ignition timing. This practice will only increase plug temperature and the chances of detonation or pre-ignition with relatively little performance gain.

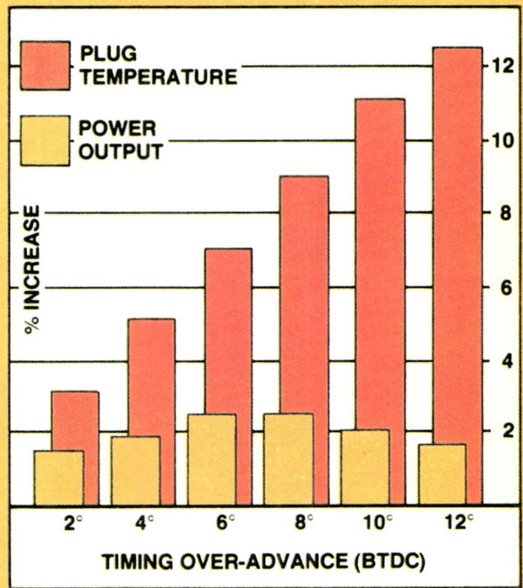

Increasing ignition timing past the specified setting results in a drastic increase in spark plug temperature with increased chance of detonation or preignition. Performance increase is considerably less. (Photo courtesy Champion Spark Plug Co.)

that form in the engine should be flushed out to allow the engine to operate at peak efficiency.

35. Clean the radiator of debris that can decrease cooling efficiency.

36. Install a flex-type or electric cooling fan, if you don't have a clutch type fan. Flex fans use curved plastic blades to push more air at low speeds when more cooling is needed; at high speeds the blades flatten out for less resistance. Electric fans only run when the engine temperature reaches a predetermined level.

37. Check the radiator cap for a worn or cracked gasket. If the cap does not seal properly, the cooling system will not function properly.

42. Be sure the front end is correctly aligned. A misaligned front end actually has wheels going in different directions. The increased drag can reduce fuel economy by .3 mpg.

43. Correctly adjust the wheel bearings. Wheel bearings that are adjusted too tight increase rolling resistance.

Check tire pressures regularly with a reliable pocket type gauge. Be sure to check the pressure on a cold tire.

GENERAL MAINTENANCE

Check the fluid levels (particularly engine oil) on a regular basis. Be sure to check the oil for grit, water or other contamination.

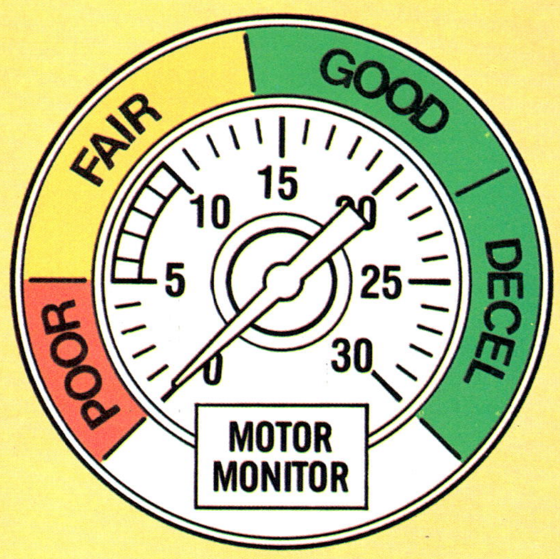

A vacuum gauge is another excellent indicator of internal engine condition and can also be installed in the dash as a mileage indicator.

44. Periodically check the fluid levels in the engine, power steering pump, master cylinder, automatic transmission and drive axle.

45. Change the oil at the recommended interval and change the filter at every oil change. Dirty oil is thick and causes extra friction between moving parts, cutting efficiency and increasing wear. A worn engine requires more frequent tune-ups and gets progressively worse fuel economy. In general, use the lightest viscosity oil for the driving conditions you will encounter.

46. Use the recommended viscosity fluids in the transmission and axle.

47. Be sure the battery is fully charged for fast starts. A slow starting engine wastes fuel.

48. Be sure battery terminals are clean and tight.

49. Check the battery electrolyte level and add distilled water if necessary.

50. Check the exhaust system for crushed pipes, blockages and leaks.

51. Adjust the brakes. Dragging brakes or brakes that are not releasing create increased drag on the engine.

52. Install a vacuum gauge or miles-per-gallon gauge. These gauges visually indicate engine vacuum in the intake manifold. High vacuum = good mileage and low vacuum = poorer mileage. The gauge can also be an excellent indicator of internal engine conditions.

53. Be sure the clutch is properly adjusted. A slipping clutch wastes fuel.

54. Check and periodically lubricate the heat control valve in the exhaust manifold. A sticking or inoperative valve prevents engine warm-up and wastes gas.

55. Keep accurate records to check fuel economy over a period of time. A sudden drop in fuel economy may signal a need for tune-up or other maintenance.

ever it has been disassembled and assembled, when the plunger or plunger barrel have been replaced or when any of the component parts have been replaced. Use nozzle tube 57805-002 and test nozzle 5000-101, starting pressure 1422.3 psi. Clean No. 2 fuel should be used. Rotating direction, from drive side, is clockwise. Sequence is 1-4-3-2

Starting Timing

1. Remove the fuel feed pump and cover plate from the injection pump.

2. Install the injection pump on the tester and holding fixture.

3. Connect the test coupling to the tester drive shaft with the coupling disc.

4. Remove the cap and position the tester dial to measure the camshaft rotating angle.

5. Install a tappet lift gauge on the #1 tappet.

6. Bottom the tappet and set the dial gauge to 0.

7. Bleed the pump at the bleeder screw.

8. Loosen the nozzle holder ball valve.

9. Feed fuel to the pump inlet while

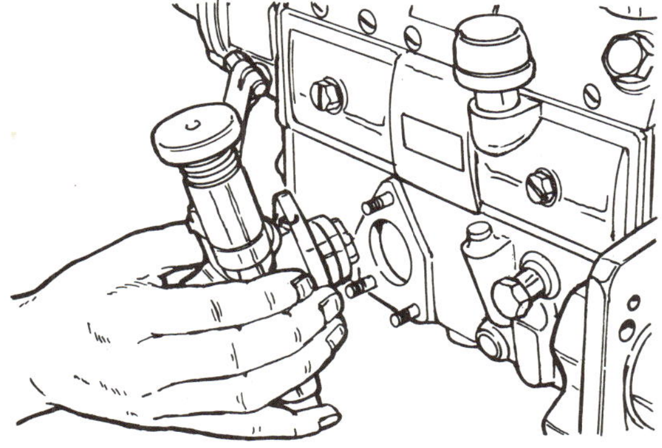

Fuel feed pump removal

Checking injection start timing with a lift gauge

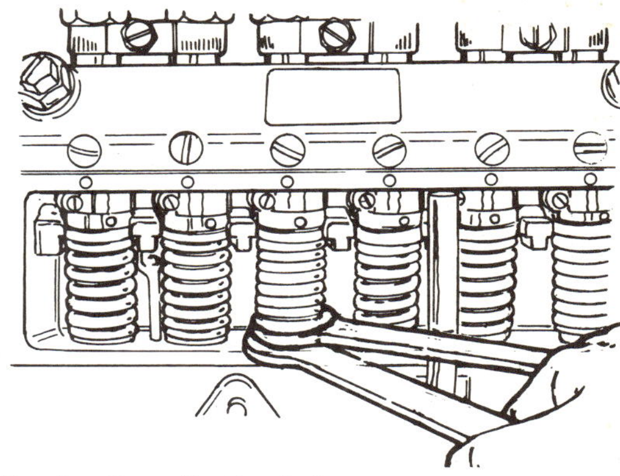

Turning the adjusting bolt

slowly turning the pump tester by hand in the normal engine rotation direction. Fuel will flow from the test nozzle. When the fuel stops flowing the injection point has been reached. The tappet, at this precise point, must be .08858−.09251″ above BDC.

10. If the fuel does not stop flowing after .0926″, turn the adjusting bolt counterclockwise to raise the position of the plunger.

11. If the fuel stops flowing before .0886″, turn the adjusting bolt clockwise to lower the plunger position.

12. When adjustment is made, torque the locknut to 43−50 ft. lb.

13. With the pump set at initial injection, set the angle scale mark on the tester flywheel at 0 or 180°.

14. If adjustment is correct, fuel should stop flowing at #4 cylinder when the tester has been turned 60° in normal rotation. If fuel does not stop flowing at the correct time, adjust as above.

15. Check and adjust each cylinder in turn.

16. When timing for each cylinder is correct, position the cam at TDC, check the

Checking tappet clearance

plunger piston pin-to-plunger barrel clearance and make sure that the tappet vertical clearance is at least .0118″ for each tappet.

Standard Fuel Injection Volume Adjustment

1. Determine the zero position of the control by attaching the measuring device to the pump and pushing the index all the way to governor side. Match the scale on the left end of the index and set the 0 (zero) position of the scale at the position at which the measuring device index stops. On RSV mechanical governors, loosen the stop bolt to align this index.

2. Remove the rack guide screw from the rear of the pump housing and apply the lock screw attached to the tester. Secure the control rack in the standard position for adjustment.

NOTE: *The lock screw should be tightened by hand; overtightening will bend the rack.*

3. Start the tester and run the pump at rated speed.

4. Set the pump feed pressure at 21.3–

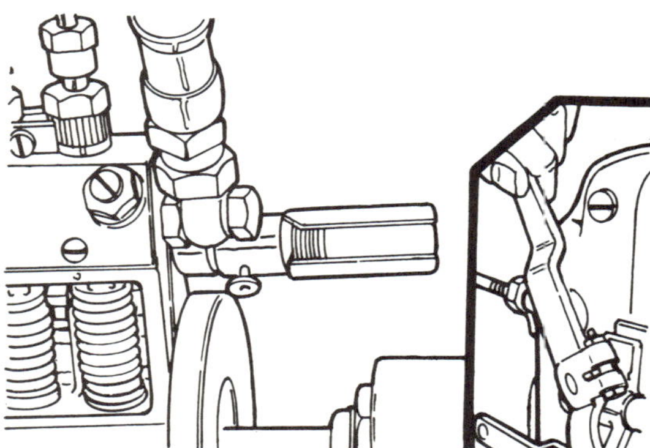

Attaching a measuring device to the pump

Installing the tester lock screw

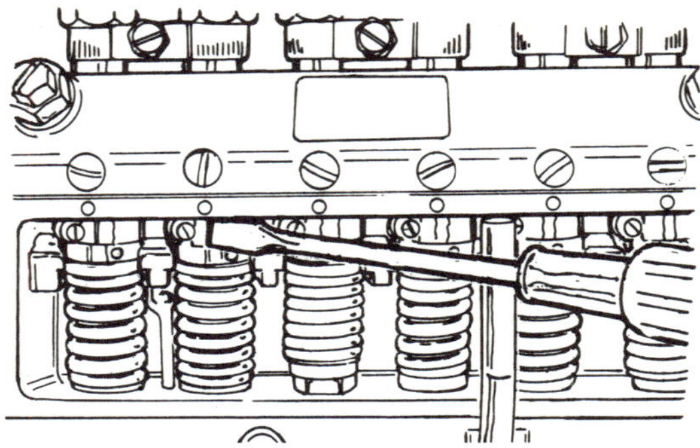

Loosening the pinion set screw

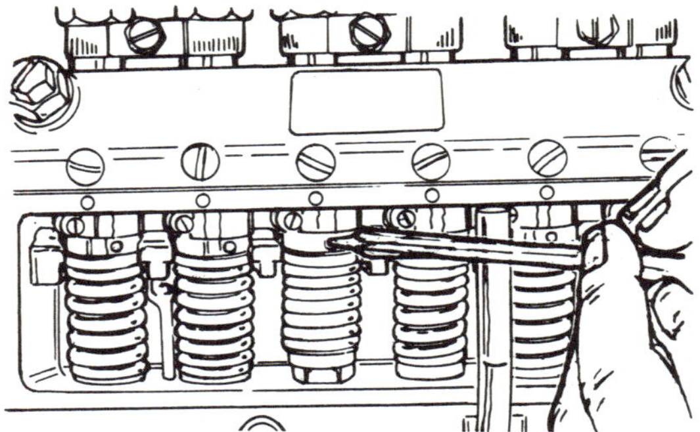

Installing the appropriate sized pin

22.75 psi and measure the injection volume at the rated stroke of the female cylinder.

5. In the same manner, measure the injection volume at rated speed and standard rack position. Compute the rate of unevenness of the injection.

6. If the results show that the mean injection volume and rate of unevenness are not within the limits, adjust by changing the relative position of the control pinion and control sleeve. This may be done by:

    a. loosen the pinion set screw.

    b. place a pin in the hole in the control sleeve and adjust by moving the control sleeve along the control rack a little at a time.

    c. when adjustment is completed, secure the pinion set screw.

    d. remove the lock screw from the control rack and reinstall the guide screw.

Testing and Adjusting the Governor

RSV AND RAD MECHANICAL GOVERNOR

1. Match the adjusting device index to the zero point on the scale and set the control rack to the zero position.

2. Operate the control lever and make certain the full stroke of the rack is .827″.

3. Make certain that the rack moves freely in the direction for maximum fuel injection by the spring force of the starting spring.

4. Set the stop bolt to remove any significant load on the governor linkage.

5. Set the stop bolt to give a control rack setting of .0197–.03937dp.

High Speed Adjustment

6. Remove the governor rear cover.

7. Loosen the full load stop lock nut and adjust the full load stop so that its setting corrsponds to an A rack position between "K" rotation and BC. See the accompanying chart.

$$\text{Uneveness} = \frac{\dfrac{\text{Max or min injection}}{\text{volume for each plunger}} - \dfrac{\text{Main injection}}{\text{volume}}}{\dfrac{\text{Mean injection}}{\text{volume}}} \times 100$$

8. To increase the volume, turn the stop to the right; left to decrease.

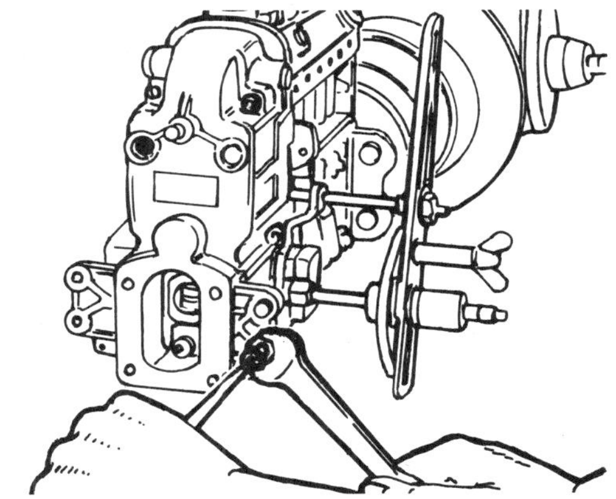

Adjusting the full load stopper bolt (RAD shown)

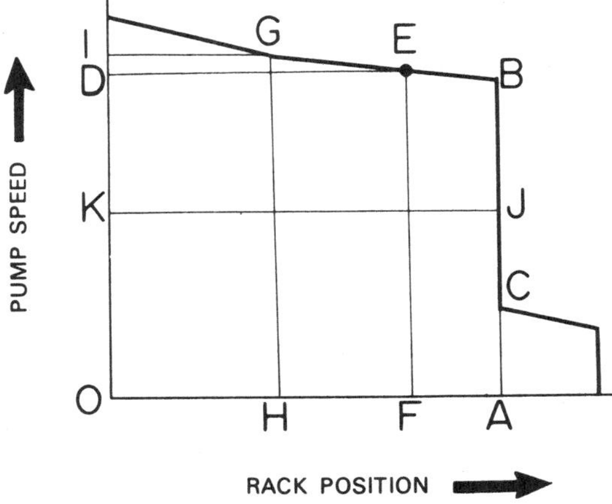

Maximum Speed Adjustment

9. Operate the pump at speed G and adjust the maximum speed stop so that the control rack position is G mm.

Speed Fluctuation Adjustment

10. Speed fluctuation can be controlled by varying the spring rate on RSV governors.

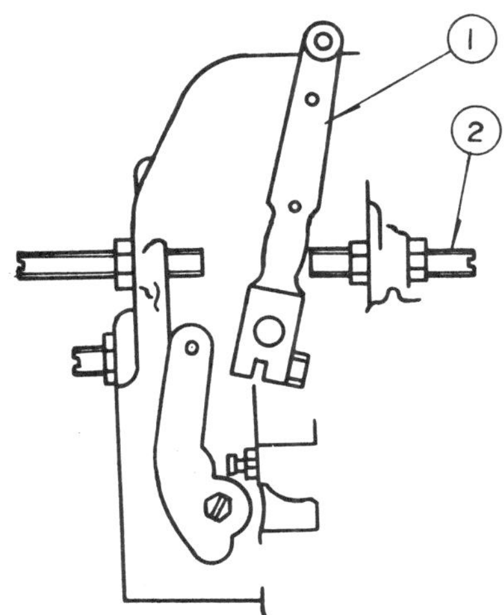

1. Speed control lever
2. Speed adjusting bolt (stopper bolt)

Adjusting the speed adjustment bolt (RSV shown)

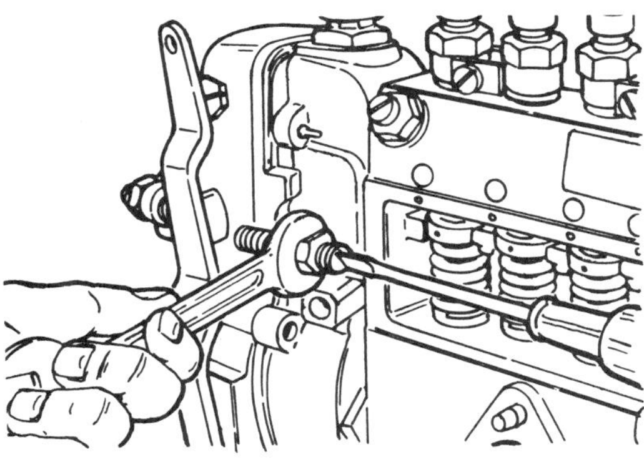

Maximum speed stopper adjustment

Balance Spring Adjustment

11. Set the control lever to the point where it contacts the maximum speed stop and operate the pump at a rate of F, between A and B.

12. Install the balance spring assembly under the tension lever.

13. Tighten the balance spring assembly with tool 57916-212 until the control rack position G is H mm. and secure with the lock nut.

14. Gradually accelerate the engine from

speed D and make certain that the control rack is at G mm when the action of the spring ends at speed E.

Idle Speed Adjustment

15. Set the control lever at the stop position so that the control rack is at point B.

16. Set the speed to D and tighten the idle auxiliary spring so that the position of the rack is C mm. Secure with the lock nut.

NOTE: *Do not tighten the auxiliary idle spring too much or overspeeding will result.*

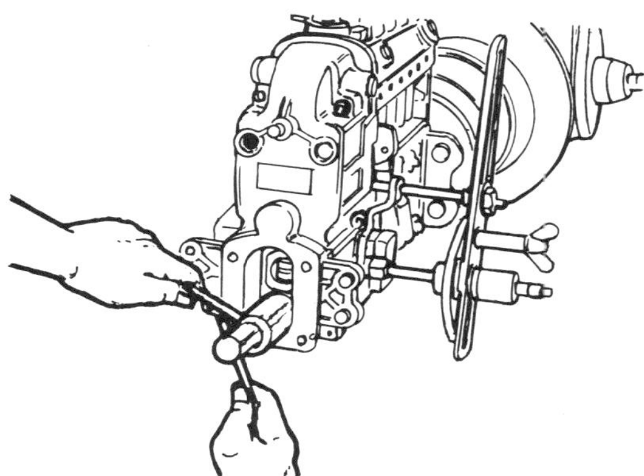

Adjusting the idling spring

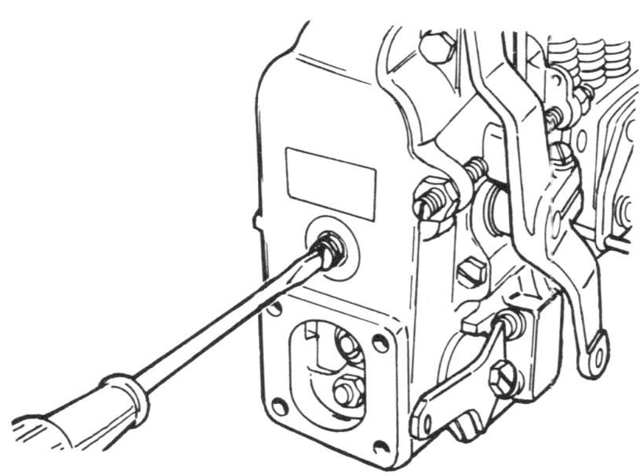

Auxiliary idle spring adjustment

Testing and Adjusting the Timer

1. Install a timing light, such as tool 5783-001 on the tester using the cover plate attaching bolts, so that the synchronizer lever attachment is applied to the tappet.

2. Start the pump and turn on the timing light.

3. Direct the light at the angle scale on the flywheel and measure the angular change based on variations in pump speed.

Pump installed on the tester, with synchronizer and strobe light

4. If the tester does not have an angle scale:

   a. Attach an angle scale to the timer coupling and mount a pointer on the tester drive shaft.

   b. Operate the pump and direct the light on the scale.

5. If the angular change is not within limits, disassemble the timer and adjust the spring force by increasing or decreasing the shims, or if necessary, replace the spring.

Fuel Tank

The fuel tank is located under the floor of the bed on the right side, directly behind the cab.

REMOVAL AND INSTALLATION

1. Disconnect the negative cable from the battery.

2. Remove the drain plug at the bottom of the tank and drain the fuel into a suitable container.

3. Disconnect the filler tube from the filler hose.

4. Disconnect the ventilation hoses, the fuel return hose and fuel outlet hose from the tank. Disconnect the gauge unit wires at the electrical connector.

5. Remove the mounting bolts and remove the fuel tank.

6. Installation is the reverse. Install the clamps securely, but do not crimp any of the lines. Install the clips holding the fuel tube to the underbody securely. Do not attach the filler hose to the tube until the tank is in place. Failure to do this will cause leaks around the connection.

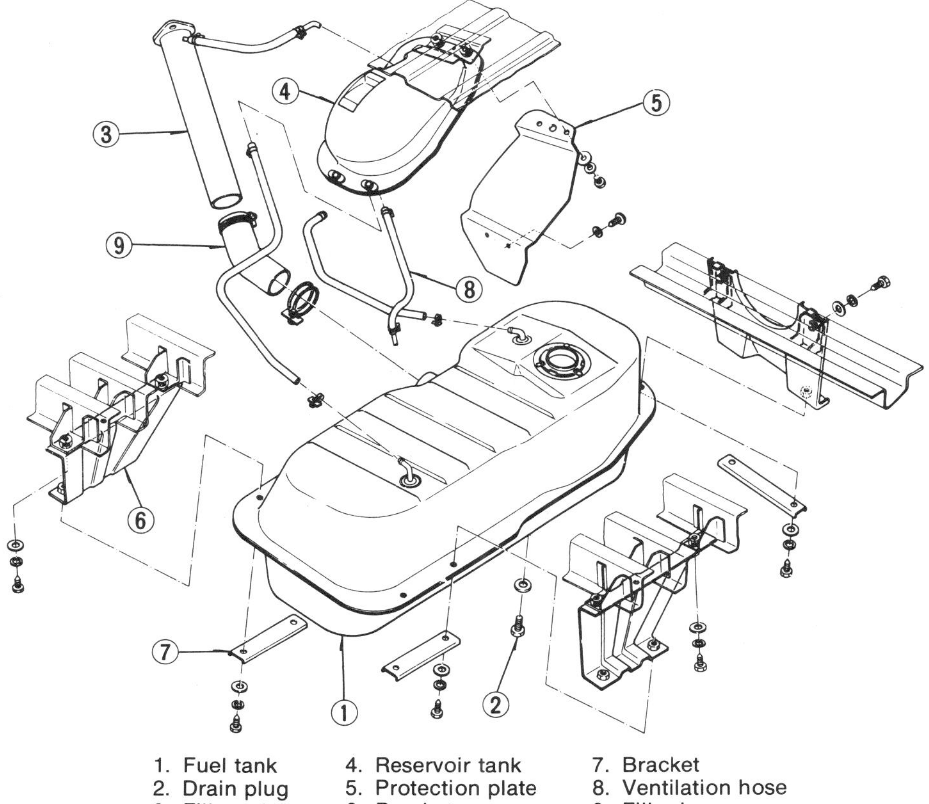

1. Fuel tank
2. Drain plug
3. Filler tube
4. Reservoir tank
5. Protection plate
6. Bracket
7. Bracket
8. Ventilation hose
9. Filler hose

Fuel tank and lines

GAUGE UNIT

The fuel tank must be removed for access to the gauge unit. The unit is installed into the tank with a bayonet-type of mount. Turn it counterclockwise with a screwdriver to remove. Use a new O-ring when installing, aligning the tab in the unit with the notch in the tank.

Chassis Electrical

HEATER

Heater Assembly
REMOVAL AND INSTALLATION
1970–79

1. Disconnect the battery ground cable.
2. Drain the engine coolant.
3. Remove the defroster hoses.
4. Remove the three cable retaining clips and disconnect the control cables from the valves and water cock.
5. Disconnect the two fan motor leads from each connector.
6. Disconnect the two resistor lead wires from each connector.
7. Disconnect the water hoses from the heater core and water cock.
8. Remove the three heater housing mounting bolts and remove the heater assembly from the vehicle.
9. Install the heater assembly in the reverse order of removal.

1980–81

1. Disconnect the battery ground cable.
2. Drain the cooling system.
3. On models with air conditioning, disconnect the heater hose from the engine.

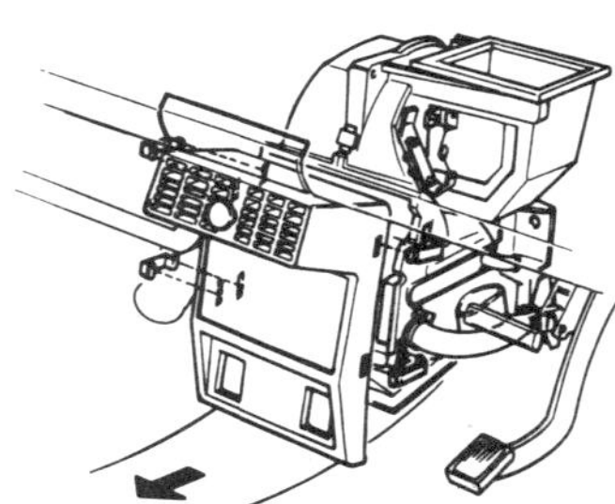

Removing the heater assembly front cover on 1970–79 models

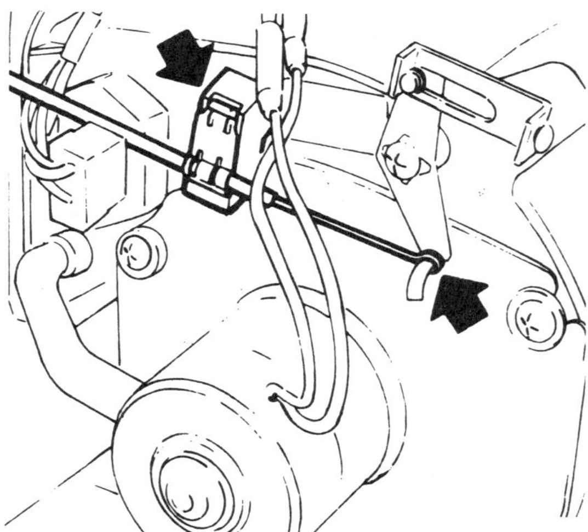

Control cable positioning

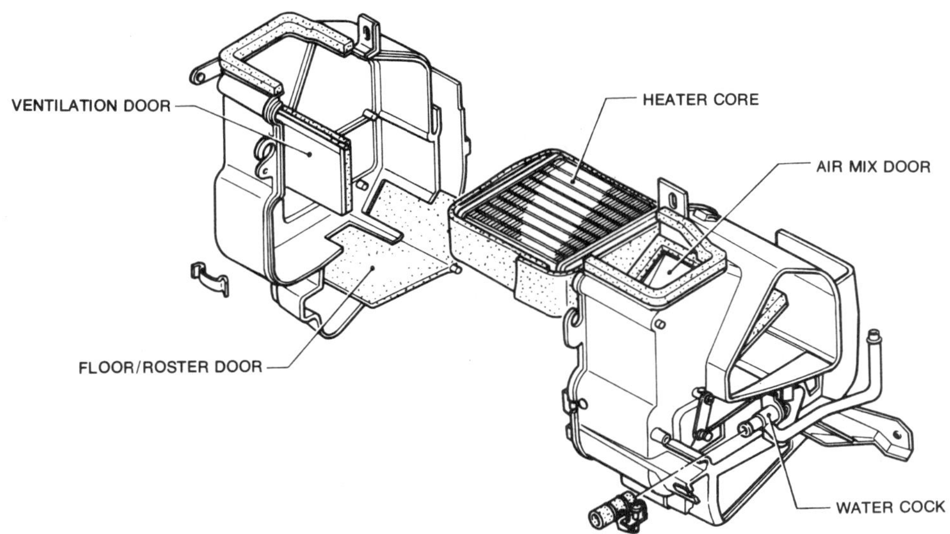

1980–81 heater assembly

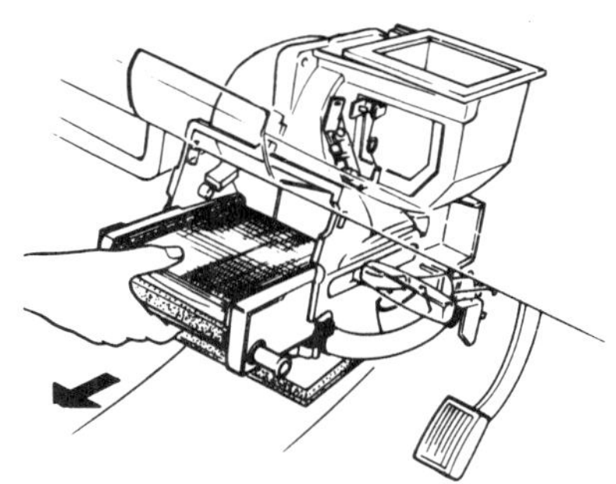

Removing the heater core from 1970–79 models

4. On models without air conditioning, remove the heater duct and disconnect the heater hose at the heater.

5. Remove the console box and instrument panel assembly.

6. Disconnect the air intake control cable from the blower.

7. On models equipped with a/c, remove the blower. Remove the evaporator unit nuts and bolts, but do not remove the evaporator unit.

8. Remove the heater assembly.

9. Installation is the reverse of removal. Adjust the control cable for proper operation.

Blower

REMOVAL AND INSTALLATION

1970–79

1. Remove the heater assembly from the vehicle as previously outlined.

2. Remove the nine spring clips and disassemble the heater housing.

3. Remove the fan from the electric motor.

4. Remove the fan motor retaining screws and remove the motor.

5. Install the blower motor and heater assembly in the reverse order.

1980–81

1. Disconnect the battery ground.

2. Remove the package tray.

3. Remove the heater duct on models without air conditioning.

4. Remove the resistor connector and disconnect the control cable.

5. Remove the blower.

6. Installation is the reverse of removal. Adjust the control cable for proper operation.

Heater Core

REMOVAL AND INSTALLATION

1. Drain the engine coolant.

2. Remove the defroster hoses.

3. Disconnect the water hoses from the inlet and outlet pipes of the heater core.

4. Remove the four clips and front cover.

5. Remove the heater core from the heater housing.

6. Install the heater core in the reverse order of removal.

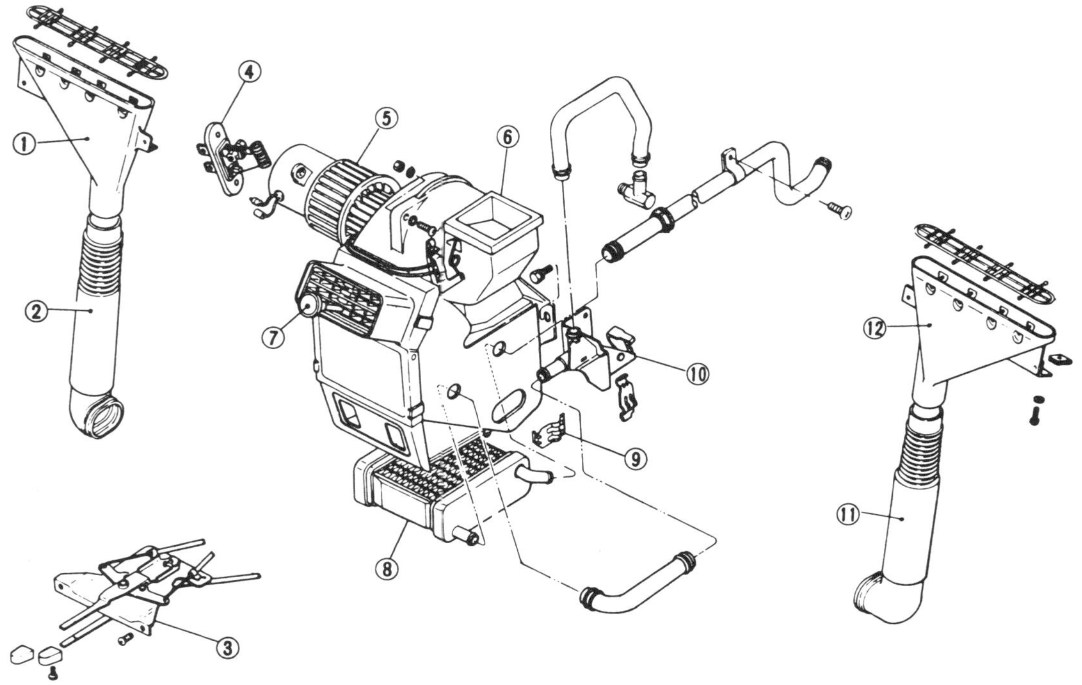

1. Defroster nozzle (L.H.)
2. Defroster duct (L.H.)
3. Heater control
4. Resistor
5. Heater motor
6. Heater case
7. Ventilator knob
8. Heater core
9. Control cable clip
10. Heater cock
11. Defroster duct (R.H.)
12. Defroster nozzle (R.H.)

Heater assembly, 1972–79

RADIO

Observe the following cautions when working on the radio:

1. Always observe the proper polarity of the connections (positive to positive and negative to negative).

2. Never operate the radio without a speaker, to prevent damage to the output transistors. If a new speaker is installed, make sure it has the correct impedance (ohms) for the radio.

If a new antenna or antenna cable is used, or if poor AM reception is noted, the antenna trimmer can be adjusted. Tune the radio to a weak station around 1400kc. Adjust the trimmer screw until best reception and maximum volume are obtained. The trimmer screw for the factory-installed radio is located on the bottom in the left rear corner on 1970–71 models, in the lower left corner of the rear of the radio case 1972–78, and above the left knob on the front of the radio in 1979. For best FM reception, raise the antenna to 31 inches. For best AM reception, raise the antenna to its full height.

REMOVAL AND INSTALLATION

1970–79

1. Pull the knobs off the radio control shafts.

2. Remove the radio retaining nuts and washer from the radio control shafts.

3. Remove the bezel plate from the front of the radio.

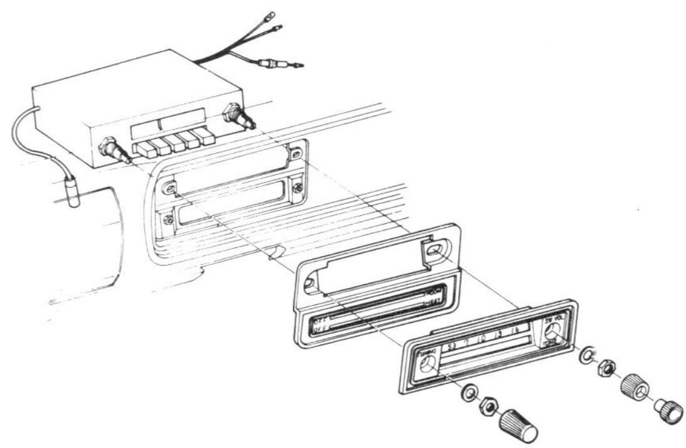

1970–79 radio removal

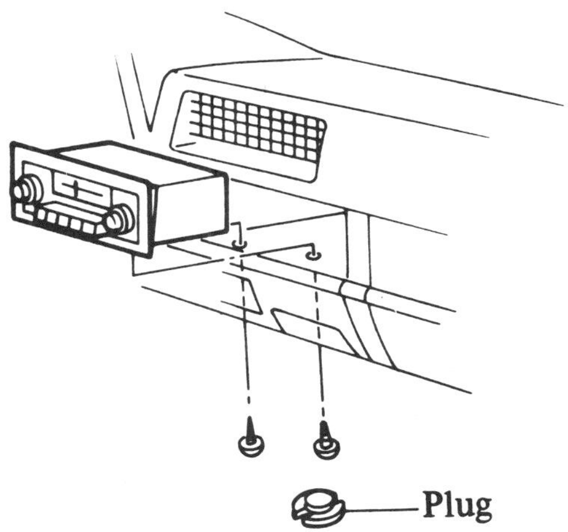

1980–81 radio removal

4. Disconnect the antenna cable and the power and speaker wires from under the instrument panel.

5. Remove the radio from the instrument panel.

6. Install the radio in the reverse order of removal.

1980–81

1. Disconnect the battery ground.

2. Remove the ash tray and heater/air conditioner control panel.

3. Disconnect the wiring plug at the back of the radio.

4. Remove the plug covering the mounting screws, remove the screws and pull the radio from the dash.

5. Disconnect the wiring harness and the antenna.

6. Installation is the reverse of removal.

WINDSHIELD WIPERS

Motor

REMOVAL AND INSTALLATION

1. Remove the wiper blades and arms as an assembly from the pivots. The arms are retained to the pivots by nuts. Remove the nuts and pull the arms straight off.

2. Remove the cowl top grille. It is retained by four screws at its front edge. Remove the screws and pull the grille forward to disengage the tabs at the rear.

3. Remove the stop ring which connects the wiper motor arm to the connecting rod.

4. Disconnect the wiper motor harness at the connector on the wiper motor body from under the instrument panel.

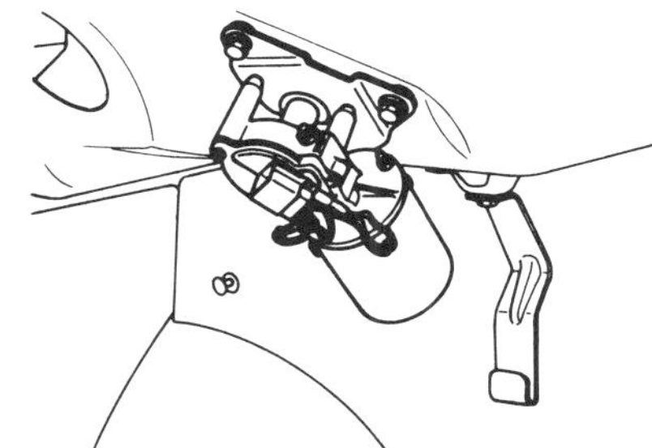

Windshield wiper motor removal

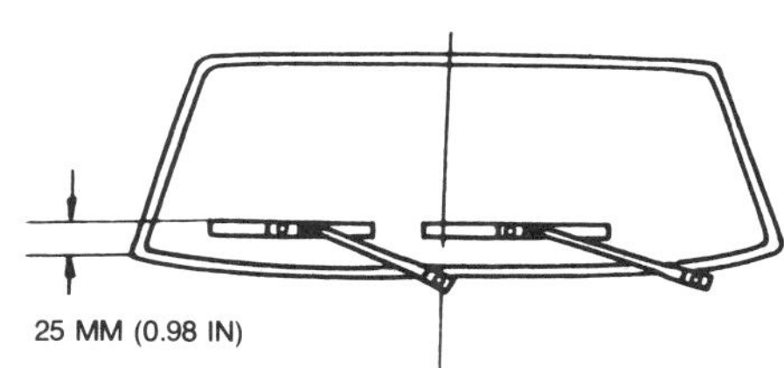

Wiper arm installation

5. Remove the three retaining screws and pull the wiper motor outward and remove the motor from the vehicle.

6. Install the wiper motor in the reverse order of removal. The wiper arms should be installed so that the blades are .98 in. (25 mm) abover, and parallel to, the windshield molding. If the motor has been run, be sure the motor and linkage is in its parked position before installing the wiper arms. To do this, turn the ignition switch on, and cycle the motor three or four times. Shut off the motor with the wiper switch (not the ignition switch), and allow the motor to return to the park position.

Linkage

REMOVAL AND INSTALLATION

1. Remove the wiper blade and arm from the pivot. See the preceding section.

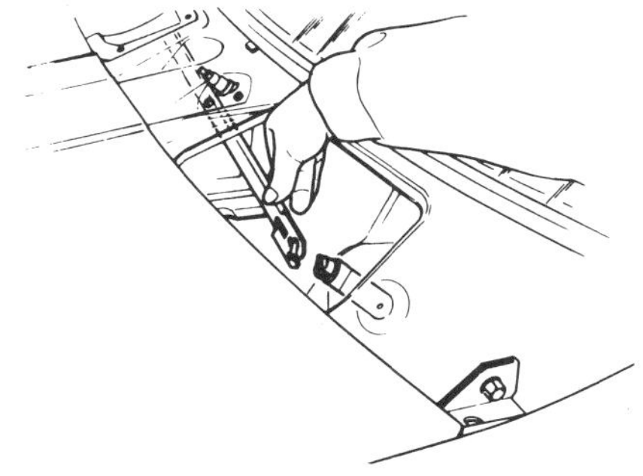

Windshield wiper linkage removal

Wiper linkage

2. Remove the cowl top grille.

3. Remove the two flange nuts retaining the wiper linkage pivot to the cowl top.

4. Remove the stop ring which retains the connecting rod to the wiper motor arm.

5. Remove the wiper motor linkage assembly from the truck.

6. Install the linkage in the reverse order of removal.

INSTRUMENT CLUSTER

REMOVAL AND INSTALLATION

1970-79

1. Disconnect the battery ground (negative) cable.

2. Working through the openings of the instrument cluster cover, remove the three screws retaining the cluster cover to the instrument panel and remove cover.

3. From underneath the instrument panel, remove the one screw retaining the cluster assembly to the lower instrument panel.

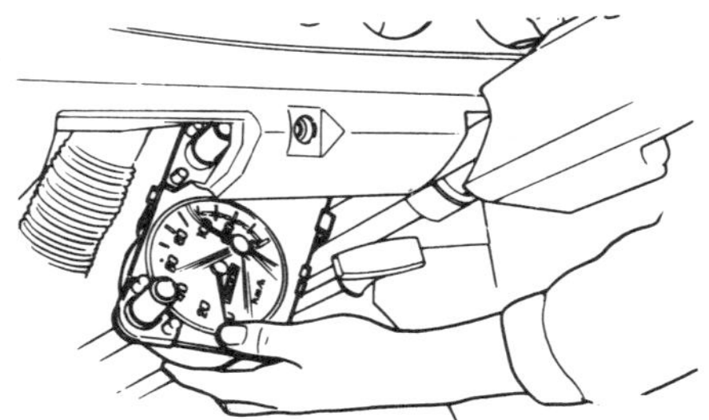

Removing the instrument cluster from under the instrument panel

4. Withdraw the cluster lid slightly. Press the windshield wiper control knob in, turn it counterclockwise and pull it off the switch. Remove the headlight/parking light switch knob in the same manner.

5. From behind the instrument cluster, disconnect the speedometer cable at the speedometer head and the multiple connector from the printed circuit.

6. On vehicles with a clock, disconnect the wires at each connection on the instrument panel printed circuit.

7. Remove the four screws retaining the cluster assembly to the cluster lid.

8. Remove the instrument cluster assembly from under the instrument panel.

9. Install the instrument cluster in the reverse order of removal.

1980-81

1. Disconnect the battery ground.

2. Remove the cluster lid.

3. Remove the cluster assembly.

4. Remove the gauges from the cluster, individually.

5. Installation is the reverse of removal.

1970-79 cluster cover removal

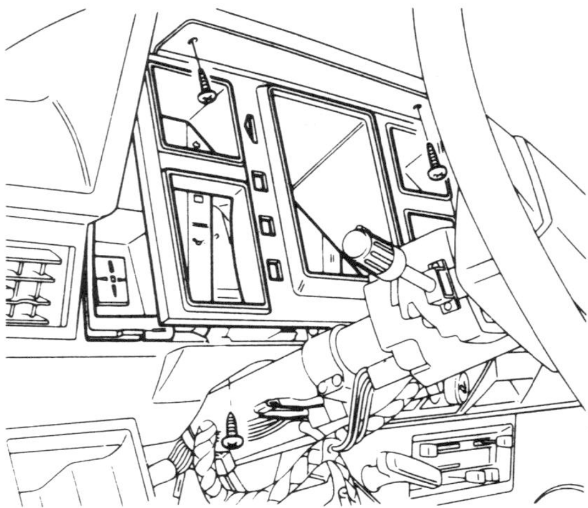

1980-81 cluster lid removal

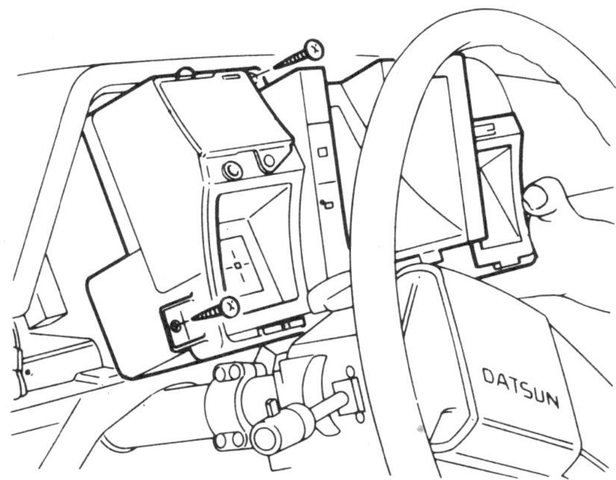

1980–81 cluster removal

SPEEDOMETER CABLE REPLACEMENT

1. Reach up under the instrument panel and disconnect the cable housing from the back of the speedometer. It is attached by a knurled knob which simply unscrews.

2. Pull the cable from the cable housing. If the cable is broken, the other half of the cable will have to be removed from the transmission end. Unscrew the retaining knob and remove the cable from the transmission extension housing.

3. Lubricate the cable with graphite powder (sold as speedometer cable lubricant, curiously enough) and feed the cable into the housing. It is best to start at the speedometer end and feed the cable down towards the transmission. It is also usually necessary to unscrew the transmission connection and install the cable end to the gear, then reconnect the housing to the transmission. Slip the cable end into the speedometer, and reconnect the cable housing.

HEADLIGHTS

REMOVAL AND INSTALLATION

1. Remove the radiator grille retaining screws and remove the radiator grille.

2. Loosen and remove, if necessary, the three retaining ring screws. Do not disturb the aiming adjusting screws.

3. Remove the retaining ring by rotating it clockwise.

4. Remove the headlight from the mounting ring and disconnect the electrical connector from behind the light.

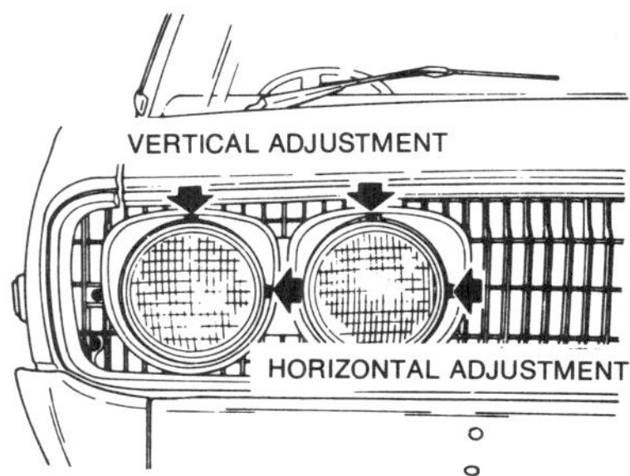

Headlight adjustment

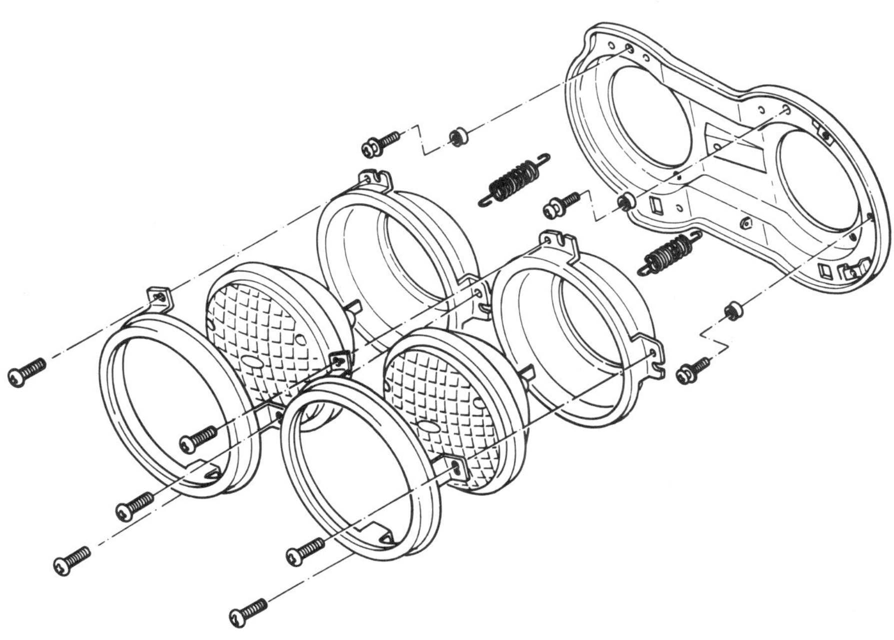

Typical headlight assembly

5. Change the headlight and connect the wiring connector to the new light.

6. Place the headlight in position so that the three locating tabs behind the light fit in the three holes on the mounting ring.

7. Install the headlight retaining ring and tighten the retaining screws.

8. Install the radiator grille.

CIRCUIT PROTECTION

Fusible Links

A fusible link is a protective device used in an electrical circuit. When current increases beyond a certain amperage, the fusible metal wire of the link melts, thus breaking the electrical circuit and preventing further damage to other components and wiring. Whenever a fusible link is melted because of a short circuit, correct the cause before installing a new fusible link.

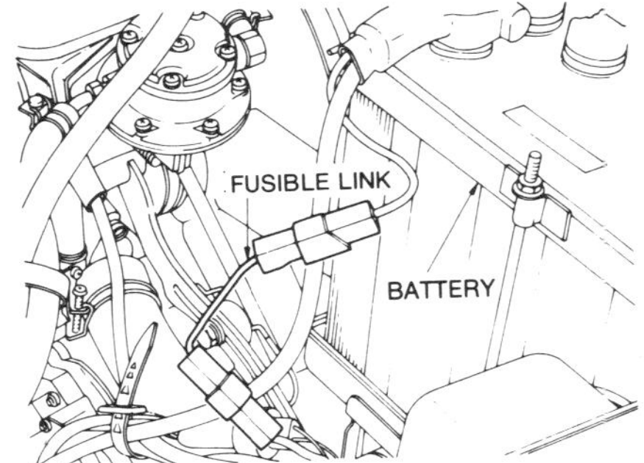

Fusible link

There is only one fusible link in Datsun pick-ups installed in the thinner of the two wires connected to the positive battery terminal. Replacements are simply plugged into the connectors in this wire.

CAUTION: *Use only replacements of the same electrical capacity as the original, available from your dealer. Replacements of a different electrical value will not provide adequate system protection.*

NOTE: *Model 521 pick-ups (1970–71) do not have a fusible link.*

Fuses

Fuses protect all the major electrical systems in the truck. In case of an electrical overload, the fuse melts, breaking the circuit and stopping the flow of electricity.

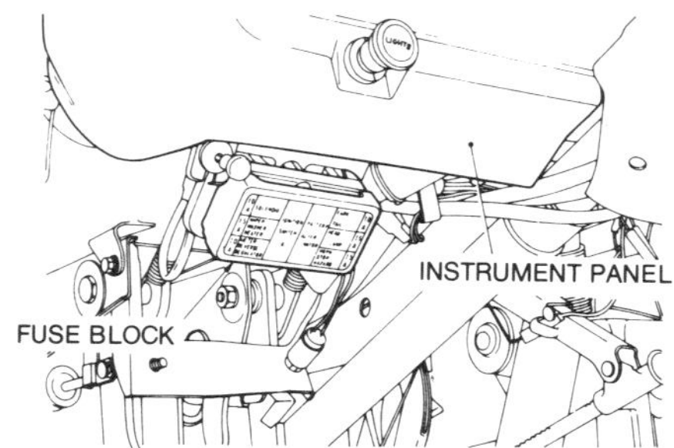

Fuse box location

If a fuse blows, the cause should be investigated and corrected before the installation of a new fuse. This, however, is easier to say than to do. Because each fuse protects a limited number of components, your job is narrowed down somewhat. Begin your investigation by looking for obvious fraying, loose connections, breaks in insulation, etc. Use the techniques outlined at the beginning of this chapter. Electrical problems are almost always a real headache to solve, but patience

Light Bulb Specifications

Light	SAE Number	Wattage
Headlight		
High beam	4001	37.5
Low beam	4002	37.5–50
Front combination light		
Turn signal	1034	8–23
Parking		
Side marker	67	8
License plate	89	7.5
Rear combination light		
Turn signal and stop	1073	23
Tail/turn signal and stop	1034	8–23
Tail	67	8
Back-up	1073	23
Cab light		5
Underhood		6
Wiper/washer illumination lamp	158	3.4
Heater control illumination lamp	(1970–76) 57	3.4
	(1977–81) 158	3.4
Combination meter, turn signal and high beam indicators, oil press, and charge warning	(1970–71) 158	3.4
	(1972–81) 161	1.7
Brake warning clock or tach. illumination	(1972–76) 161	1.7
	(1977–81) 158	3.4

and persistence, coupled with logic, usually provide a solution.

The amperage of each fuse and the circuit it protects are marked on the cover of the fusebox, which is located under the instrument panel next to the steering column, on 1972 and later trucks. On model 521 pickups (1970–71) the fusebox is in the engine compartment on the firewall.

Flashers and Relays

The turn signal and four-way hazard flashers are located under the instrument panel on opposite sides of the steering column. The turn signal flasher is the larger of the two. Replacement is made by unplugging the old flasher and plugging in the new one.

Relays are used for the horn, headlights, wiper, heater, choke heater, catalyst floor sensor, air conditioner compressor, and transmission switches, although obviously not all relays are used on all models. All relays used are grouped together, and mounted on the right fender in the engine compartment.

WIRING DIAGRAMS

Wiring diagrams have been left out of this book. As trucks have become more complex, and available with longer and longer option lists, wiring diagrams have grown in size and complexity also. It has become virtually impossible to provide a readable reproduction in a reasonable number of pages. Information on ordering wiring diagrams from the vehicle manufacturer can be found in the owners manual.

Clutch and Transmission

MANUAL TRANSMISSION

The Datsun pick-up uses an F4W63 four speed transmission through 1974, an F4W71B four speed 1975–81, or an optional FS5W71B five speed transmission from 1977 to 1981. The F4W63 is a bottom cover unit, with an extension housing for the shift rail, while the two later models have a one piece case, an adapter plate which supports the mainshaft and countershaft, and an extension housing. All units have internal shift rails; no linkage adjustments are necessary.

REMOVAL AND INSTALLATION

1. Disconnect the battery ground cable from the battery.

2. On 1975 and later models only, remove the shift lever from inside the cab. It is retained to the shift rail by a c-clip, accessible under the boot. Remove the c-clip and retaining pin, and remove the lever.

3. Jack up the vehicle and support it with jackstands.

4. On 1970–74 models, unscrew the nut securing the bottom of the shifter to the transmission shifting mechanism.

5. Disconnect the exhaust pipe from the exhaust manifold. On trucks with a catalytic converter, also remove the exhaust pipe bracket next to the speedometer cable by unscrewing the two mounting bolts.

6. Remove the clutch slave cylinder from the transmission case.

7. Disconnect the speedometer cable from the transmission extension housing and the back-up light and transmission switch wires at the switch(es).

8. Remove the bracket holding the center bearing of the driveshaft on the third crossmember of the frame.

9. Remove the driveshaft(s). See driveshaft removal in chapter 7.

NOTE: *On 4-wheel drive models, this requires removal of the transfer case, which should be performed at this time.*

10. Support the engine with a jack located under the oil pan. Place a block of wood between the jack and the oil pan to prevent damage to the oil pan. Support the transmission with a jack.

11. Remove the rear engine mount securing bolts and the crossmember mounting bolts. 521 models (1970–71) do not have a removable crossmember. Only the rear engine/transmission extension housing mount is removable from the crossmember.

12. Remove the starter motor.

13. Remove the bolts securing the transmission to the engine, pull the transmission toward the rear until the transmission main-

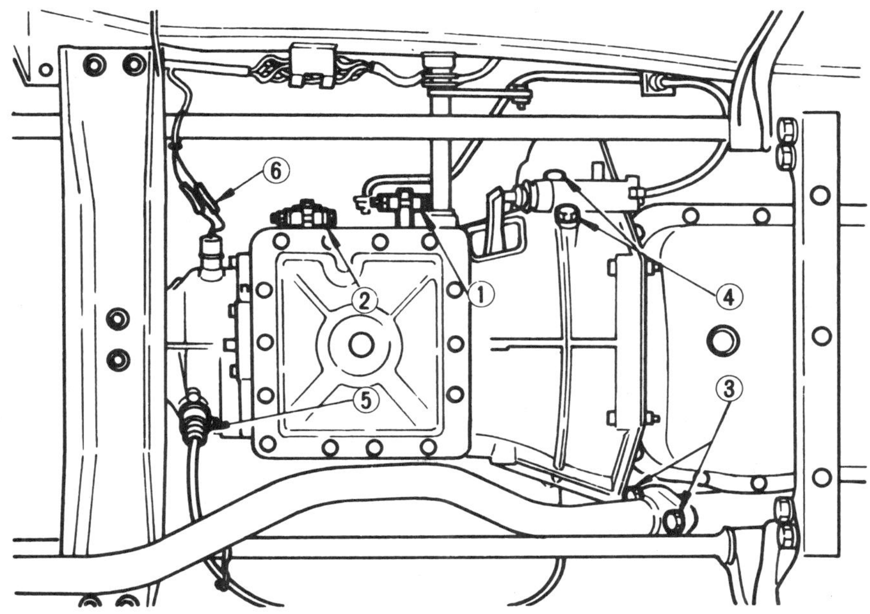

1. Neutral safety switch
2. Downshift solenoid
3. Exhaust pipe-to-flange attaching nuts
4. Clutch slave cylinder attaching bolts
5. Speedometer cable
6. Back-up light switch wires

Bottom view of the transmission 1972–74; others similar

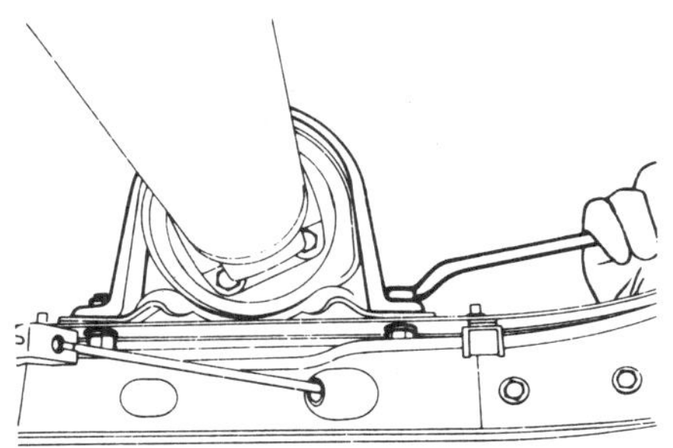

Removing the driveshaft center bearing bracket

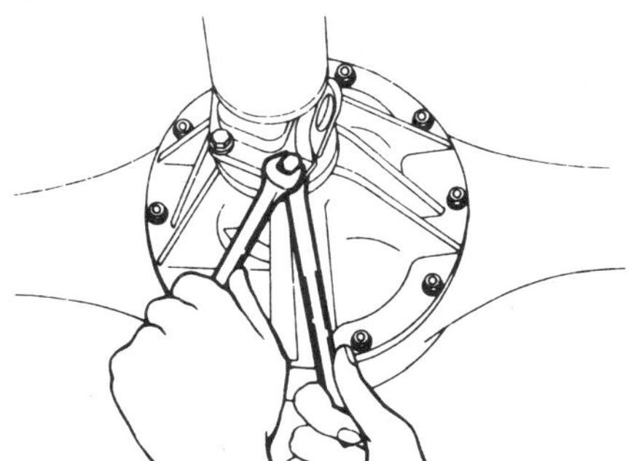

Disconnecting the driveshaft from the differential flange

shaft is free of the back of the engine. For 1970–71 (521) models, place the rear of the transmission on the crossmember and then pull the transmission down toward the front

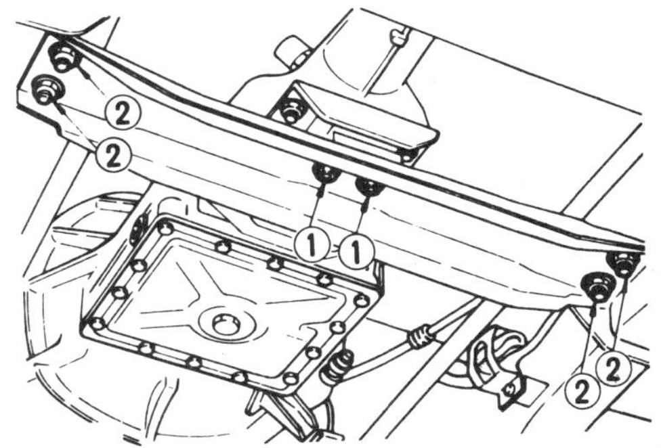

1. Transmission-to-crossmember attaching nuts and bolts
2. Crossmember-to-frame attaching nuts and bolts

The crossmember supporting the transmission, 1972 and later

of the truck and out from under the vehicle. For 1972 and later models, separate the transmission from the engine, then lower the transmission out from under the truck.

14. Install the transmission in the reverse order of removal. Before installing, clean the mating surfaces of the engine and transmission thoroughly. Lightly coat the input shaft splines with grease. Tighten the engine-to-transmission bolts to 17–20 ft. lbs., 1970–73; 29–36 ft. lbs. 1974–76; or 32–43 ft. lbs., 1977 and later. Tighten the crossmember to chassis bolts to 20–27 ft. lbs. and the clutch

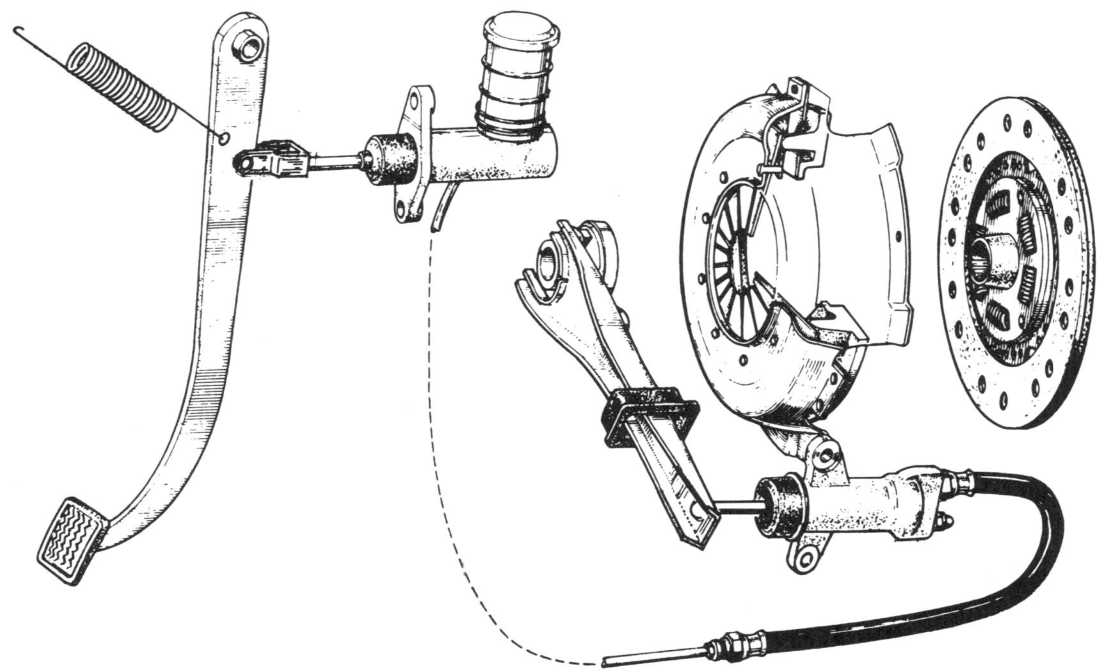

The clutch operating system

slave cylinder mounting bolts to 18–22 ft. lbs. Be sure to align the marks made earlier on the U-joint and differential flange when installing the driveshaft, to maintain driveline balance.

CLUTCH

The clutch is a hydraulically-operated single-plate, dry friction disc, diaphragm spring type.

The clutch is operated by a clutch pedal which is mechanically connected to a clutch master cylinder. When the pedal is depressed, the piston in the master cylinder is moved in the master cylinder bore. This movement compresses the fluid in the master cylinder causing hydraulic pressure which is transferred through a tube to the slave cylinder. The slave cylinder is mounted to the clutch housing with its piston connected to the clutch release lever. The hydraulic pressure in the slave cylinder forces the slave cylinder piston to travel out the cylinder bore and move the clutch release lever, disengaging the clutch.

PEDAL HEIGHT AND FREE-PLAY ADJUSTMENT

1. Disconnect the pushrod clevis from the clutch pedal and adjust the clutch pedal height to 5.91 in., 1970–73; 6.02 in., 1974–77; 6.42 in., 1978–79, by loosening the locknut on the pedal stopper. Turn the pedal stopper in or out as necessary and retightening the locknut.

2. Loosen the locknut on the pushrod and adjust the position of the clevis so that 0.039–0.118 in. (through 1977) or 0.04–0.20 in. (1978–79) of free-play exists between the clevis and the clutch pedal. Retighten the locknut and install the clevis pin and cotter pin.

3. After making the adjustments, make sure that the clutch pedal travels its full stroke of: 4.61 in., 1970–71; 5;08 in., 1972–

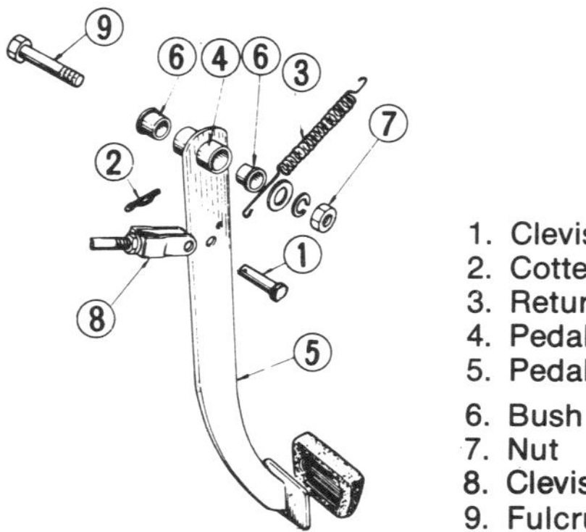

1. Clevis pin
2. Cotter pin
3. Return spring
4. Pedal boss
5. Pedal assembly
6. Bush
7. Nut
8. Clevis
9. Fulcrum pin

Exploded view of the clutch pedal assembly

73; 4.61–4.84 in., 1974–77; 4.69–4.92 in., 1978–79. The clutch linkage should operate smoothly and completely disengage and engage the clutch.

REMOVAL AND INSTALLATION

1. Raise the vehicle on a hoist.

2. Remove the transmission. On 1970–71 521 models with the non-removable crossmember, you can move the transmission back and rest it on the crossmember while working on the clutch, if you prefer. You will not have much working area, but it will save you the trouble of getting the transmission out from under the truck.

3. Mark the clutch assembly-to-flywheel relationship with paint or a center punch so that the clutch assembly can be reassembled in the same position from which it is removed. Insert a clutch aligning tool (dummy shaft) into the hub. This tool is available from your dealer or an auto parts store. It is important to support the weight of the clutch while the retaining bolts are being removed.

4. Loosen the six clutch cover-to-flywheel attaching bolts, one turn at a time in an alternating sequence, until the spring tension is relieved to avoid distorting or bending the clutch cover. Remove the clutch assembly.

5. Inspect the flywheel for scoring, roughness, or signs of overheating. Light scoring may be cleaned up with emery cloth, but any deep grooves or scoring warrant replacement or refacing (if possible) of the flywheel. If the clutch facings or flywheel are

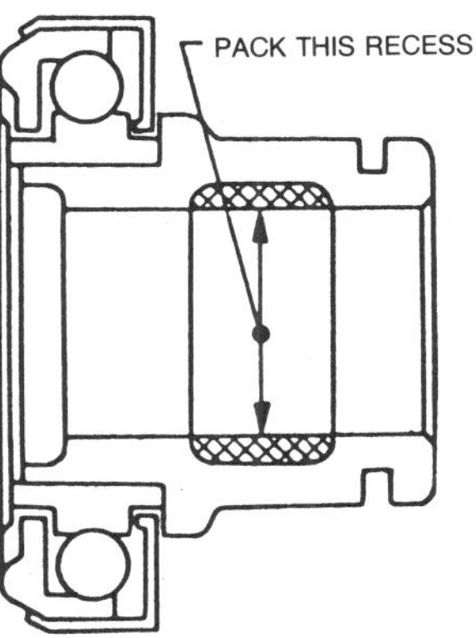

Coat the area indicated in the bearing sleeve with grease

oily, inspect the transmission front cover oil seal, the pilot bushing, and engine rear seals, etc. for leakage, and correct before replacing the clutch. If the pilot bushing in the crankshaft is worn, replace it. Install it using a soft hammer. The factory-supplied part does not have to be oiled, but check the procedure if you are using an aftermarket part. Inspect the clutch cover for wear or scoring, and replace as necessary. The pressure plate and spring cannot be disassembled; you must replace the clutch cover as an assembly.

6. Inspect the clutch release bearing. If it is rough or noisy, it should be replaced. The bearing can be removed from the sleeve with a puller; this requires a press to install the new bearing. After installation, coat the groove in the sleeve, the contact surfaces of the release lever, pivot pin and sleeve, and the release bearing contact surfaces on the transmission front cover with a light coat of grease. Be careful not to use too much grease, which will run at high temperatures and get onto the clutch facings. Reinstall the release bearing on the lever.

7. Apply a thin coat of grease to the pressure plate wire ring, diaphragm spring, clutch cover grooves and the drive bosses on the pressure plate.

8. Apply a thin coat of Lubriplate® to the splines in the driven plate. Slide the clutch disc onto the splines, and move it back and forth several times. Remove the disc and wipe off the excess lubricant. Be very careful not to get any grease on the clutch facings.

9. Assemble the clutch cover and the clutch plate on the clutch alignment arbor.

10. Align the marks made on the clutch cover and the flywheel (if the old cover is being used) and install the six clutch cover-

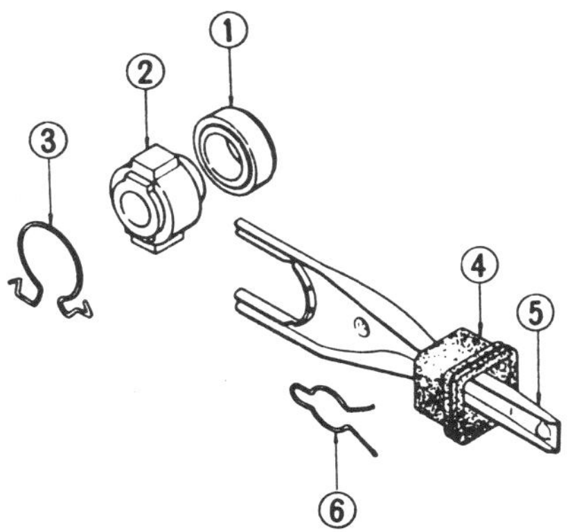

1. Release (throwout) bearing
2. Bearing sleeve
3. Sleeve spring
4. Boot
5. Release lever
6. Retaining spring

Clutch release mechanism

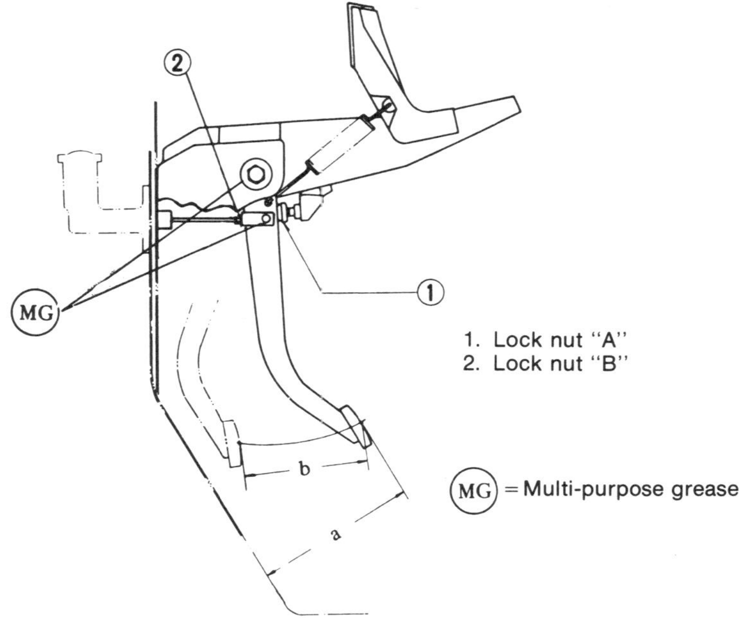

Clutch pedal adjustment: "a" is pedal height, "b" is pedal stroke

to-flywheel attaching bolts. Three dowels are used to locate the clutch cover on the flywheel properly. Tighten the bolts in an alternating sequence one turn at a time to 12–15 ft. lbs. Remove the aligning arbor.

11. Install the transmission.

Clutch Master Cylinder

REMOVAL AND INSTALLATION

1. Disconnect the clutch pedal arm from the pushrod clevis. Remove the dust cover (boot) from the master cylinder body and pushrod. It will not go through the firewall without tearing.

2. Disconnect the clutch hydraulic line from the master cylinder.

NOTE: *Take precautions to keep brake fluid from coming in contact with any painted surfaces.*

3. Remove the nuts attaching the master cylinder and remove the master cylinder and pushrod toward the engine compartment side.

4. Install the master cylinder in the reverse order of removal and bleed the clutch hydraulic system.

OVERHAUL

NOTE: *Datsun obtains its clutch master cylinder parts from two suppliers: Nabco and Tokico. There is no interchangeability*

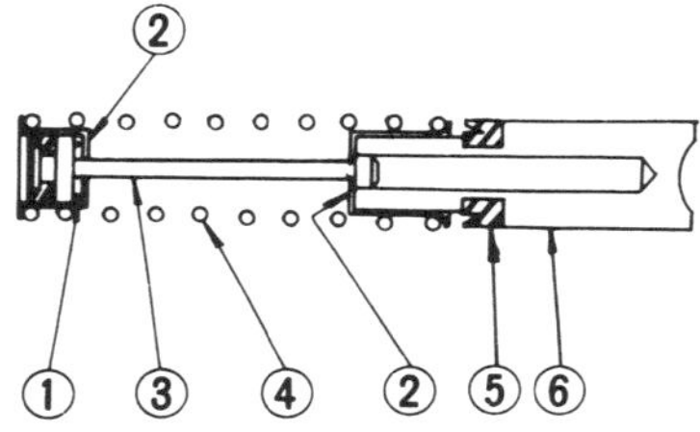

1. Valve spring
2. Spring seat
3. Valve assembly
4. Return spring
5. Piston cup
6. Piston

Cutaway view of the clutch master cylinder piston

between the parts. Be absolutely certain that you get the correct parts for the master cylinder installed on your truck. The manufacturer's name is clearly written on the cylinder.

1. Remove the master cylinder.

2. Drain the clutch fluid from the master cylinder reservoir.

3. Remove the circlip and remove the pushrod.

4. Remove the stopper, piston, cup, and return spring.

5. Clean all of the parts in clean brake fluid.

6. Check the master cylinder and piston for wear, corrosion and scores, and replace the parts as necessary. Light scoring and glaze can be removed with crocus cloth

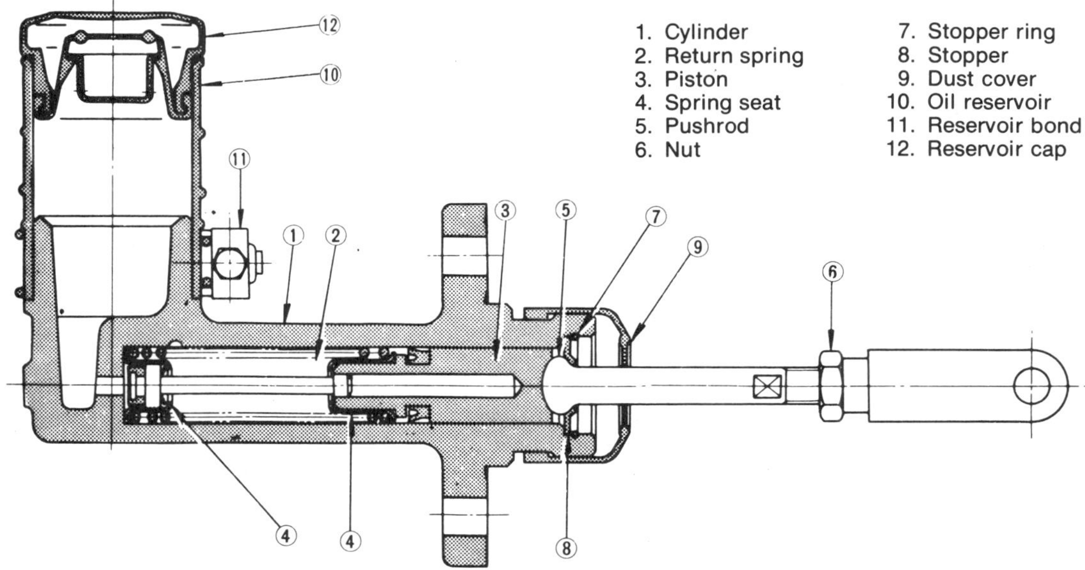

Cutaway view of the clutch master cylinder

soaked in brake fluid. Move the crocus cloth in a circular motion; not in and out.

7. Generally, the cup seal should be replaced each time the master cylinder is disassembled. Check the cup and replace it if it is worn, fatigued, or damaged.

8. Check the clutch fluid reservoir, filler cap, dust cover, and the pipe for distortion and damage and replace the parts as necessary.

9. Lubricate all new parts with clean brake fluid.

10. Reassemble the master cylinder parts in the reverse order of disassembly, taking note of the following:

    a. Reinstall the cup seal carefully to prevent damaging the lipped portions;

    b. Adjust the height of the clutch pedal after installing the master cylinder in position on the vehicle;

    c. Fill the master cylinder and clutch fluid reservoir and then bleed the clutch hydraulic system.

Clutch Slave Cylinder

REMOVAL AND INSTALLATION

1. Remove the slave cylinder attaching bolts and the pushrod from the shift fork.

2. Disconnect the flexible fluid hose from the slave cylinder and remove the unit from the vehicle.

3. Install the slave cylinder in the reverse order of removal and bleed the clutch hy-

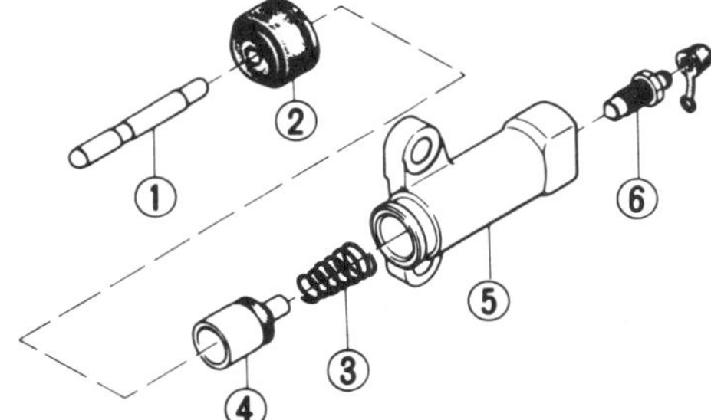

Exploded view of the clutch slave cylinder

draulic system. Tighten the attaching bolts to 18–25 ft. lbs.

OVERHAUL

NOTE: *Datsun obtains its slave cylinders from two manufacturers: Nabco and Tokico. Parts are not interchangeable. Be sure you get the correct rebuilding parts for the model on your truck.*

1. Remove the slave cylinder from the vehicle.

2. Remove the pushrod and boot.

3. Force out the piston by blowing compressed air into the slave cylinder at the hose connection.

NOTE: *Be careful not to apply excess air pressure to avoid possible injury.*

4. Clean all of the parts in clean brake fluid.

5. Check and replace the slave cylinder bore and piston if wear or severe scoring exists. Light scoring and glaze can be removed with crocus cloth soaked in brake fluid. Move the crocus cloth in a circular motion; not in and out.

6. Normally the piston cup should be replaced when the slave cylinder is disassembled. Check the piston cup and replace it if it is found to be worn, fatigued or scored.

7. Replace the rubber boot if it is cracked or broken.

8. Lubricate all of the new parts in clean brake fluid and reassemble in the reverse order of disassembly, taking note of the following:

    a. Use care when reassembling the piston cup to prevent damaging the lipped portion of the piston cup;

    b. Fill the master cylinder with brake fluid and bleed the clutch hydraulic system;

    c. Adjust the clearance between the pushrod and the shift fork to $5/64$ in. Only 1970–71 models are adjustable.

Bleeding the Clutch Hydraulic System

1. Check and fill the clutch fluid reservoir to the full mark if necessary. During the bleeding process, continue to check and replenish the reservoir to prevent the fluid level from getting lower than ½ full.

2. Connect a clear vinyl hose to the bleeder screw on the slave cylinder. Immerse the other end of the hose in a clear jar half filled with brake fluid.

NOTE: *Don't drip brake fluid on any painted surfaces.*

3. Have an assistant pump the clutch pedal several times and hold it down. Loosen the bleeder screw slowly.

4. Tighten the bleeder screw and release the clutch pedal gradually. Repeat this operation until air bubbles disappear from the brake fluid being expelled out through the bleeder screw.

5. When the air is completely removed, securely tighten the bleeder screw and replace the dust cap.

6. Check and refill the master cylinder reservoir as necessary.

7. Depress the clutch pedal several times to check the operation of the clutch and check for leaks.

AUTOMATIC TRANSMISSION

The optional automatic transmission is a JATCO (Japan Automatic Transmission Co., Ltd.) model 3N71B. It is a fully automatic unit, with a three element torque converter and two planetary gear sets. The transmission shifts gears in response to signals of both engine speed and manifold vacuum.

While it is unlikely that you will ever disassemble the transmission yourself, there are a few adjustments you can perform which will prolong the transmission's life if performed accurately. The most important thing is to change the fluid regularly, which is covered in Chapter One. There is one internal adjustment which can be made to the brake band; however, this requires a certain amount of expertise, and so is purposely not covered here.

PAN REMOVAL

Loosen the automatic transmission pan attaching bolts more at one corner than the other three corners. Allow the fluid to drain out the one corner. Remove all of the pan attaching bolts and remove the pan. Install the pan in the reverse order of removal. Always use a new gasket.

SHIFT LINKAGE ADJUSTMENT

Adjustment of the shift linkage is a critical operation. If the adjustment is made sloppily, the result will be partial application of the band or clutches, and will eventually burn up the transmission.

1. If the control knob is removed, prior to installation set dimension "A" in the illustration to 0.43–0.47 in. (11–12 mm).

2. Install the control knob, adjusting dimension "B" to 0.004–0.043 in. (0.1–1.1 mm) by turning the pushing rod (2).

3. Loosen the adjusting nuts ("H"). Set both the shift lever (3) and the transmission selector lever (4) into the neutral positions. Set clearance "C" to 0.04 in. (1 mm) by turning the adjusting nuts which connect to the selector rod (6).

After making the adjustments, check for proper engagement in each gear, and that the mechanism operates without binding. Readjust as necessary.

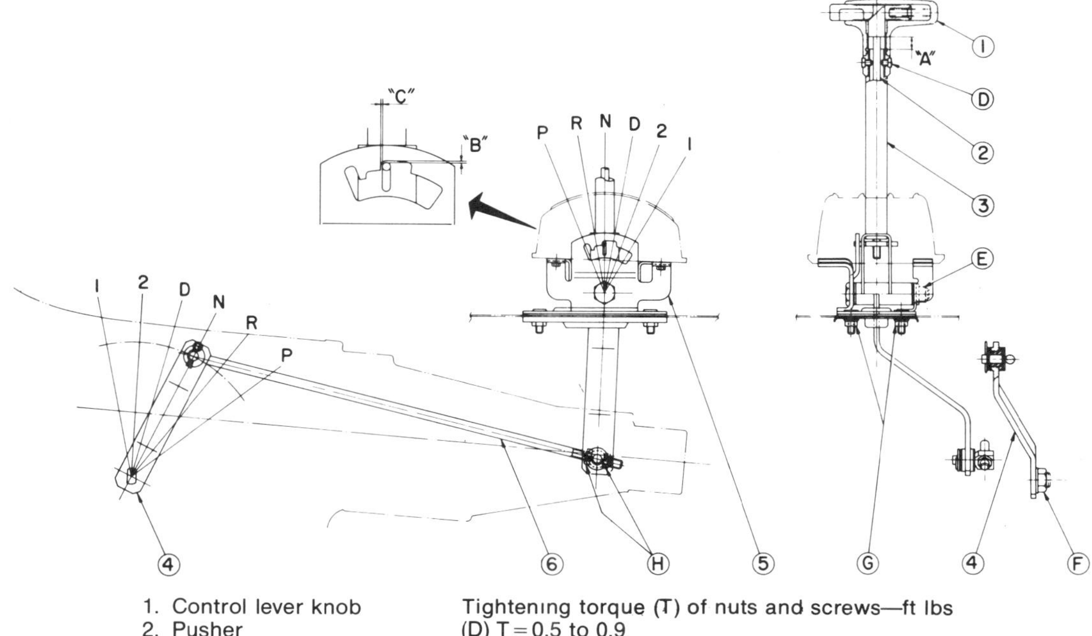

1. Control lever knob	Tightening torque (T) of nuts and screws—ft lbs
2. Pusher	(D) T = 0.5 to 0.9
3. Control lever assembly	(E) T = 12 to 16
4. Selector range lever	(F) T = 22 to 29
5. Control lever bracket	(G) T = 2.5 to 3.3
6. Selector rod	(H) T = 5.8 to 8.0

Automatic transmission shift linkage

KICK-DOWN SWITCH AND DOWNSHIFT SOLENOID

With the ignition switch in the ON position and the engine off, when the accelerator pedal is depressed fully, the kick-down switch contacts should be closed and the downshift solenoid activated, emitting a clicking sound. If the components fail to operate in this manner, check for continuity first at the switch and then at the solenoid if the switch checks out as being satisfactory. Replace either of the components as necessary.

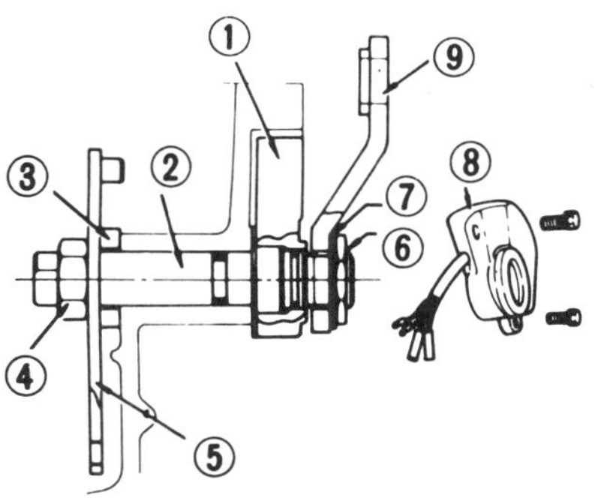

1. Inhibitor switch	6. Washer
2. Manual shaft	7. Nut
3. Washer	8. Inhibitor switch
4. Nut	9. Selector lever
5. Manual plate	

The neutral safety switch

NEUTRAL SAFETY SWITCH

The neutral safety switch is located on the transmission range selector lever. The switch operates the back-up lights and controls the operation of the starter. The starter should only operate when the transmission is in Park or Neutral.

To adjust the neutral safety switch, unscrew the securing nut of the range selector

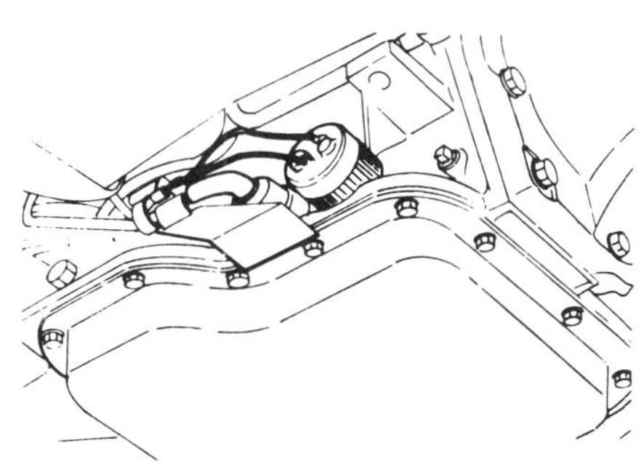

The downshift solenoid

lever and the two bolts securing the switch body. Remove the machine screw under the switch body. Adjust the shift selector to the Neutral position (in vertical position and detent clicks).

Move the switch slightly aside so that the screw hole will be aligned with the pin hole of the internal rotor combined with the manual shaft and check their alignment by inserting a 0.0591 in. (1.5 mm) diameter pin into the holes. A #53 drill bit will work for this. Fasten the switch body with the bolts, pull out the pin, and tighten the screw into the hole. Connect the selector lever.

If the neutral safety switch does not perform satisfactorily after adjustment, replace it with a new one.

REMOVAL AND INSTALLATION

1. Disconnect the negative battery cable.
2. Disconnect the shaft from the accelerator linkage.
3. Raise and support the truck.
4. Matchmark the U-joint and differential flange and disconnect them. Remove the center bearing mounting bolts and remove the driveshaft. Plug the transmission extension housing.
5. Disconnect the exhaust pipe from the manifold and discard the gasket. Use a new gasket upon assembly. On trucks with a catalytic converter, disconnect the exhaust pipe bracket.
6. Disconnect the shift linkage at the transmission.
7. Disconnect the neutral switch wires. Disconnect the vacuum hose from the diaphragm, and the wire from the downshift solenoid. Disconnect the speedometer cable from the extension housing.
8. Remove the fluid filler tube.
9. Disconnect the fluid cooler lines at the transmission. Use a flare nut wrench if one is available.
10. Support the engine with a jack under the oil pan, placing a wooden block between the pan and the jack as a buffer. Also support the transmission with a jack.
11. Remove the torque converter cover. Matchmark the converter and the drive plate for reassembly; they were balanced as a unit at the factory. Remove the bolts attaching the converter to the drive plate (flywheel). You will have to rotate the engine to do this, using a wrench on the crankshaft pulley bolt.
12. Remove the bolts for the rear engine mount and the crossmember. Remove the crossmember.
13. Remove the starter.
14. Remove the transmission-to-engine bolts. Lower the transmission back and down, out from under the truck.
15. Before installing the transmission, check the drive plate runout with a dial indicator. Turn the crankshaft one full turn. Maximum allowable runout is 0.012 in. (0.3 mm). Replace the drive plate if runout exceeds 0.020 in. (0.5 mm); otherwise, reface it.
16. When installing the torque converter, be sure to line up the notch in the converter with the projection on the oil pump. Align the marks made during removal and bolt the converter to the drive plate, tightening the bolts to 29–36 ft. lbs. Then rotate the engine a few turns to make sure the transmission rotates freely without binding. The engine-to-transmission bolt torque is 29–36 ft. lbs. Adjust the shift linkage and neutral switch after installation.

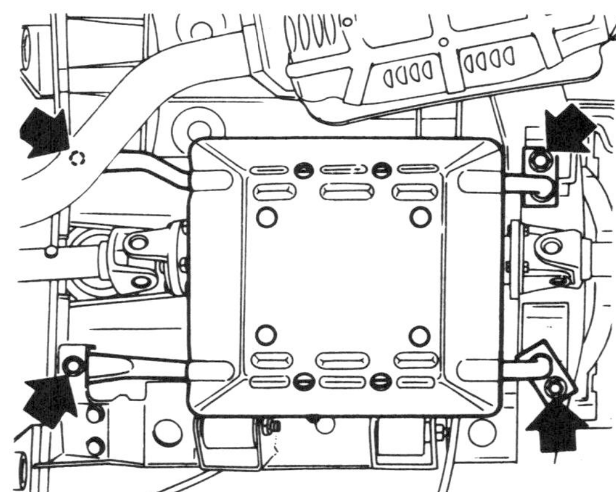

Transfer case shield bolts

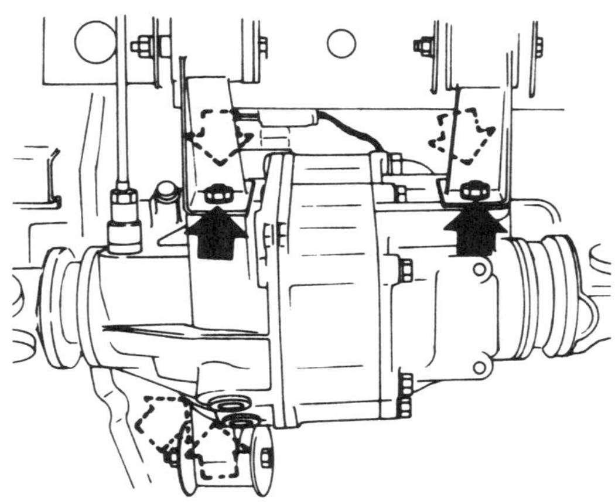

Transfer case mounting bolts

TRANSFER CASE

REMOVAL AND INSTALLATION

1. Disconnect the battery ground.
2. Raise the vehicle and support it on stands.
3. Remove the transfer case shield.
4. Remove the primary driveshaft securing nuts.
5. Remove the front and rear driveshafts.
6. Disconnect the 4-wd switch wire.
7. Disconnect the speedometer cable.
8. Remove the exhaust pipe.
9. Support the transfer case with a jack.
10. Temporarily loosen the transfer case insulator bolts.
11. Remove the shift lever rubber boot from the floor.

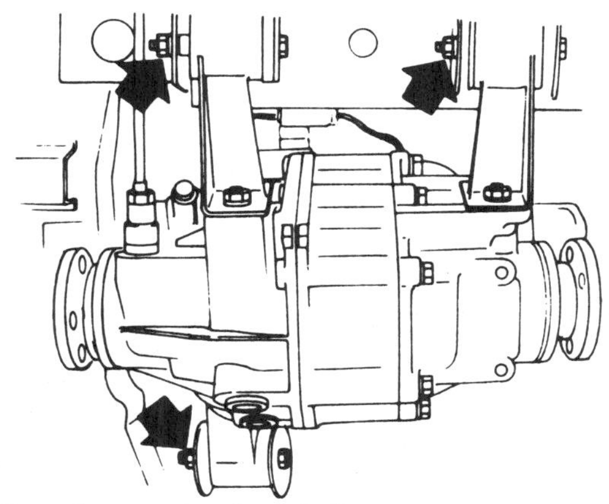

Transfer case insulator bolts

12. Unbolt and remove the transfer case and primary driveshaft from the vehicle.
13. Installation is the reverse of removal. Torque all mounting bolts to 20–26 ft. lb.

DRIVELINE

2-Wheel Drive 1970–81
REMOVAL AND INSTALLATION
Driveshaft and U-Joints

1. Raise the truck on a hoist. Mark the relationship of the driveshaft to the companion flange at the differential housing so that the driveshaft can be reinstalled in the same position. All 521 models (1970–71) have a flange at the transmission output shaft, instead of a sliding spline yoke. In this case also mark the front flange. It is important to reassemble the parts in their correct relationship, because they were balanced as an assembly at the factory.

2. Remove the bolts retaining the center bearing bracket.

3. Remove the bolts connecting the driveshaft to the companion flange at the differential housing (and at the front flange on early models).

4. Move the driveshaft assembly toward the rear of the truck, passing it under the rear axle, removing the sleeve yoke from the transmission. Watch for oil leaking out the end of the transmission. Plug if necessary.

5. Install the driveshaft in the reverse order of removal. Be careful not to bang the sleeve yoke (620 models) into the seal inside the transmission extension housing, which will damage the seal, causing leakage. Align the driveshaft with the differential housing companion flange in their original position. Tighten the U-joint nuts to 15–20 ft. lbs. (17–24 ft. lbs., 1974 and later) and the nuts retaining the center bearing bracket to 12–16 ft. lbs.

4-Wheel Drive 1981
REMOVAL AND INSTALLATION
Primary Driveshaft

1. Match-mark the flanges and separate the primary driveshaft from the transfer case.

2. Remove the transfer case.

3. Pull the primary shaft from the transmission and plug the opening.

4. Installation is the reverse of removal. Align your match-marks.

Front Driveshaft

1. Match-mark the flanges and unbolt the front shaft from the front differential.

2. Pull the shaft from the transfer case and plug the opening.

3. Installation is the reverse of removal. Align your match-marks.

Rear Driveshaft

1. Match-mark the flanges and unbolt the shaft from the rear differential.

2. Remove the center bearing bracket.

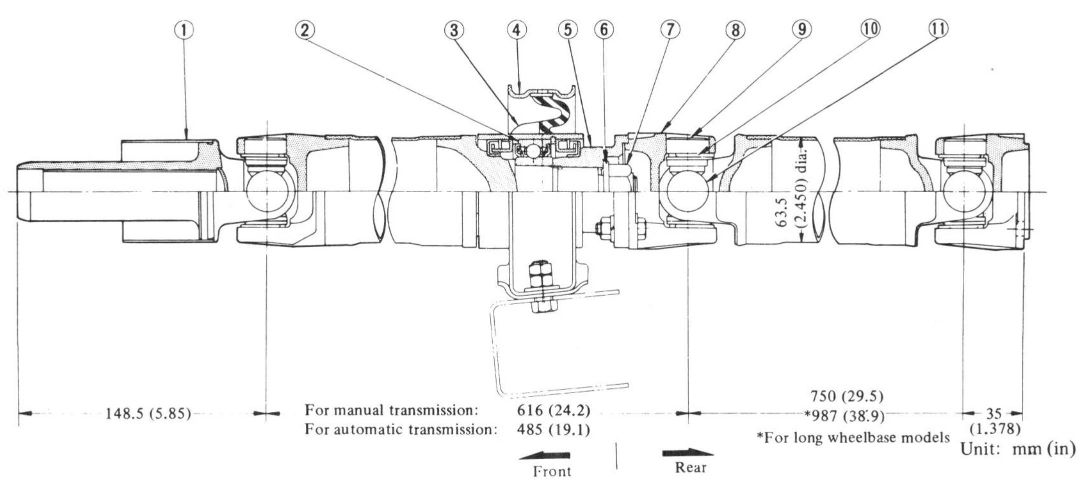

Front axle propeller shaft; the left insert shows the primary driveshaft which connects the transfer case with the transmission

3. Pull the shaft from the transmission and plug the opening.

4. Installation is the reverse of removal. Align your match-marks.

U-JOINT OVERHAUL

1. Remove the driveshaft.

2. Punch mating marks on both the yokes at either end of the driveshaft and the driveshaft itself so that the driveshaft assembly can be reassembled in the same position.

3. Remove the snap-rings from the bearing hole of the yokes with a screwdriver.

4. Place the yoke in a vise with a small socket positioned against one of the bearing cups and a larger socket placed against the yoke on the opposite side. The larger socket must be able to receive the bearing cap when it is pressed out of the yoke.

5. Tighten the vise until the bearing caps are free of the yoke.

6. Remove the two remaining bearings from the opposite yoke in the same manner and remove the spider bearing journal.

1. Sleeve yoke assembly
2. Center bearing
3. Center bearing insulator
4. Center bearing bracket
5. Companion flange
6. Plain washer
7. Self-locking nut
8. Flange yoke
9. Bearing race assembly
10. Snap-ring
11. Journal assembly

A cutaway view of the driveshaft assembly, 1972 and later

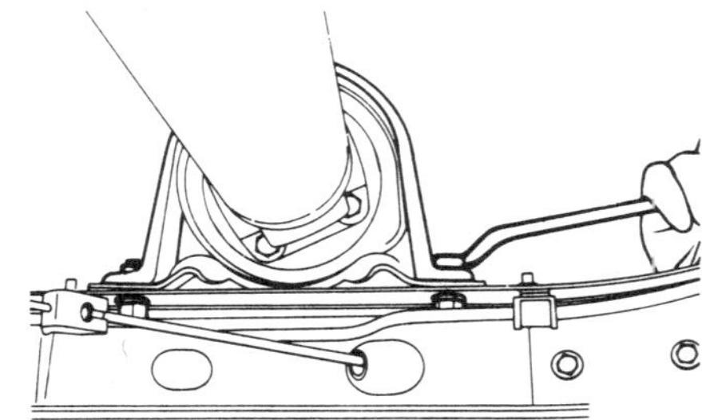

Removing the center bearing bracket

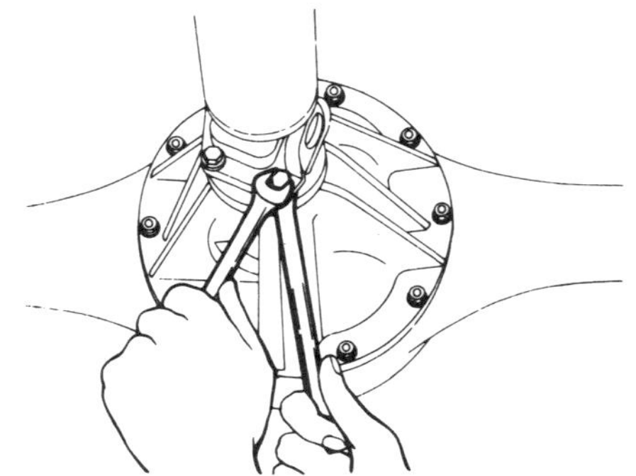

Disconnecting the driveshaft from the differential flange

7. Make sure that the new spiders and needle bearings in the bearing caps are well lubricated.

8. Assemble the universal joint spider and bearing caps to the yoke in the reverse manner of removal, using the smaller socket to press the bearing caps into the yoke and the larger socket to bear against the yoke bearing cap hole at the opposite end. Use a vise to press the bearing caps in place.

9. Install the hole snap-ring to secure the bearing caps. Use snap rings of the same thickness on both sides of the U-joint to maintain balance. After installing the snap rings, check the end play of the spider within the yoke. End play should be under 0.02 mm (0.0008 in.). If it is not, different snap rings should be used to bring the end play to within tolerance. Snap rings are available in seven thicknesses, from 2.00 mm to 2.12 mm, in 0.02 mm increments. Be sure to use the same size snap rings on each side of the spider.

10. Assemble the yoke to the driveshaft, aligning the marks made prior to disassembly.

11. Install the driveshaft.

CENTER BEARING REPLACEMENT

The center bearing is a sealed unit which must be replaced as an assembly if defective.

1. Remove the driveshaft.

2. Matchmark the flange yoke (#8 in the figure) and the companion flange (#5 in the figure) which connect the front half of the driveshaft to the rear. Also matchmark the companion flange and the front driveshaft. Remove the bolts and separate the shafts.

3. You must devise a way to hold the driveshaft while unbolting the companion flange from the front driveshaft. Do not place the front driveshaft tube in a vise, because the chances are it will get crushed. The best way is to grip the flange somehow while loosening the nut. It is going to require some strength to remove.

4. Slide the companion flange off the front driveshaft and remove the center bearing from its mount.

5. The new bearing is already lubricated. Install it into the mount, making sure that the seals and so on are facing the same way as when removed.

6. Slide the companion flange onto the front driveshaft, aligning the marks made during removal. Install the washer and locknut. Tighten the nut to 145–175 ft. lbs. Check that the bearing rotates freely around the driveshaft.

7. Connect the companion flange to the flange yoke, aligning the marks made during disassembly. Torque to 18–23 ft. lbs.

8. Install the driveshaft, aligning the marks made at the axle flange (and the transmission flange on 521 models) during removal.

REAR AXLE

Axle Shaft, Bearing and Seal
REMOVAL AND INSTALLATION

1. Raise the rear of the vehicle and support it. Remove the rear wheel and tire.

2. Disconnect the rear parking brake cable by removing the adjusting nut and clamps.

3. Disconnect the brake tube at the rear brake backing plate. Plug the end of the brake tube to prevent loss of brake fluid.

4. Remove the brake drum.

5. Remove the nuts securing the wheel bearing retainer to the brake backing plate.

6. Pull out the axle shaft assembly together with the brake backing plate using a slide hammer.

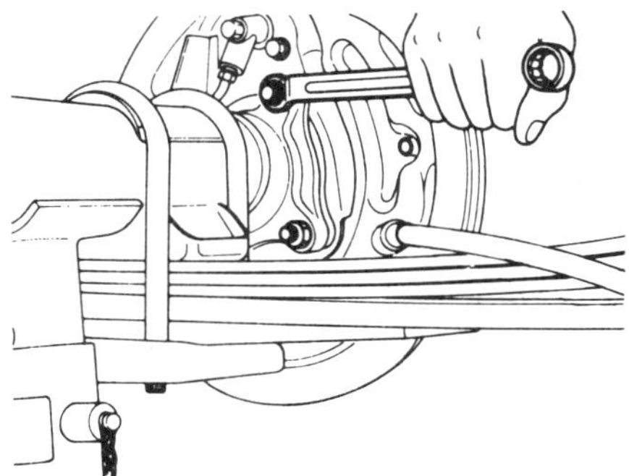

Disconnecting the brake backing plate from the axle housing

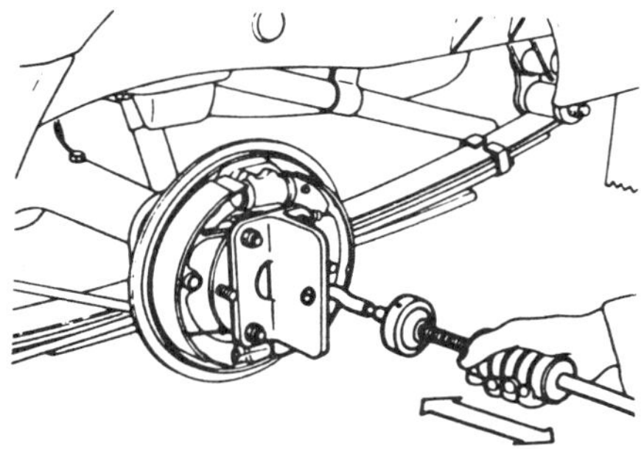

Remove the axle shaft with a slide hammer

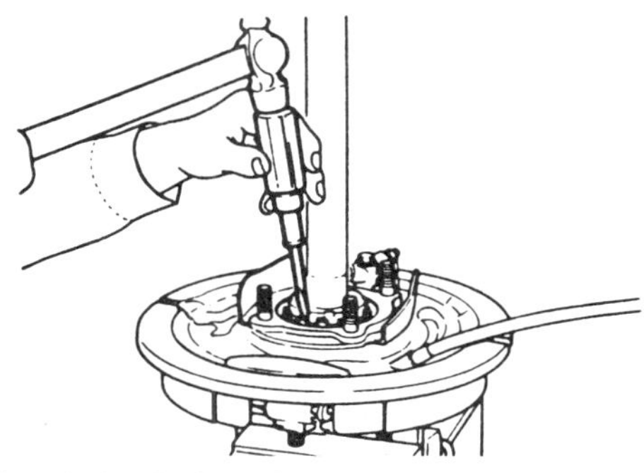

Unbend the lockwasher to remove the bearing locknut; use a new one for installation

7. Remove the oil seal in the axle housing if necessary. It can be pried out with a screwdriver. Oil the lips of the new seal and install it carefully to avoid damage to the lip.

8. To replace the bearing, unbend and discard the lockwasher. Remove the locknut with a soft drift and a hammer.

9. Press the old bearing and cage off the shaft.

10. Remove the oil seal in the cage. Use a brass drift to remove the bearing cup after the seal has been removed.

11. Install the new cup with a brass drift. Install a new oil seal over the bearing cup. Lubricate the area between the seal lips with grease after installation.

12. Place the bearing cage and spacer on the axle shaft, then fit the bearing, tapping it into place with a soft drift and light hammer blows.

13. Place the flat bearing lockwasher over the bearing, then the new nut lockwasher. Install the locknut, tightening to 108 ft. lbs. Continue to tighten after that until the grooves line up with the lockwasher tabs. The nut can be tightened up to 145 ft. lbs. Bend the lockwasher tabs into place.

14. Lubricate the bearing and the recess in the axle housing with wheel bearing grease. Coat the axle splines with gear oil. Coat the seal surface of the shaft with grease.

15. Install the axle shaft in the reverse order of removal. The axle end-play should be 0.012–0.035 in. The end-play is adjusted by adding or removing shims behind the brake backing plate. Tighten the backing plate attaching nuts to 27–35 ft. lbs. through 1973, or 39–46 ft. lbs., 1974 and later.

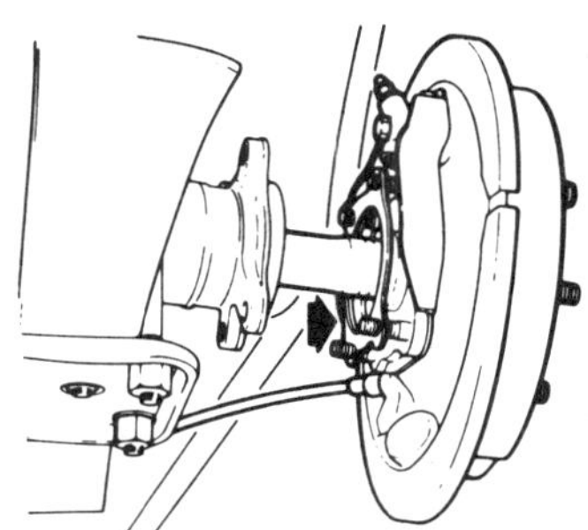

Axle shaft end-play is adjusted by the addition or subtraction of shims behind the brake backing plate

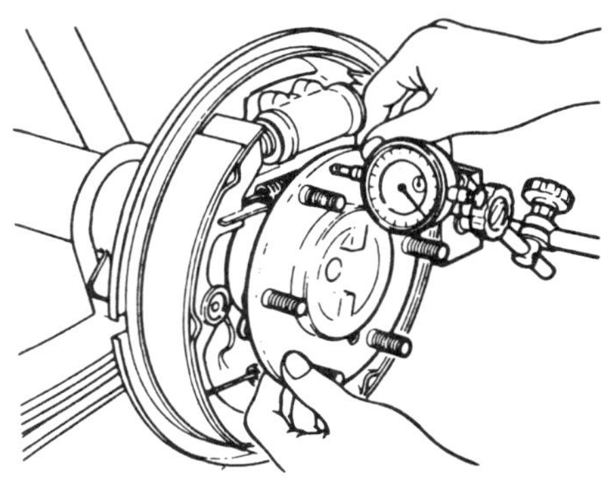

Measure the axial end play with a dial indicator

FRONT AXLE

Free Running Hub

REMOVAL AND INSTALLATION

1. Raise and support the front axle on stands.

2. Set the hub in the lock position.

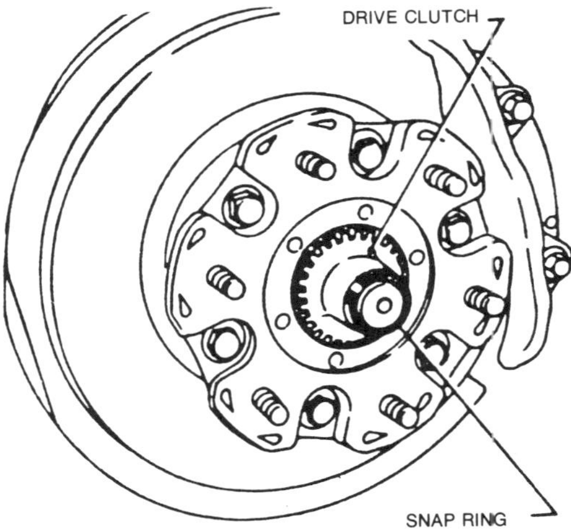

Free running hub snap ring and driven clutch location

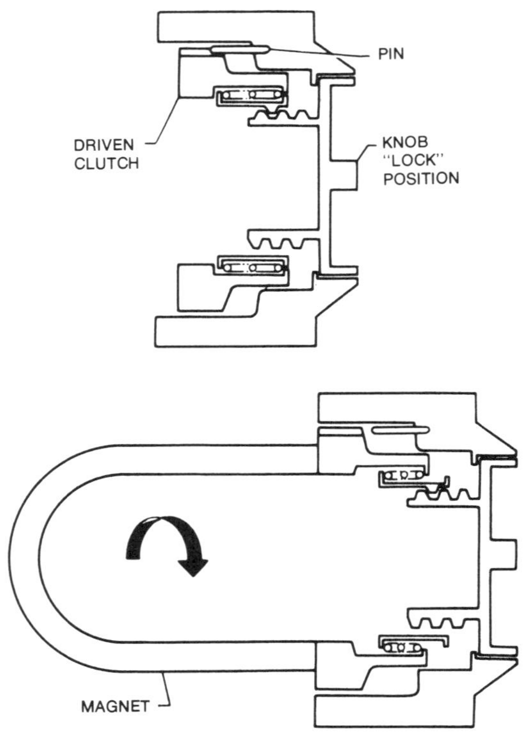

Hub removal sequence

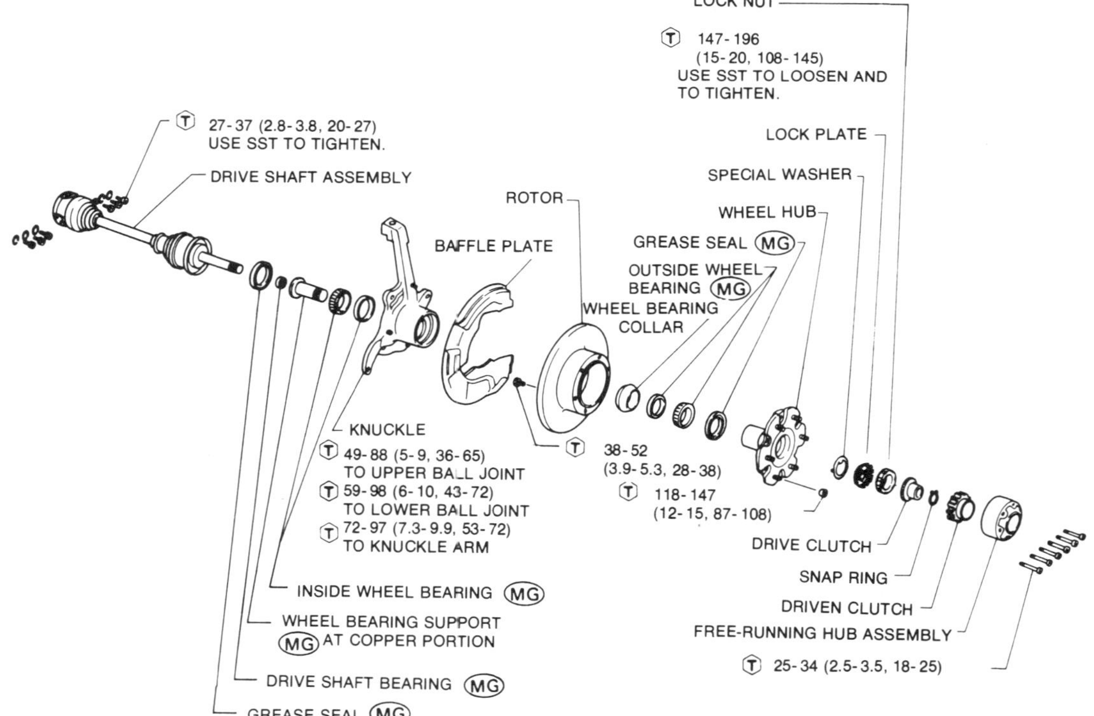

The 4-wheel drive front axle

3. Remove the driven clutch by turning it clockwise.

NOTE: *A pin is located inside the hub case to lock the driven clutch. Pull and turn the clutch while attracting the pin with a magnet.*

4. Remove the lock pin.

5. Set the hub at the free position.

6. Screw the driven clutch into place by turning it counterclockwise until it bottoms.

7. Turn it clockwise until it aligns with the bolt hole.

8. Install the lock pin.

Axle Shaft

REMOVAL AND INSTALLATION

1. Remove the free running hub.

2. Remove the snap ring and remove the driven clutch.

3. Remove the rebound bumper.

4. Disconnect the stabilizer bar at the lower link.

5. Remove the bolts attaching the axle shaft to the carrier. DO NOT REMOVE THE RUBBER BOOTS!

6. Pull the axle shaft from the suspension.

7. Installation is the reverse of removal. Observe the following points:

   a. Apply multi-purpose wheel bearing grease to the copper portion of the wheel bearing support.

   b. Adjust the axle shaft axial end play by installing the proper thickness of snap rings on the end of the shaft.

   c. Torque the axle shaft flange bolts to 20–27 ft. lb.; the free running hub bolts to 18–25 ft. lb.; the stabilizer bar-to-frame bolts to 12–16 ft. lb.; the stabilizer bar-to-lower link bolts to 12–16 ft. lb. and the wheel nuts to 87–108 ft. lb.

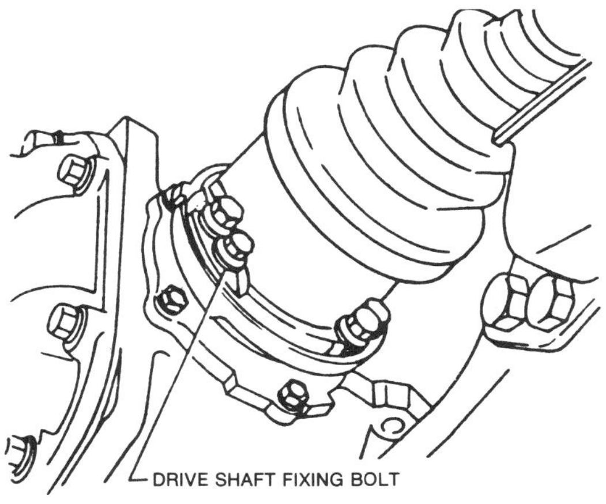

Axle shaft-to-carrier bolts

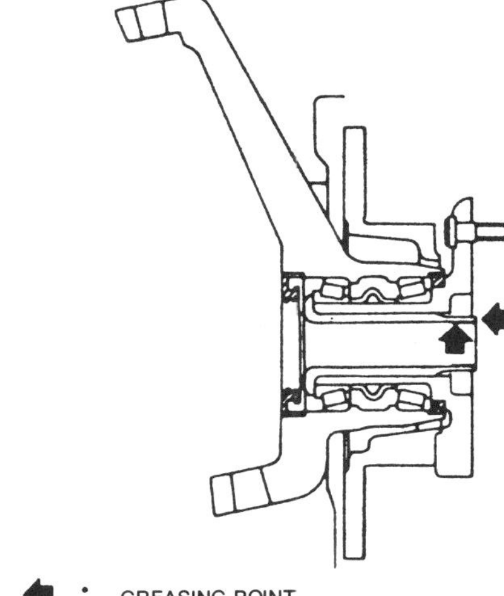

Copper wheel bearing support greasing locations

DIFFERENTIAL

Differential service is best left to those extremely familiar with their vagaries and idiosyncrasies. A great many specialized tools are required as well as a good deal of experience.

Introduction

The axle must transmit power through a 90° bend. To accomplish this, straight cut bevel gears or spiral bevel gears were originally used. This type of gear is satisfactory for differential side gears, but since the centerline of the gears must intersect, they rapidly became unsuited for ring and pinion gears. The lowering of the driveshaft brought about a variation of bevel gears called the hypoid gear. This type of gear does not require a meeting of the gear centerlines and can therefore be underslung, relative to the centerline of the ring gear.

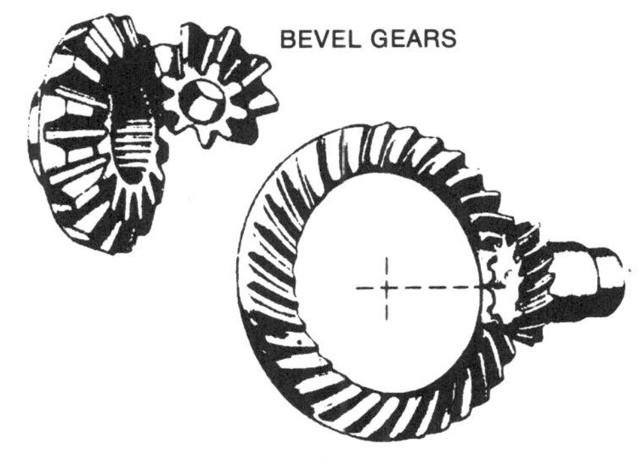

Bevel gears

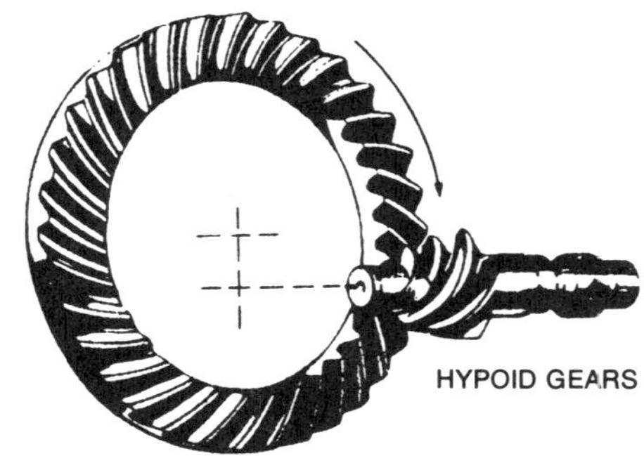

Hypoid gears

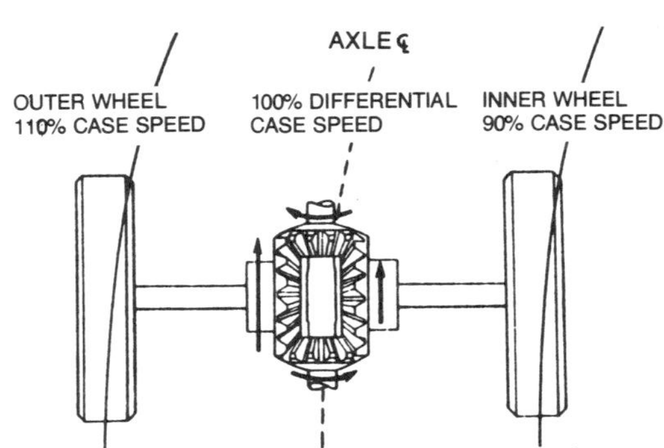

Differential action during cornering

Operation

The differential is an arrangement of gears which permits the rear wheels to turn at different speeds when cornering and divides the torque between the axle shafts. The differential gears are mounted on a pinion shaft. The pinion shaft is fitted in a bore in the differential case and is at right angles to the axle shafts.

Power flow through the differential is as follows. The drive pinion, which is turned by the driveshaft, turns the ring gear. The ring gear, which is bolted to the differential case, rotates the case. The differential pinion forces the pinion gears against the side gears. In cases where both wheels have equal traction, the pinion gears do not rotate on the pinion shaft, because the input force of the pinion gear is divded equally between the two side gears. Consequently the pinion gears revolve with the pinion shaft, although they do not rotate on the pinion shaft itself. The side gears, which are splined to the axle shafts, and meshed with the pinion gears, rotate the axle shafts.

When it becomes necessary to turn a corner, the differential becomes effective and allows the axle shafts to rotate at different speeds. As the inner wheel slows down, the side gear splined to the inner wheel axle shaft also slows down. The pinion gears act as balancing levers by maintaining equal tooth loads to both gears while allowing unequal speeds of rotation at the axle shafts. If the vehicle speed remains constant, and the inner wheel slows down to 90 percent of vehicle speed, the outer wheel will speed up to 110 percent.

DETERMINING GEAR RATIO

Determining the axle ratio of any given axle is an esoteric subject, relatively useless until you have to know. But, as a "junkyard art," it is invaluable.

The rear axle ratio is said to have a certain ratio, say, 4.11. It is called a 4.11 rear although the 4.11 actually means 4.11:1. This means that the driveshaft will turn 4.11 times for every turn of the rear wheel. The number 4.11 is determined by dividing the number of teeth on the pinion gear into the number of teeth on the ring gear. In the case of a 4.11, there could be 9 teeth on the pinion and 37 teeth on the ring gear ($37 \div 9 = 4.11$). This provides a sure way (although troublesome—except to those who are really interested) of determining your rear axle ratio. You must drain the rear axle and remove the rear cover, if it has one, and count the teeth on the ring and pinion.

An easier method is to jack and support the vehicle so that BOTH rear wheels are off the ground. Make a chalk mark on the rear wheel and the driveshaft. Block the front wheels, set the parking brake and put the transmission in Neutral. Turn the rear wheel one complete revolution and count the number of turns that the driveshaft makes. The number of turns that the driveshaft makes in one complete revolution of the rear wheel is an *approximation* of the rear axle ratio.

Suspension and Steering

2-WHEEL DRIVE FRONT SUSPENSION

The front suspension on Datsun pick-ups is of an independent, unequal length control arm type with torsion bar springs.

The control arms are attached to a bracket which is welded to the frame at their inner pivot points and to the steering knuckle/spindle support at their outer points.

The torsion bar has splines on each end; the front end is installed to the torque arm which is attached to the lower control arm and the rear end is installed to the torsion bar anchor which is secured to the chassis frame.

Fore-and-aft movement of the front suspension is controlled by strut bars connected to the lower control arms at one end and mounted to the chassis frame at the forward end.

An optional torsion bar type stabilizer is connected to the lower control arms through vertical stabilizer links.

The steering knuckle/spindle is attached to the steering knuckle/spindle support by a kingpin through 1977. Ball joints are used in 1978 and 1979.

Torsion Bars

REMOVAL AND INSTALLATION

1. Jack up the front of the vehicle and support it with jackstands. Remove the wheel. On trucks with a catalytic converter, the converter must be removed if the left torsion bar is being removed.

2. Loosen the ride height adjusting nuts at the anchor (rear) end of the torsion bar, allowing the anchor arm to hang down.

3. Remove the dust cover at the rear end of the torsion bar and remove the snap-ring.

4. Pull the anchor arm rearward and off the torsion bar. Withdraw the torsion bar from the lower control arm and remove it from under the vehicle.

5. Before installing the torsion bar, apply a light coat of grease to the splines. Install the torsion bar to the lower control arm.

NOTE: *The torsion bars are marked on the end with an "L" (left) or an "R" (right). The torsion bars must be installed on the same side from which they are removed.*

6. Install the anchor arm on the rear end of the torsion bar. For 521 models (1970–71), with the lower arm against the rebound

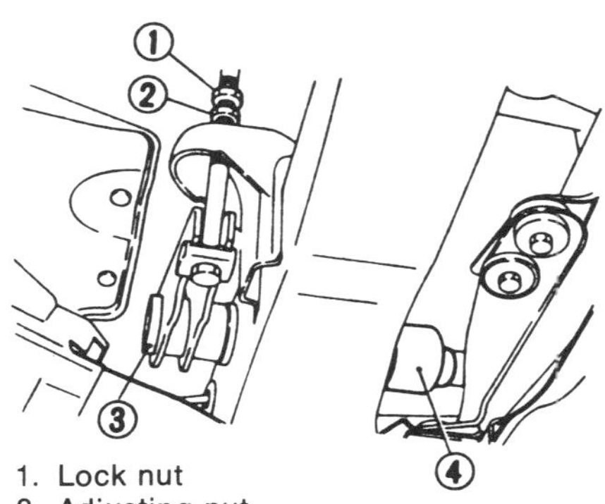

1. Lock nut
2. Adjusting nut
3. Anchor arm
4. Dust cover

2-wheel drive torsion bar anchor removal

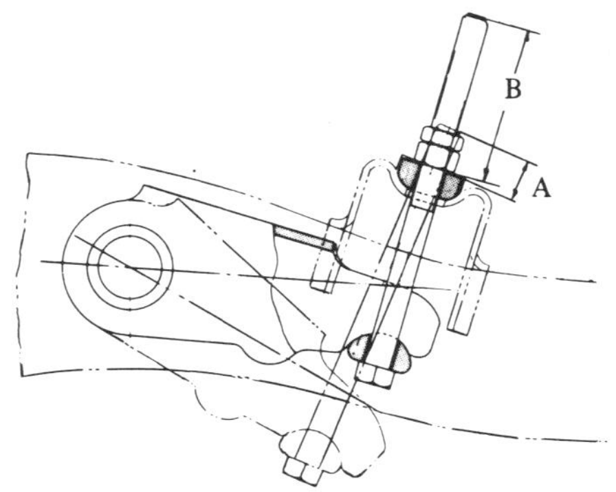

Installing the torsion bar anchor end on 1972 and later 2-wheel drive models

bumper, dimension "A" must be 2.697 in. (68.5 mm). For 620 models (1972 and later), dimension "A" must be:

- 1972–76; 0.20–0.60 in. (5–15 mm), standard bed,
  0.59–0.98 in. (15–25 mm), long bed;

- 1977:   0.59–0.98 in. (15–25 mm), all models;
- 1978–79: 0.28–0.67 in. (7–17 mm), all models.

Note that there are two different methods used for measuring this distance. Be sure you are using the correct illustration for your truck.

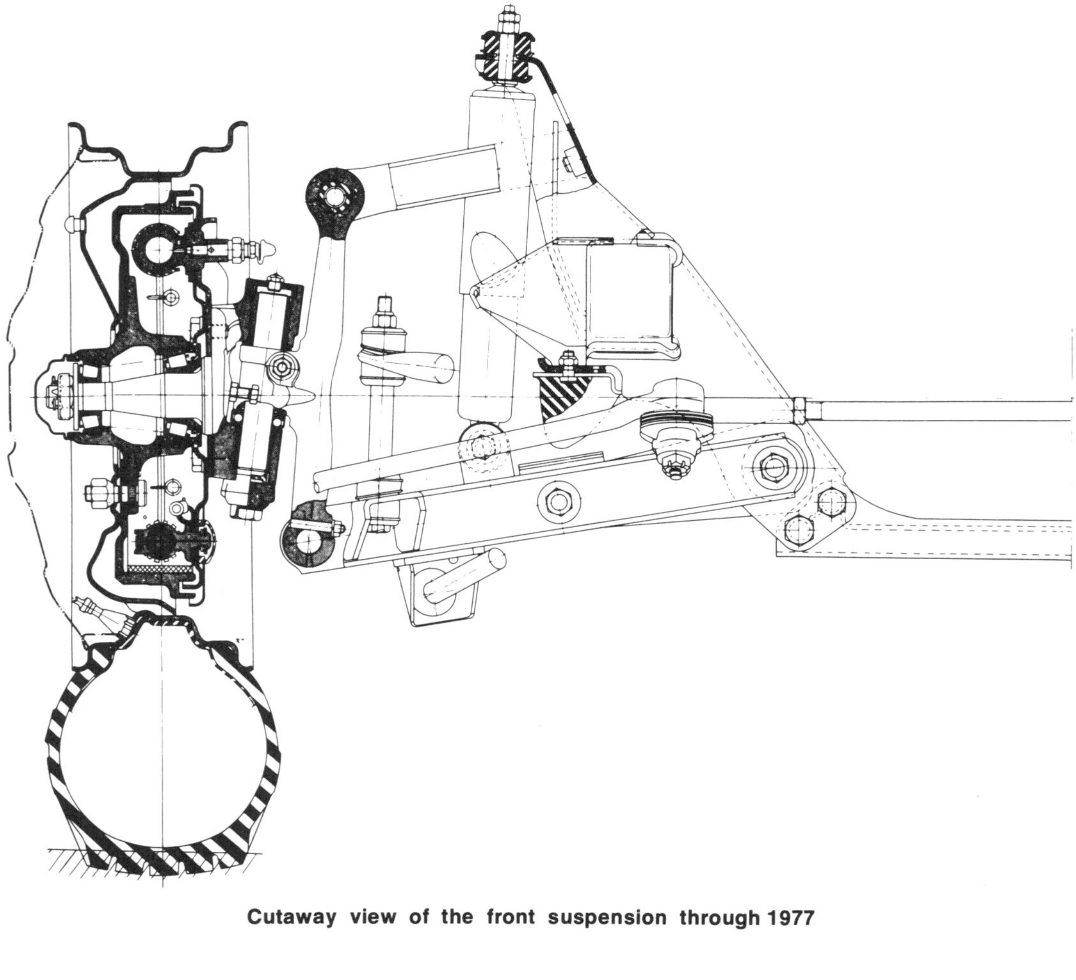

Cutaway view of the front suspension through 1977

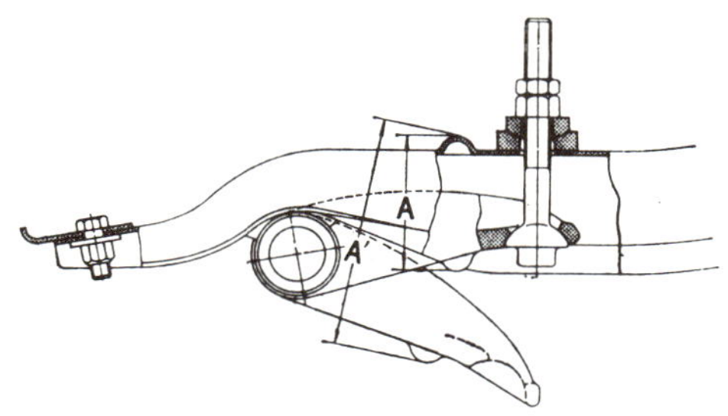

Installing the torsion bar anchor end—1970–71

7. Install a new retaining snap-ring and dust boot to the anchor arm end of the torsion bar.

NOTE: *Always use a new snap-ring. Never reinstall the old one.*

8. For 521 models, tighten the adjusting nut until dimension "A" is reached: 3.86 in. (98 mm) for standard bed models, 3.62 in. (92 mm) for long bed models. For 620 models, tighten the adjusting nut until the link protrudes above the support bracket 2.36–2.76 in. (60–70 mm) on all models (dimension "B"). Again, make sure you are looking at the correct illustration for your truck.

9. Install the wheel and lower the vehicle.

10. Adjust the vehicle ride height with the truck at curb weight (full tank of gas and no passengers). Refer to "Ride Height Adjustment." Tighten the locknut to 23–30 ft. lbs.

Shock Abosrbers
TESTING

Visually inspect the shock absorber. If there is evidence of leakage and the shock absorber is covered with oil, the shock is defective and must be replaced.

If there is no sign of excessive leakage (a small amount of weeping is natural) bounce the truck at one corner by pressing up and down on the fender or bumper. When you have the truck bouncing as much as you can, stop bouncing it, and release the fender or bumper. The truck should stop bouncing after the first rebound. If the bouncing continues past the center point of the bounce more than once, the shock absorbers are worn and should be replaced.

REMOVAL AND INSTALLATION

1. Jack up the vehicle and support it. Remove the wheel.

2. Hold the upper stem of the shock absorber and remove the nuts, washer, and rubber bushing.

3. Remove the bolt from the lower end of the shock absorber and remove the shock absorber from the vehicle.

4. Install the shock absorber in the reverse order of removal. Replace all of the rubber bushings with new ones if a new shock absorber is being installed. Install the lower retaining bolt from the front of the truck. Tighten the upper attaching nut to 12–16 ft. lbs. and the lower nut to 23–30 ft. lbs.

Kingpins
INSPECTION

1. Jack the vehicle so that the tire on the side to be checked is off the ground. Adjust wheel bearing preload to the proper specification.

2. Grasp the top and bottom of the tire and try to move the top and bottom of the tire alternately in and out. If there is noticeable play between the steering knuckle/spindle and the spindle support then it can be assumed that the kingpins or bushings are worn and should be replaced.

NOTE: *Before performing this test make sure that the wheel bearings are properly adjusted.*

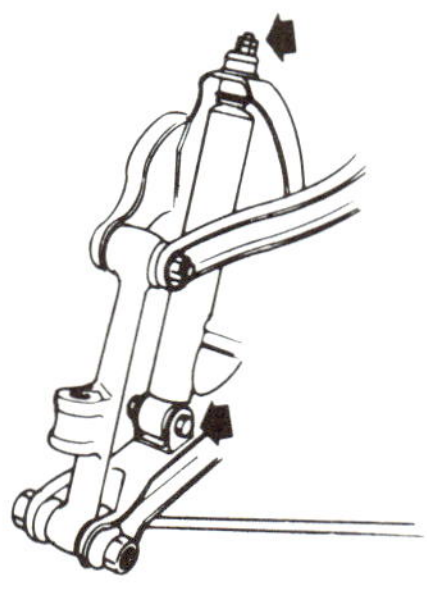

Front shock absorber attaching bolts

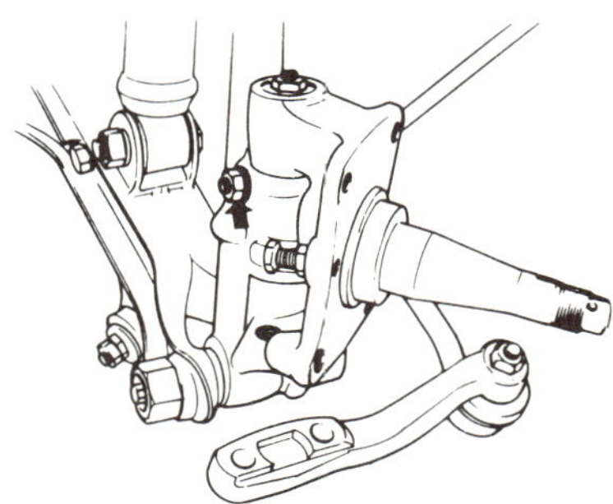

The kingpin lockbolt (arrow); steering arm disconnected

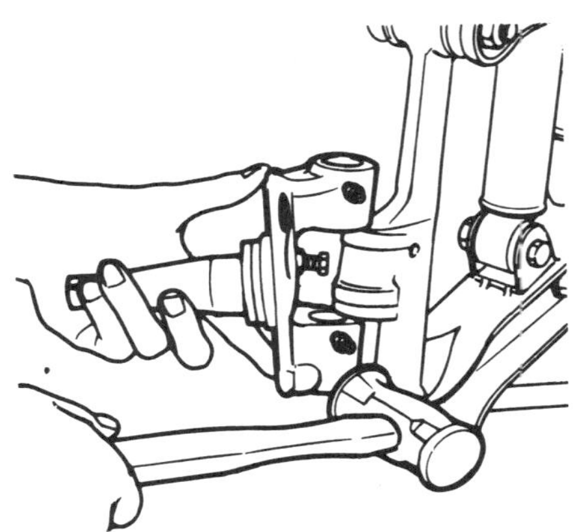

Removing the steering knuckle/spindle

REMOVAL AND INSTALLATION

1. Jack up the front of the vehicle and support it by placing jackstands under the frame.

2. Remove the front wheel.

3. Remove the brake hose and connector from the wheel cylinder.

> NOTE: *It is not absolutely necessary to remove the drum, hub, brake, and backing plate from the spindle, although it will make working with the spindle a great deal easier. If you choose to leave the parts in place, skip down to Step 8.*

4. Remove the brake drum.

5. Remove the hub dust cap, cotter pin, adjusting cap, and spindle nut from the spindle.

6. Remove the wheel hub, inner and outer wheel bearings, bearing washer, and grease seal from the spindle.

7. Remove the brake backing plate from the knuckle/spindle flange.

8. Remove the knuckle/spindle steering arm from the knuckle/spindle.

9. Remove the kingpin lockbolt.

10. Remove the plug from the top of the

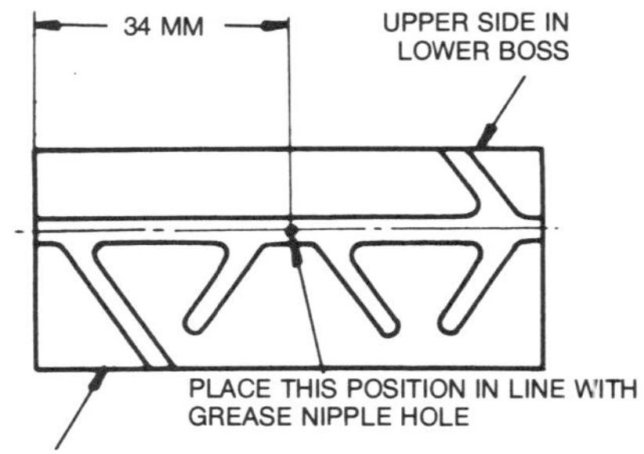

Kingpin bushing to be aligned with the grease nipple hole in the spindle

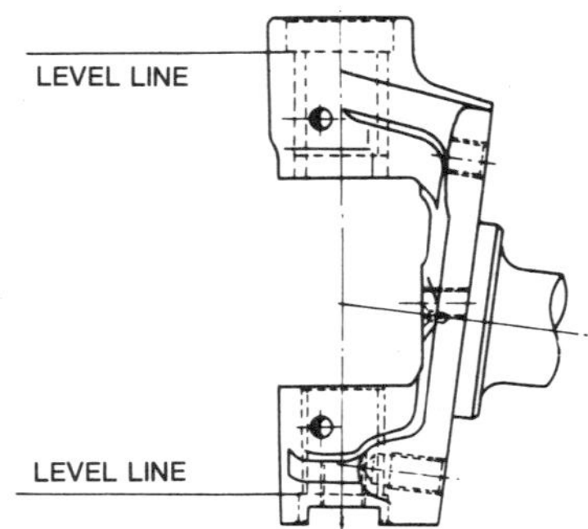

Installation of the kingpin bushings flush with the plug counterbores

kingpin by drilling a small hole in the plug, screwing a sheet metal screw into the hole, and pulling out the plug.

11. Drive out the kingpin together with the lower plug with a suitable drift and a hammer.

12. Tap the steering knuckle/spindle lightly with a hammer and detach it from knuckle/spindle support. Be careful not to drop the thrust bearing.

13. Drive the steering knuckle/spindle bushing and grease seal out of the kingpin bore with a kingpin bushing driver. Do not try to use a drift, because it will probably score the inner wall of the knuckle spindle.

14. After cleaning the pin bores thoroughly, install the new bushing carefully by using a bushing driver. Position the bushings in the spindle so that they are flush with the counterbores for the plugs. The bushings in the factory rebuilding kit have a grease gallery which should align with the grease nipple hole in the spindle (see the illustration).

15. Remove the grease nipples and drill grease holes through the bushings, placing the drill through the threaded grease nipple hole. The grease hole should be ⅛ in. or less. Remove *all* metal filings and burrs after drilling the grease holes. This step is not necessary if you have used the factory bushings and installed them as outlined in Step 14.

16. Ream the inside of the kingpin bushings, if necessary, to 0.7878–0.7888 in. The fit should be such that the kingpin, when greased, can be turned or pushed in or out with thumb pressure. Use the lower bore as a guide for reaming the upper bore, and vice versa, to keep the bores aligned.

17. Press fit the grease seal on the upper bushing. Take care not to damage the grease seal lip.

18. Install the steering knuckle/spindle to

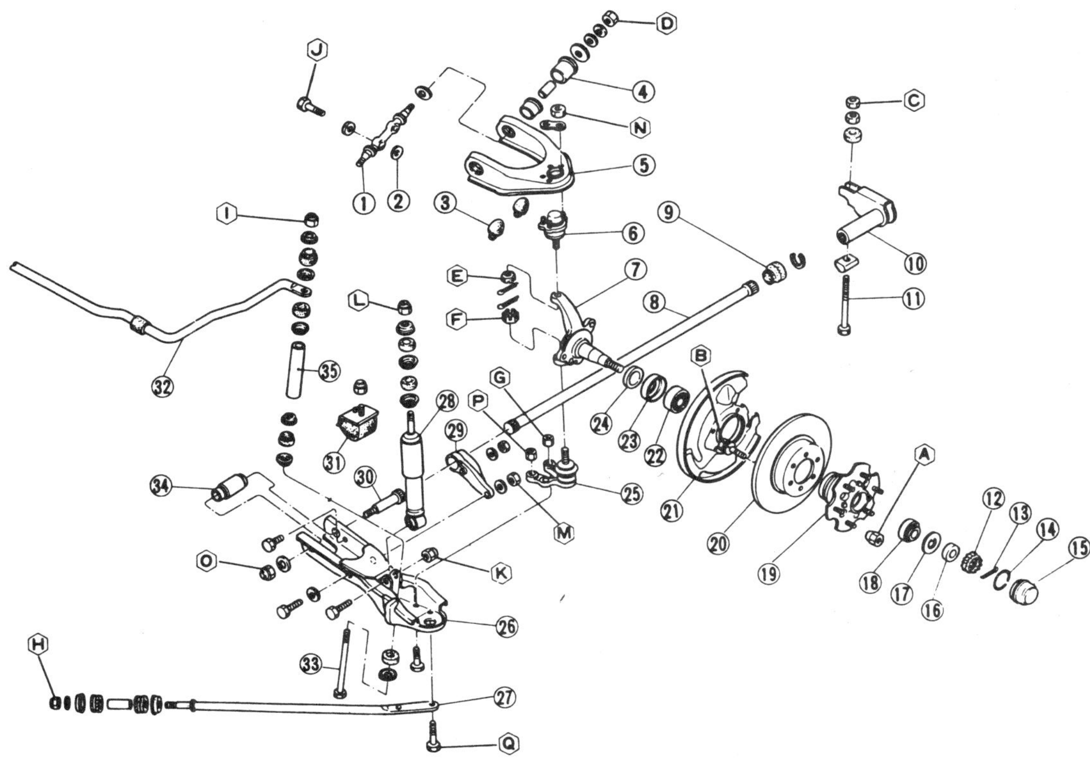

1. Upper arm pivot shaft	19. Wheel hub	Tightening torque kg-m (ft-lb)
2. Camber adjusting shim	20. Rotor	A. 8.0 to 10.0 (58 to 72)
3. Rebound bumper	21. Backing plate	B. 3.9 to 5.3 (28 to 38)
4. Bushing	22. Inner wheel bearing	C. 3.1 to 4.1 (22 to 30)
5. Upper arm	23. Grease seal	D. 7.7 to 10.5 (56 to 76)
6. Upper ball joint	24. Spacer	E. 8.0 to 10.0 (58 to 72)
7. Knuckle spindle	25. Lower ball joint	F. 17.2 to 19.5 (124 to 141)
8. Torsion bar	26. Lower arm	G. 3.9 to 5.3 (28 to 38)
9. Boot	27. Tension rod (strut)	H. 3.0 to 4.2 (22 to 30)
10. Anchor arm	28. Shock absorber	I. 1.6 to 2.2 (12 to 16)
11. Anchor arm adjusting bolt	29. Torque arm	J. 11.1 to 15.0 (80 to 108)
12. Adjusting nut	30. Lower arm pivot shaft	K. 3.1 to 4.1 (22 to 30)
13. Cotter pin	31. Bumper stop	L. 1.6 to 2.2 (12 to 16)
14. O-ring	32. Stabilizer (optional)	M. 2.7 to 3.7 (20 to 27)
15. Hub cap	33. Stabilizer connecting bolt	N. 1.7 to 2.2 (12 to 16)
16. Spindle nut	34. Lower arm bushing	O. 11.1 to 15.0 (80 to 108)
17. Washer	35. Stabilizer collar	P. 3.6 to 4.6 (26 to 33)
18. Outer wheel bearing		Q. 3.9 to 5.3 (28 to 38)

1978–79 front suspension

the steering knuckle/spindle support in the reverse order of removal as follows:

19. Insert the O-ring in the lower end of the knuckle/spindle support. Install the thrust bearing and shim together with the steering knuckle/spindle to the knuckle/spindle support.

Select the spindle shims which will obtain 0.004 in. or less clearance between the steering knuckle/spindle and the support. To measure this clearance with a feeler gauge, jack up the bottom of the spindle slightly.

NOTE: *The thrust bearing is installed so that the covered side faces upward. You will probably be able to use the same shims that were originally installed.*

20. Drive the new kingpin into the kingpin bores, securing the steering knuckle/spindle to the support.

21. Align the locking bolt hole of the knuckle/spindle support with the notch in the kingpin and install the lockbolt.

22. Install the upper and lower kingpin plugs.

23. Install the steering knuckle/spindle arm to the steering knuckle/spindle together

with a new lockplate. Tighten the steering knuckle arm attaching bolts to 75–88 ft. lbs. Bend the tabs of the lockplate to engage the flats on the bolt head.

24. Install the brake backing plate, hub and wheel bearings, brake drum, and wheel. Bleed the brake hydraulic system and grease the newly-installed bushings until grease is visible around the upper and lower grease seals.

Ball Joints

Ball joints are used on 1978 and later trucks, replacing the kingpins used previously.

INSPECTION

The ball joint should be replaced when play becomes excessive. Datsun does not publish specifications on just what constitutes excessive play, relying instead on a method of determining the force (in inch pounds) required to keep the ball joint turning. This method is not very helpful to the backyard mechanic since it involves removing the ball joint, which is what we are trying to avoid in the first place. An effective way to determine ball joint play is to jack up the truck until the wheel is just a couple of inches off the ground and the ball joint is unloaded, which means that you can't jack directly under the ball joint. Place a long bar under the tire and move the wheel and tire assembly up and down. Keep one hand on top of the tire while you are doing this. If there is over ¼ inch of play at the top of the tire, the ball joint is probably bad. This assuming that the wheel bearings are in good shape and properly adjusted. As a double check, have someone watch the ball joint while you move the tire up and down with the bar. If considerable play is seen, besides feeling play at the top of the wheel, the ball joints need to be replaced.

REMOVAL AND INSTALLATION

Upper

1. Raise and support the truck on stands placed on the frame rails.
2. Remove the wheels.
3. Loosen the torsion bar anchor lock and adjusting nuts to relieve spring tension.
4. Remove and discard the cotter pin from the ball joint stud and remove the nut. Separate the stud from the knuckle spindle with a ball joint removal tool.

5. Loosen the bolts retaining the ball joint to the control arm, and remove the joint.
6. Install the new ball joint into the control arm, tightening the bolts to 12–16 ft. lbs. Install the ball joint stud into the knuckle spindle and install the nut. Tighten the nut to 60 ft. lbs., then continue to tighten until the holes align (limit: 72 ft. lbs.). Install a new cotter pin. Install the wheel, lower the truck, and adjust the ride height. Have the alignment checked.

Lower

1. Perform Steps 1 and 2 of the upper ball joint removal procedure. Remove the lower shock absorber mounting bolt.
2. Loosen the torsion bar spring anchor lock and adjusting nuts, and remove the anchor arm bolt from the anchor arm.
3. Remove the snap ring, then move the anchor arm and torsion bar fully rearward.
4. Disconnect the stabilizer bar from the lower arm, if equipped.
5. Disconnect the strut (tension rod) from the lower arm.
6. Remove and discard the cotter pin from the ball joint stud, and remove the nut. Separate the ball joint from the knuckle spindle with a ball joint removal tool.
7. Remove the attaching bolts, and remove the ball joint from the lower arm.
8. Install the new ball joint in the arm, tightening the bolts to 28–38 ft. lbs. Install the ball joint stud into the knuckle spindle and tighten the nut to 127 ft. lbs. Continue to tighten until the holes align, then install the new cotter pin (limit: 141 ft. lbs.). The torsion bar ride height must be adjusted after assembly.

Upper and Lower Control Arms
REMOVAL AND INSTALLATION
1970–77

1. Raise the truck on a hoist or jack up the front end and support it with jackstands placed under the frame.
2. Remove the wheel and brake drum.
3. Remove the hub.
4. Remove the brake backing plate from the steering knuckle/spindle support.
5. Remove the steering knuckle/spindle arm, torsion bar, stabilizer bar, shock absorber, and strut rod, in this order.
6. Remove the upper fulcrum bolt securing the knuckle/spindle support to the upper control arm assembly and detach the two.

7. Remove the upper control arm bushings from the knuckle/spindle support.

8. Remove the screw bushings from both ends of the lower control arm fulcrum pin.

9. Loosen the nut at the lower end of the knuckle/spindle support from the inside and pull out the cotter pin retaining the fulcrum.

10. Drive the fulcrum pin out of the lower control arm and remove the knuckle/spindle support and steering knuckle/spindle from the lower control arm.

11. Remove the bolts retaining the upper control arm pivot shaft and remove the upper control arm pivot shaft with the camber adjusting shims from the body bracket.

12. Remove the nut retaining the lower control arm pivot shaft and remove the lower control arm pivot shaft. Remove the lower control arm with the torque arm from the mounting bracket.

13. The lower control arm bushing is removed with a drift and hammer.

14. Install the upper and lower control arms in the reverse order of removal. Tighten the lower control arm attaching nut to 54–58 ft. lbs. Tighten the upper control arm and camber adjusting shim bolts to 51–65 ft. lbs.

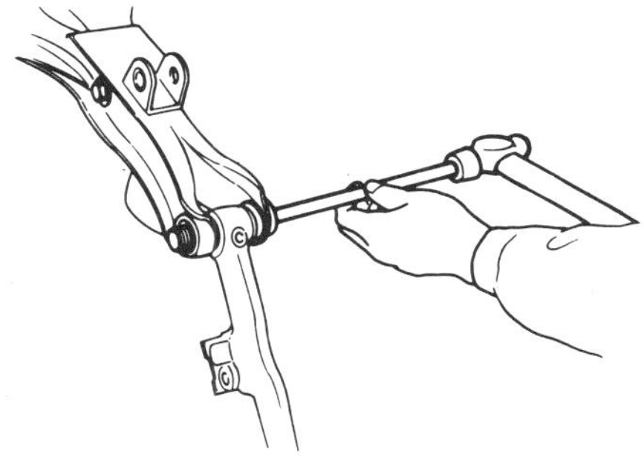

Driving the fulcrum pin out of the lower control arm in order to remove the steering knuckle/spindle support

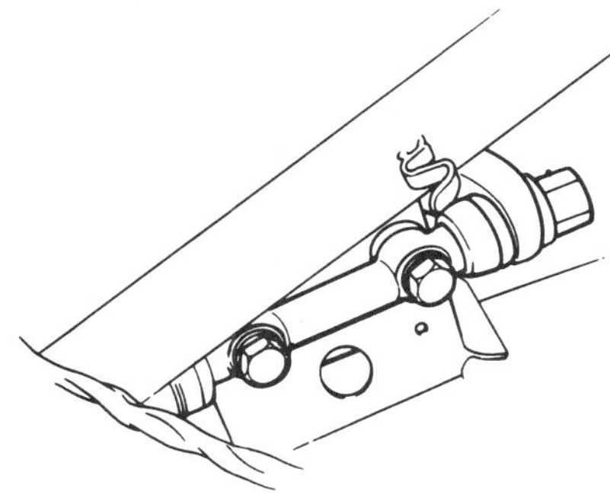

The upper control arm pivot shaft

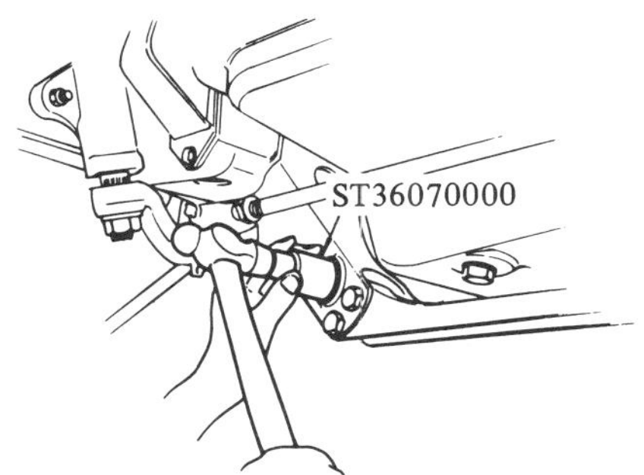

Removing the lower control arm pivot shaft bushing

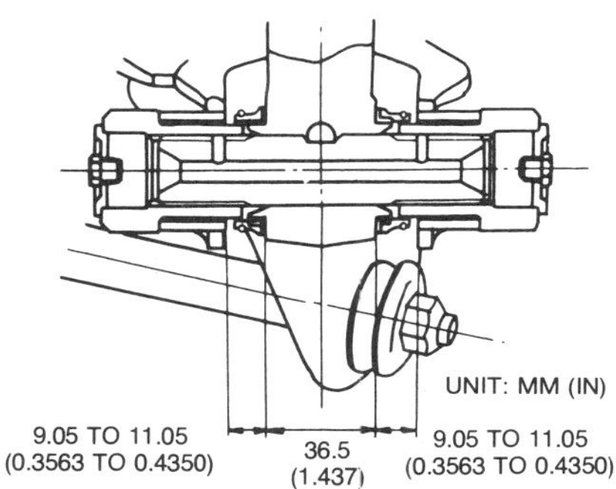

Installation of the lower control arm-to-knuckle screw bushing

15. Coat the threads of the fulcrum pin with grease and line up the notch of the fulcrum pin with the knuckle/spindle support for the insertion of the cotter pin. Install the fulcrum pin to the knuckle/spindle support with a soft hammer, attaching the support to the upper control arm. Install the cotter pin and tighten the locknut to 5.8–8.0 ft. lbs.

16. Coast the threaded portion inside of the screw bushing liberally with grease. Position the support at the center of the lower control arm and install the screw bushings. Check the dimensions of the installed screw bushing against those in the illustration, then tighten the bushing to 145–216 ft. lbs.

17. Replace the grease filler plug with a grease fitting and pump grease in until grease comes out around the dust cover. Reinstall the filler plug.

18. Install the upper control arm bushing to the knuckle/spindle support and then connect the knuckle/spindle support to the upper control arm. Insert the connecting bolt from the rear and tighten the nut to 28–38 ft. lbs.

19. Install the strut rod, shock absorber, stabilizer rod, torsion bar, and steering knuckle arm.

20. Install the brake backing plate to the steering knuckle/spindle and tighten the attaching bolts to 30–36 ft. lbs.

21. Install the brake drum and wheel and adjust the wheel bearing preload.

1978–81

UPPER

1. Perform Steps 1–4 of the upper ball joint removal procedure.

2. Remove the bolts retaining the upper arm pivot shaft and remove the shaft, arm, and camber adjusting shims from the body. Note the location of the shims so that they may be installed in their original positions during assembly.

3. To remove the upper arm shaft and bushings from the arm, remove the nuts and washers from the shaft. Use a press, first on one end of the shaft and then the other, to press out the bushings. Remove the shaft.

4. To install, coat the bushing with soapy water and press it into the upper arm. Install the washers onto the shaft and install the shaft into the arm. Be sure the chamfered side of the washer is against the shaft flange. Measure the distance between the outer collar on the bushing and the washer on the shaft (dimension "C" in the illustration); it should exceed 4.5 mm (0.177 in.).

5. Press the other bushing into the arm, and check dimension "C." Also check the distance between the outer collars of both bushings (dimension "D") and the distance between the end of the bushing and the centerline of the shaft mounting bolt bore (dimension "E"). D should be 144.6–146.6

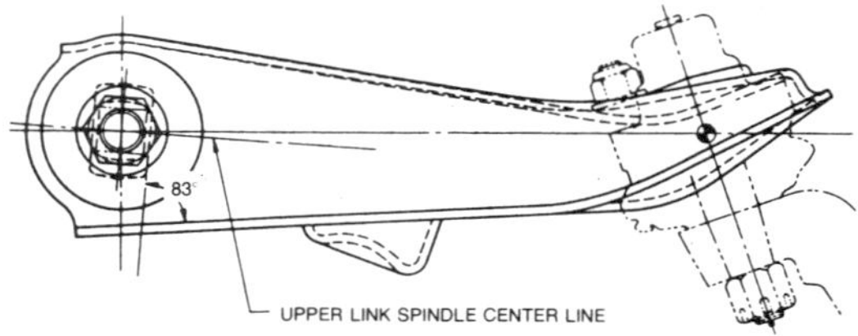

Upper arm pivot shaft installation before tightening the shaft nuts—1978–81

mm (5.69–5.77 in.); E should be 28.3 mm (1.114 in.).

6. Rotate the pivot shaft in the arm until the angle is as specified in the illustration. Install the nuts and washers on the shaft. Tighten the bushings to 56–76 ft lbs.

7. Install the upper arm and shaft assembly to the body, replacing the camber shims in their original locations. Tighten the bolts to 80–108 ft. lbs.

8. Follow Step 6 of the upper ball joint removal procedure.

LOWER

1. Perform Steps 1–6 of the lower ball joint removal procedure.

2. Remove the lower arm pivot shaft bolt and washer. Tap the pivot shaft out of the bushing. Push down on the torsion bar and remove the lower arm.

3. Use a bushing driver to tap the lower arm bushing from the frame.

4. Drive the new bushing into the frame.

5. Install the arm and pivot shaft, tightening the nut to 80–108 ft. lbs.

6. Follow Step 8 of the lower ball joint removal procedure.

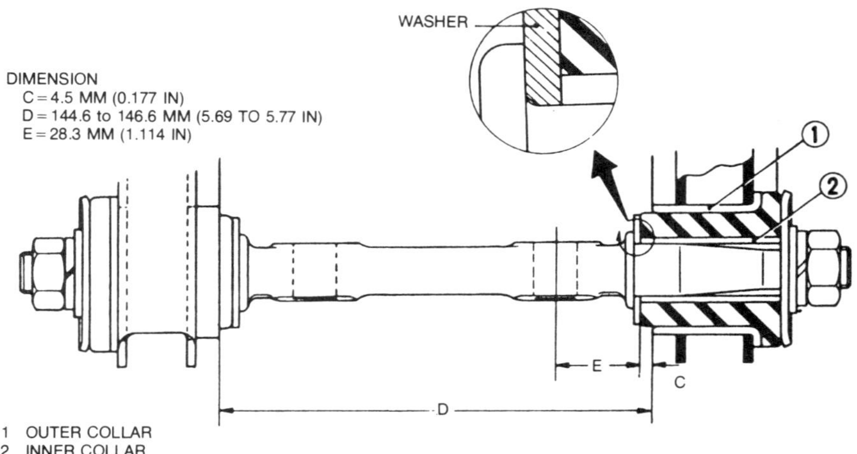

Upper arm pivot shaft and bushing installation dimensions—1978–81

4-WHEEL DRIVE FRONT SUSPENSION

Hub and Knuckle

REMOVAL AND INSTALLATION

1. Block the rear wheels, raise and support the front of the vehicle on jackstands.

2. Remove the wheels.

3. Remove the brake caliper and suspend it out of the way. DO NOT DISCONNECT THE BRAKE HOSE!

4. Remove the axle shaft.

5. Remove the bolt securing the knuckle arm to the knuckle.

6. Loosen, but do not remove the upper and lower ball joint tightening nuts.

7. Separate the ball joints from the knuckle with a ball joint removing tool.

CAUTION: *NEVER REMOVE THE BALL JOINT NUT IN THE ABOVE STEP!*

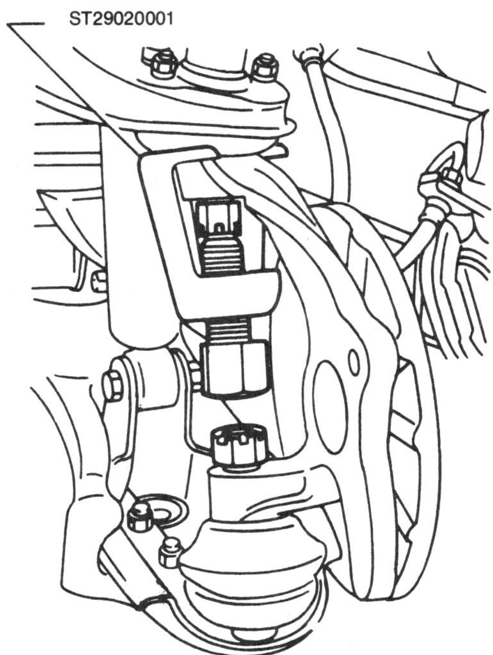

Removing the ball joints with a ball joint tool

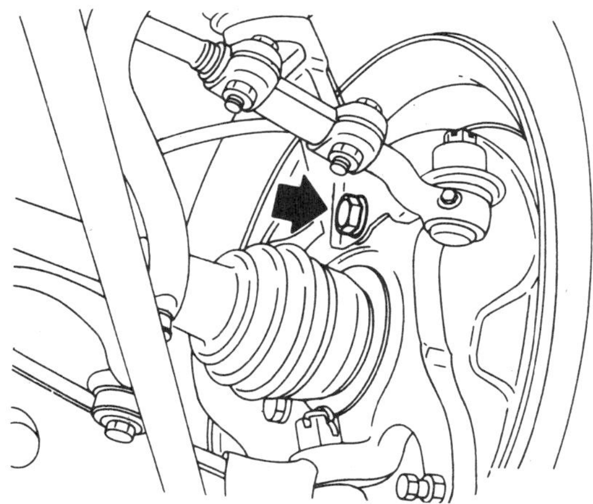

Knuckle arm attaching bolt

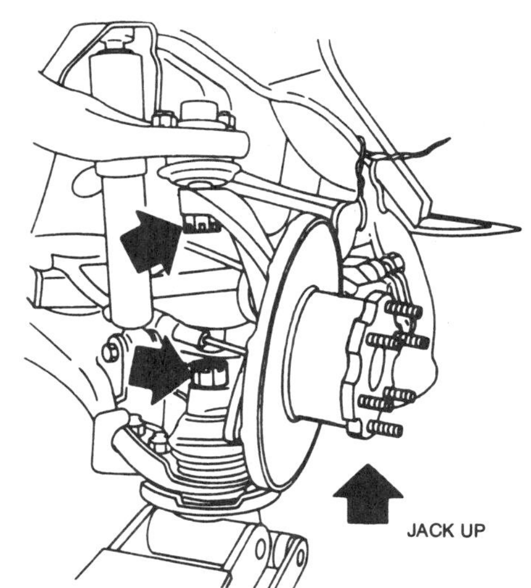

Removing the ball joint tightening nuts

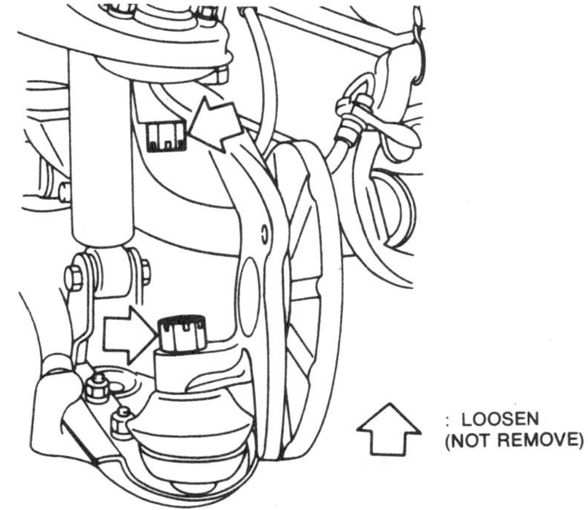

Upper and lower ball joint nuts

8. Jack up the lower link and remove the ball joint tightening nut.

9. Remove the knuckle.

10. Unbend the lockwasher with a screwdriver and remove the front hub locknut.

11. Remove the lockwasher and special washer.

12. Push the wheel bearing support out of the hub.

13. Separate the knuckle from the hub with a puller.

14. Remove the bearing collar.

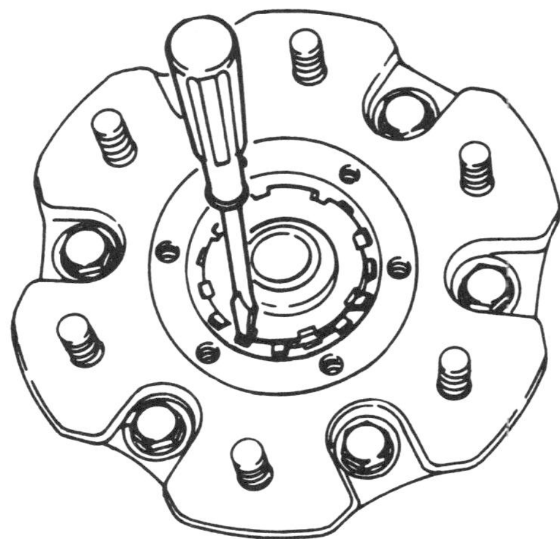

Unbending the lockwasher

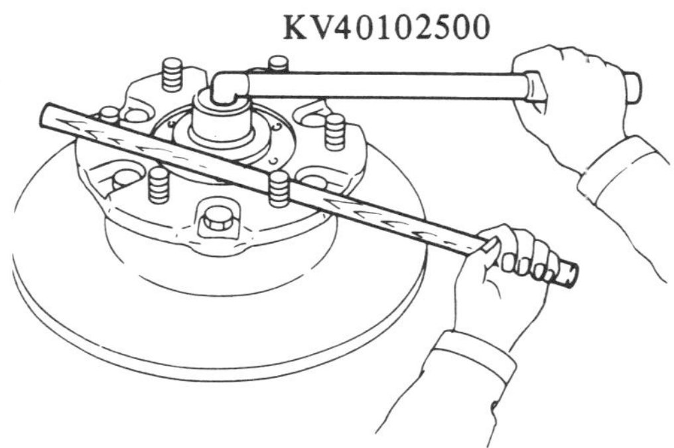

KV40102500

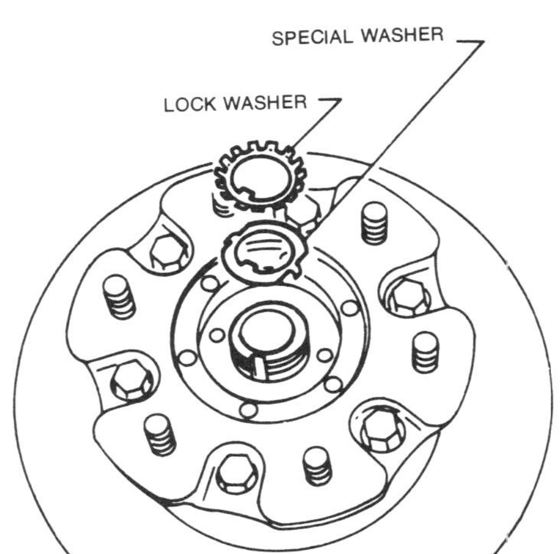

Removing the special washer and lock washer

15. Remove the inside bearing and seal. Drive the race out with a brass driver.

16. Separate the hub from the rotor.

17. Knock the hub on a wood block to move the outer bearing away from the hub surface, then pull it the rest of the way with a bearing puller. Remove the grease seal.

18. Remove the axle shaft bearing from the bearing support with a brass driver.

19. Clean and thoroughly repack the bearings.

20. Assembly is the reverse of removal. Note the following points:

a. Install the bearing outer race into each side of the knuckle with a brass driver.

b. Install the outer grease seal and bearing with a brass driver.

c. Pack the seal lip with wheel bearing grease. Be sure that the seal faces the right direction.

d. The wheel bearing collar thickness determines end play. Determine what thickness to use as follows:

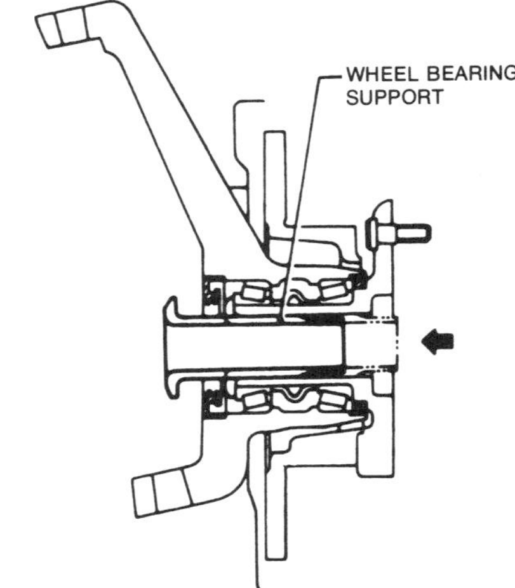

Wheel bearing support greasing location

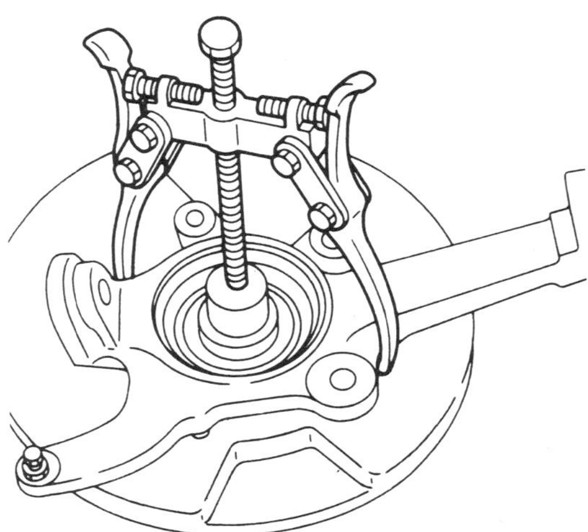

Using a puller to separate the hub from the knuckle

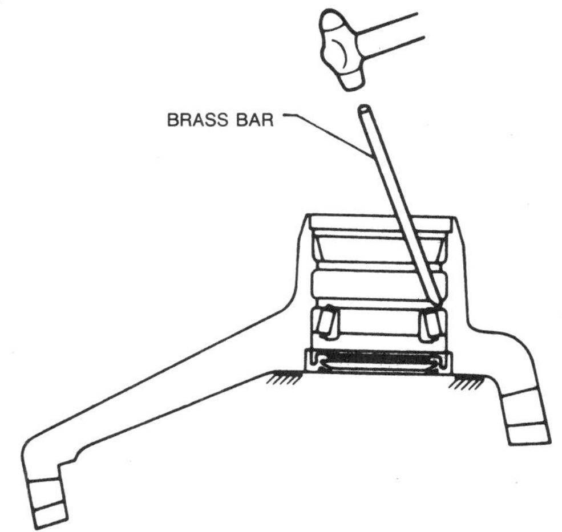

Driving out the inner bearing and seal

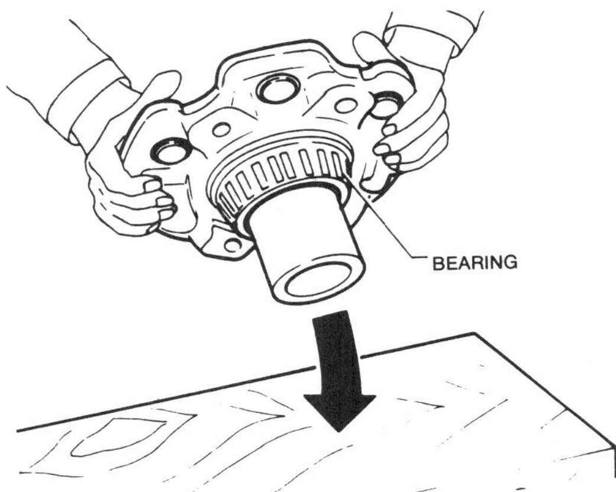

Hit the hub on a block of wood to break the outer bearing loose from the hub face

1.) Install the collar which was removed.
2.) Install the inside bearing with a brass driver.
3.) Install the special washer and lockwasher.
4.) Tighten the locknut to 108–145 ft. lb.
5.) Turn the hub several times in both directions to seat the bearings.
6.) Using a spring scale as shown, check the preload to see that it falls between 2.2 and 9.5 ft. lb.
7.) If not, adjust by replacing the collar with one of a different thickness, as shown by the number stamped on the collar. The larger the number, the thicker the collar.
8.) When preload has been correctly set, secure the nut by bending the lockwasher tip.

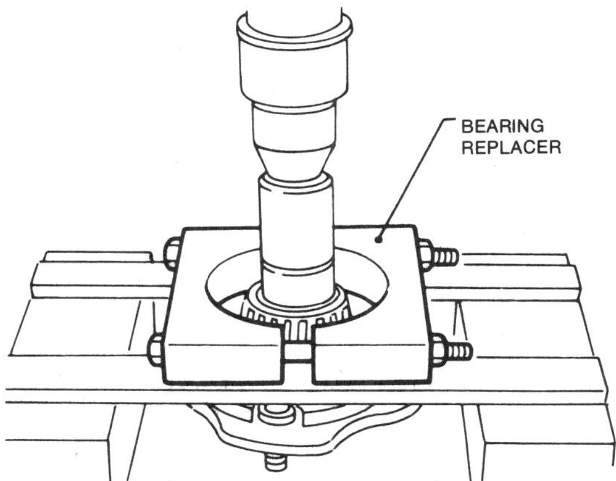

Removing the outer bearing with a puller

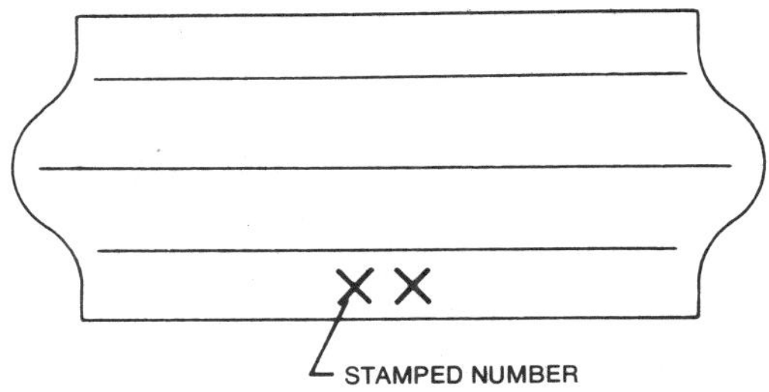

Bearing collars are stamped with a number to indicate thickness

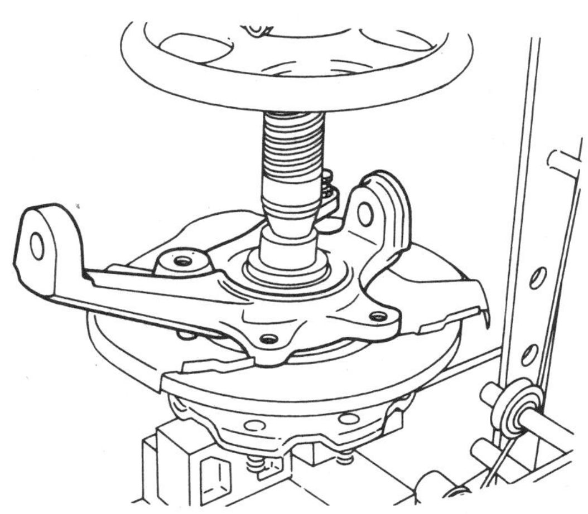

Installing the outer bearing into the knuckle

e. Observe the following torques:
Upper ball joint-to-knuckle: 36–65 ft. lb.
Lower ball joint-to-knuckle: 43–72 ft. lb.
Knuckle arm-to-knuckle: 53–72 ft. lb.
Caliper-to-knuckle: 53–72 ft. lb.
Axle shaft-to-carrier: 20–27 ft. lb.
Free running hub: 18–25 ft. lb.
Stabilizer bar: 12–16 ft. lb.
Wheel nut: 87–108 ft. lb.

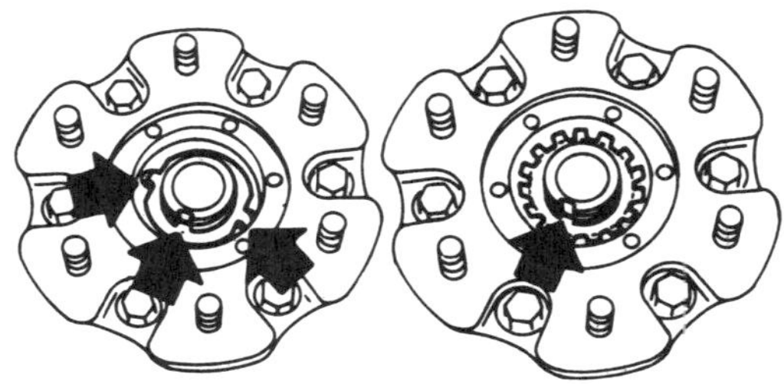

Proper positioning of the lockwasher and special washer

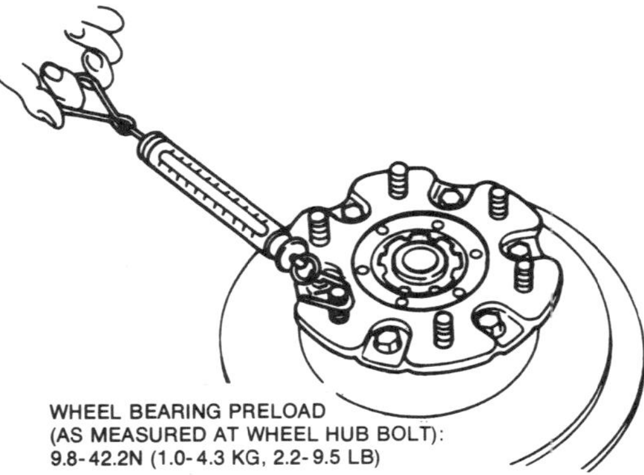

WHEEL BEARING PRELOAD
(AS MEASURED AT WHEEL HUB BOLT):
9.8-42.2N (1.0-4.3 KG, 2.2-9.5 LB)

Measuring preload with a spring scale

Shock Absorbers

See the 2-wheel drive section.

Torsion Bars

REMOVAL AND INSTALLATION

1. Raise and support the front end on jack stands.

2. Remove the torsion bar spring anchor bolt.

3. Pull the anchor arm out rearward.

4. Pull the torsion bar spring rearward.

5. Remove the torsion bar.

6. Install the torsion arm to the lower link. Torque the outer bolt to 20–27 ft. lb. and the inner to 26–33 ft. lb.

7. Place a coating of grease on the torsion bar spring serrated end and install it in the torsion arm.

8. Install the anchor arm to the serrated end of the torsion bar spring and install the anchor arm adjusting bolt. Turn the bolt until the bottom of the nut is about ½ (Dimension A) inch from the end of the bolt.

9. Install the dust cover.

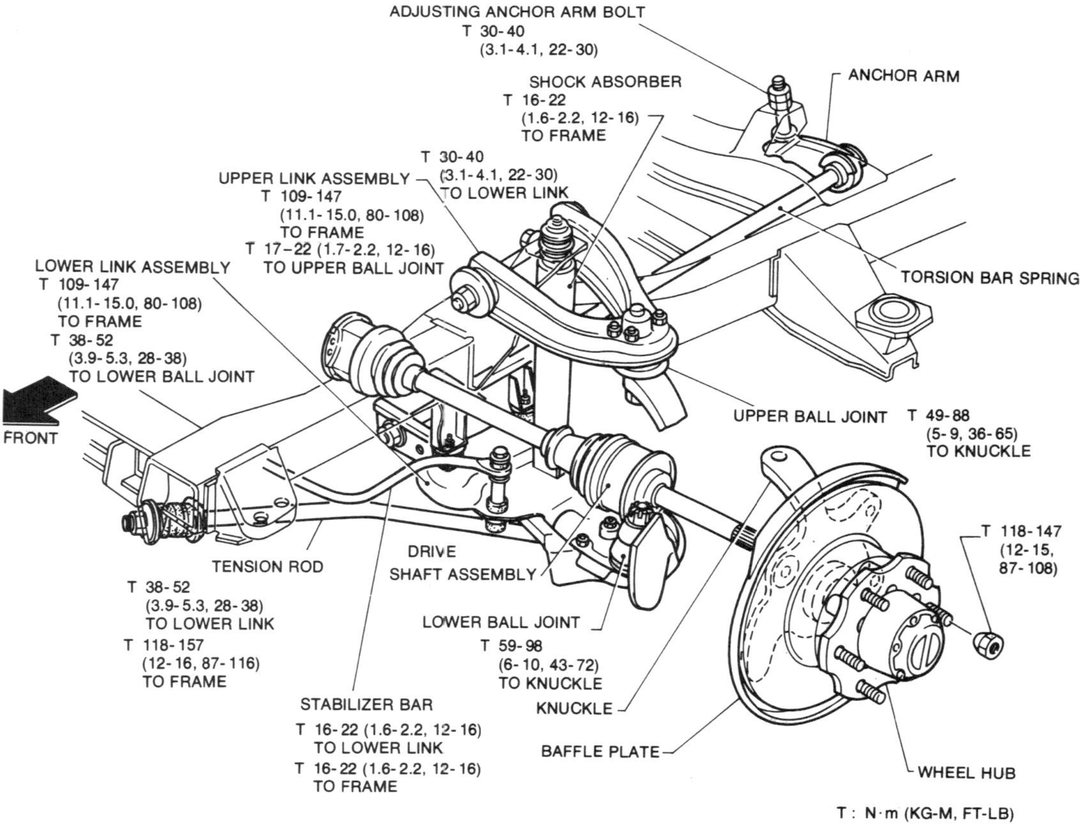

4-wheel drive front suspension

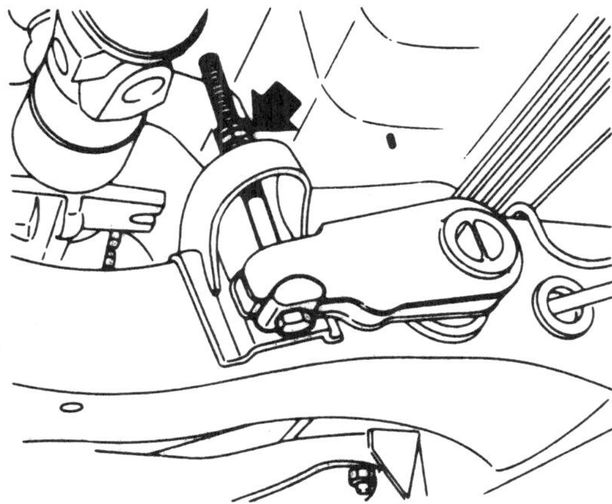

Torsion bar spring anchor bolt

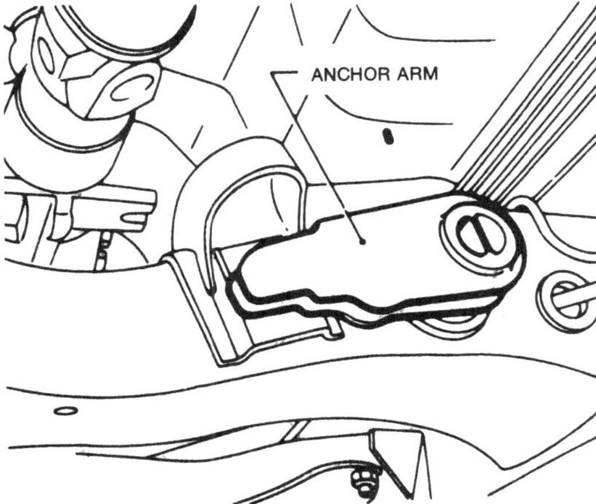

Torsion bar anchor arm

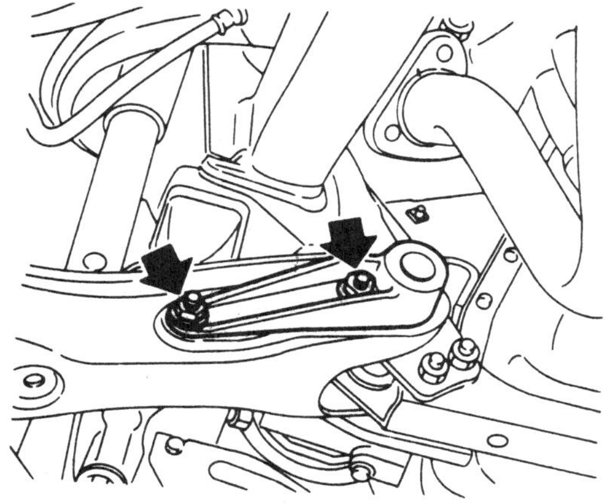

Torsion arm-to-link bolts

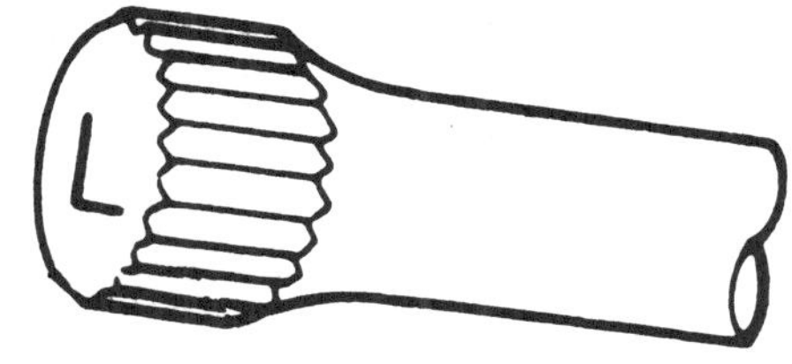

Torsion bar spring serrated ends. Note that they are marked and are not interchangeable

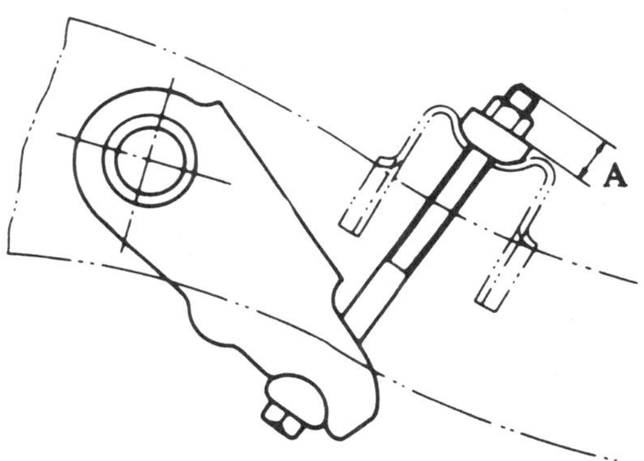

A: 7 - 17 mm (0.28 - 0.67 in)

Anchor arm adjusting bolt measurement

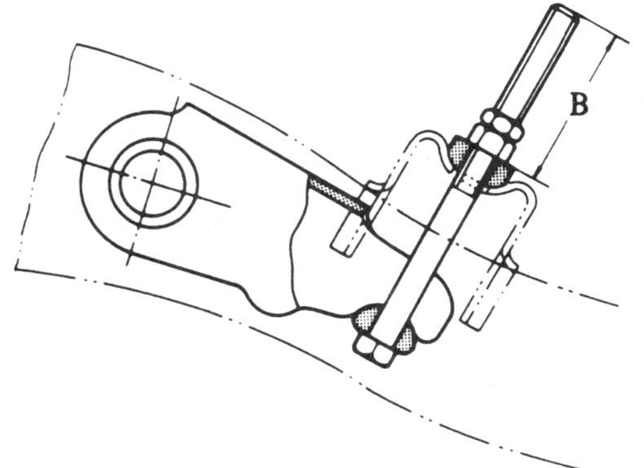

B: 60 - 70 mm (2.36 - 2.76 in)

Anchor arm position adjustment

10. Adjust the anchor arm position until the distance between the end of the dust cover and the bottom of the nut (Dimension B) is about 2½ inches.

11. Lower the vehicle.

12. Turn the anchor bolt adjusting nut until the center line of the lower link spindle is 5.28–5.47 inches above the tension rod at-

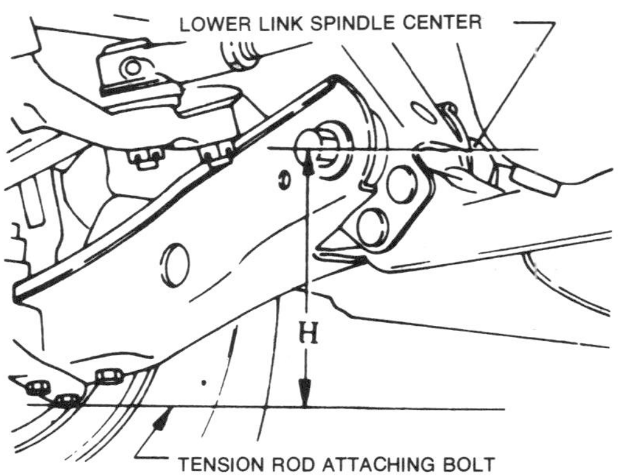

Ride height adjustment

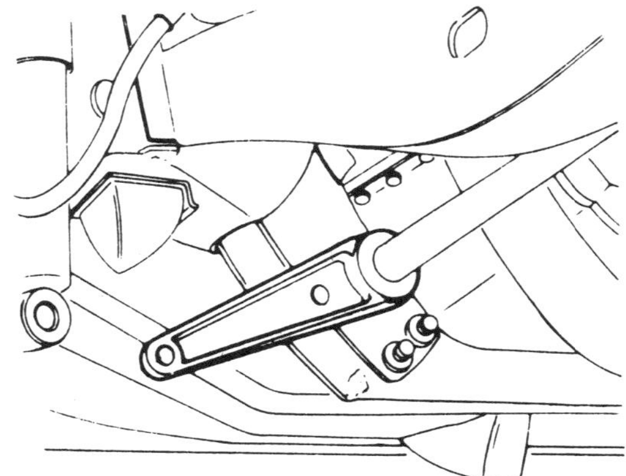

Torque arm-to-torsion bar attachment

taching bolts. This is dimension H in the accompanying figure.

Tension Bar and Stabilizer Link

See the 2-wheel drive section

Ball Joints and Control Arms

See the 2-wheel drive section with the following exceptions:

INSTALLATION

1. Install the lower ball joint to the control arm. Torque the attaching bolts to 28–38 ft. lb.

2. Install the lower control arm spindle bushing to the frame.

3. Attach to torque arm to the torsion bar.

4. Install the lower control arm.

5. Install the torque arm on the lower control arm. Torque the inner side to 26–33 ft. lb. and the outer side to 20–27 ft. lb.

6. Jack up the lower link.

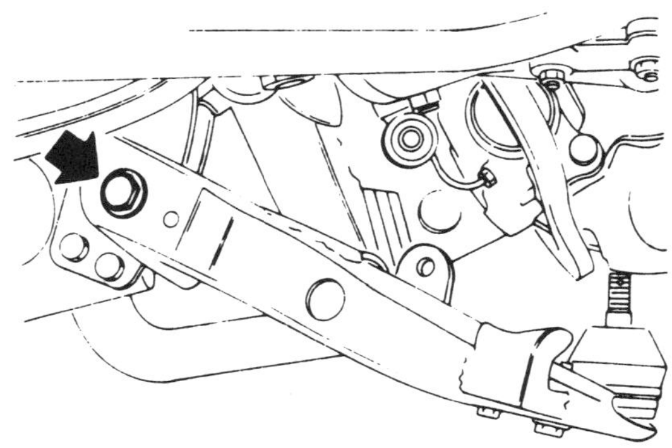

Torque arm-to-lower control arm attachment

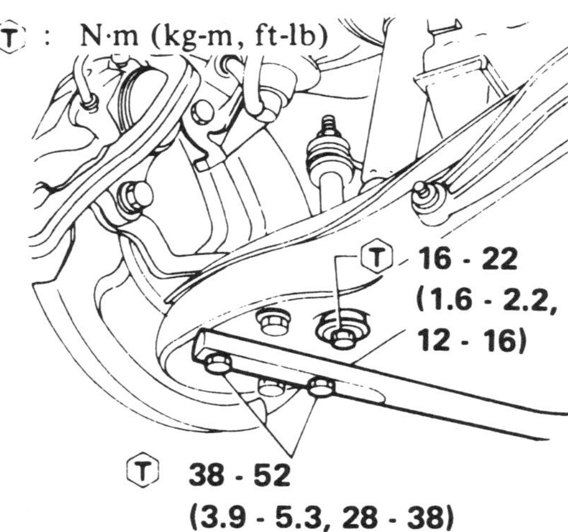

Attaching the tension rod and stabilizer bar to the lower control arm

7. Install the lower ball joint in the spindle and torque the nut to 43–72 ft. lb.

8. Install the shock absorber lower end to the control arm and torque to 22–30 ft. lb.

9. Install the tension rod and stabilizer connecting rod to the control arm.

10. Lower the vehicle.

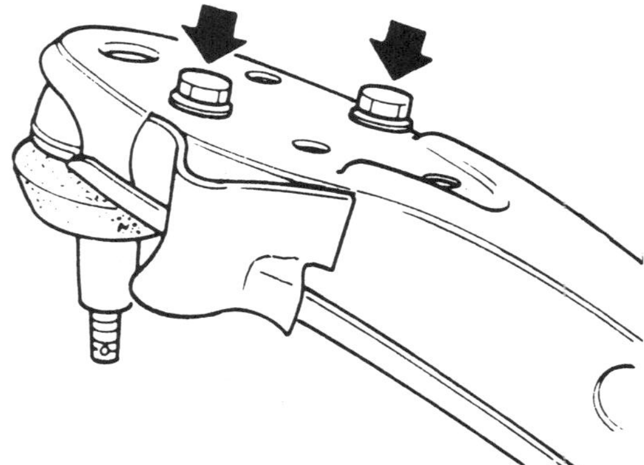

Ball joint-to-control arm bolts

11. Turn the anchor bolt adjusting nut to obtain the dimension specified in step 12 of Torsion Bar Removal and Installation.

12. After installation, check wheel alignment.

Front End Alignment

CASTER

Caster is the forward or rearward tilt of the upper end of the kingpin, or the upper ball joint, which results in a slight tilt of the steering axis forward or backward. Rearward tilt is referred to as positive caster, while forward tilt is referred to as negative caster.

Caster is adjusted by creating a difference in the total number (thickness) of shims front and rear between the upper control arm pivot shaft and its mounting bracket. Adjustment requires the use of special equipment; thus, it is not covered here.

CAMBER

Camber is the inward or outward tilt from the vertical, measured in degrees, of the front wheels at the top. An outward tilt gives the wheel positive camber. Proper camber is critical to assure even tire wear.

Camber is adjusted by adding or subtracting the same number and thickness of shims at the front and rear upper arm pivot shaft attaching bolts. Adjustment requires the use of special equipment; thus, it is not covered here.

TOE

Toe is the amount, measured in a fraction of an inch, that the wheels are closer together at one end than the other. Toe-in means that the front wheels are closer together at the front than the rear; toe-out means the rears are closer than the front. Datsun pick-ups are adjusted to have a slight amount of toe-in. Toe-in is adjusted by turning the tie-rod, which has a right-hand thread on one end and a left-hand thread on the other.

Toe-in must be checked after caster and camber have been adjusted, but it can be adjusted without disturbing the other two settings. You can make this adjustment without special equipment if you make careful measurements. The wheels must be straight ahead.

1. Toe-in can be determined by measuring the distance between the centers of the tire treads, at the front of the tire and at the rear. If the tread pattern of your truck's tires makes this impossible, you can measure between the edges of the wheel rims, but make sure to move the truck forward and measure in a couple of places to avoid errors caused by bent rims or wheel runout.

2. If the measurement is not within specifications, loosen the locknuts at both ends of the tie-rod (the driver's side locknut is left-hand threaded).

3. Turn the top of the tie-rod toward the front of the truck to reduce toe-in, or toward the rear to increase it. When the correct dimension is reached, tighten the locknuts to 58–72 ft. lbs. and check the adjustment.

STEERING ANGLE ADJUSTMENT

The maximum steering angle is adjusted by stopper bolts located on the inside of the steering knuckle/spindle. Loosen the locknut on the stopper bolt, turn the stopper bolt in or out as required to obtain the proper maximum steering angle and retighten the locknut.

Steering angle adjustment

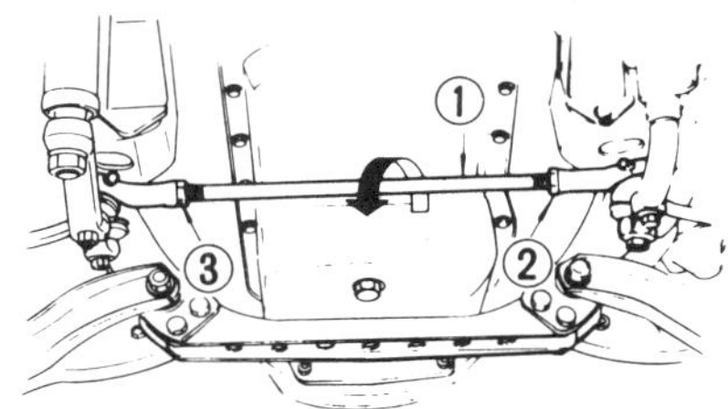

1. Tie-rod
2. Left-hand threaded locknut
3. Right-hand threaded locknut

Toe-in adjustment

RIDE HEIGHT ADJUSTMENT

The vehicle ride height is adjusted by turning the torsion bar anchor adjusting nut after loosening the locknut. The ride height adjustment should be performed with the gas

Wheel Alignment Specifications

Year	Toe-In		Camber		Caster		Kingpin Inclination (deg)	Steering Angle	
	Range (in.)	Preferred (in.)	Range (deg)	Preferred (deg)	Range (deg)	Preferred (deg)		Inner Wheel (deg)	Outer Wheel (deg)
1970–76	0.0394–0.1969	0.1182	15′–2°15′	1°15′	1°5′–2°35′	1°50′	6°15′	36° ± 1°	31° ± 1°
1977	0.079–0.118	0.100	15′–2°15′	1°15′	1°5′–2°35′	1°50′	6°15′	36° ± 1°	31° ± 1°
1978–79	0.20–0.28	0.240	15′–1°15′	1°	35′–2°05′	1°30′	—	35° ± 1°	30°30′ ± 1°
1980	0.20–0.28	0.240	0–1°	½°	50′–1°50′	1°25′	—	35° ± 1°	31° ± 1°
1981 2-wd	0.20–0.28	0.240	0–1°	½°	50′–1°50′	1°25′	—	35° ± 1°	31° ± 1°
1981 4-wd	0.20–0.28	0.240	0–1°	½°	1°10′–2°10′	1°40′	—	31° ± 2°	28° ± 1°

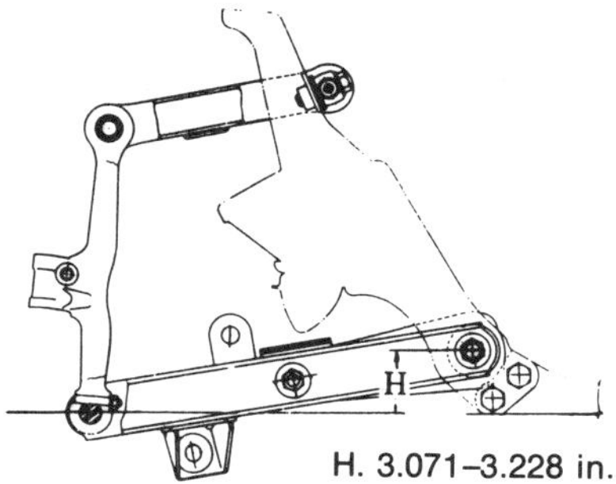

H. 3.071–3.228 in.

Ride height adjustment through 1977

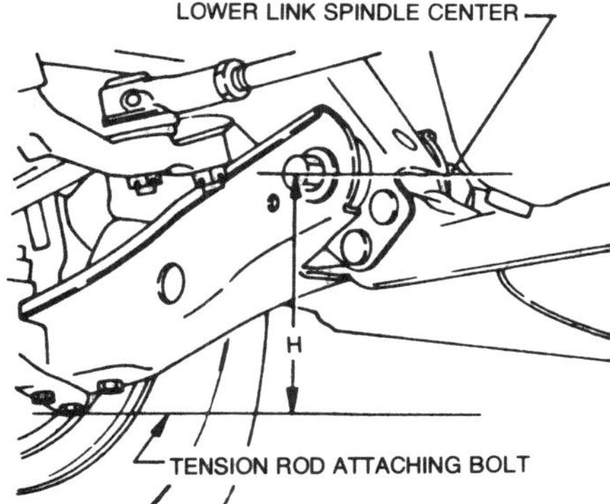

Ride height adjustment—1978–79

tank full, the radiator and engine oil at the proper level, the spare tire/wheel, jack, and jack handle in the vehicle and no passengers. Also make sure that the tires are inflated to the proper pressure.

1. Raise and support the truck under the front suspension crossmember to unload the torsion bars.

2. Turn the rear torsion bar anchor bolt to the right to lower the truck or to the left to raise it.

3. For trucks through 1976, dimension "H" in the illustration should be 3.07–3.23 in. with the truck empty and resting on its wheels. For 1977 models, the distance should be 3.11–3.31 in.

4. For 1978–79 models, measure the distance "H" between the center of the lower control arm pivot bolt and the tension rod (strut) attaching bolt, as shown in the other illustration; "H" should be 4.92 in.

REAR SUSPENSION

The rear suspension consists of semielliptic leaf springs and telescopic hydraulic shock absorbers. There are rubber bushings at

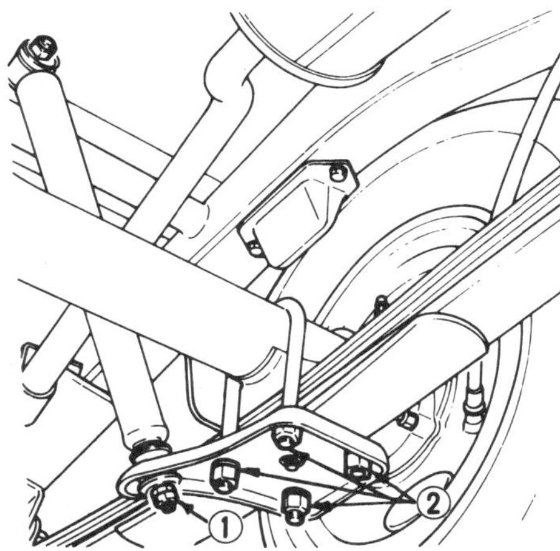

1. Shock absorber lower attaching nut
2. U-bolt attaching nut

Shock absorber lower end and U-bolt attaching nuts

either end of the leaf springs and shock absorbers to absorb vibration and noise.

Springs

REMOVAL AND INSTALLATION

CAUTION: *The leaf springs are under a considerable amount of tension. Be very careful when removing or installing them; they can exert enough force to cause serious injuries.*

1. Jack up the rear of the truck and support it with jackstands placed under the frame.

2. Disconnect the shock absorbers at their lower end.

3. Remove the nuts securing the U-bolts around the axle housing.

4. Place a jack under the rear axle housing and raise the housing to remove the weight off the springs.

5. Remove the nuts from the spring shackles, drive out the shackle pins and remove the spring from the vehicle.

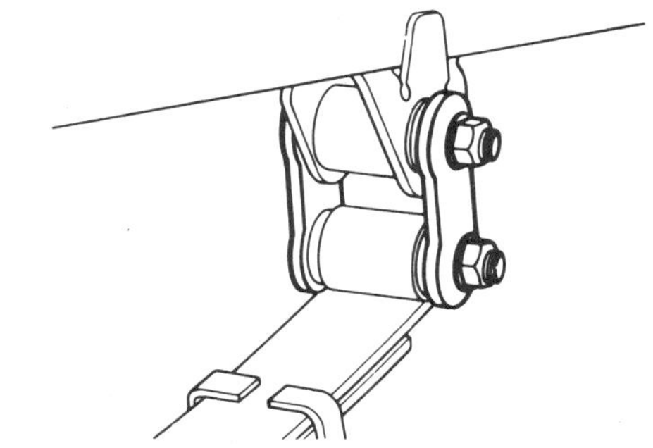

Spring shackle

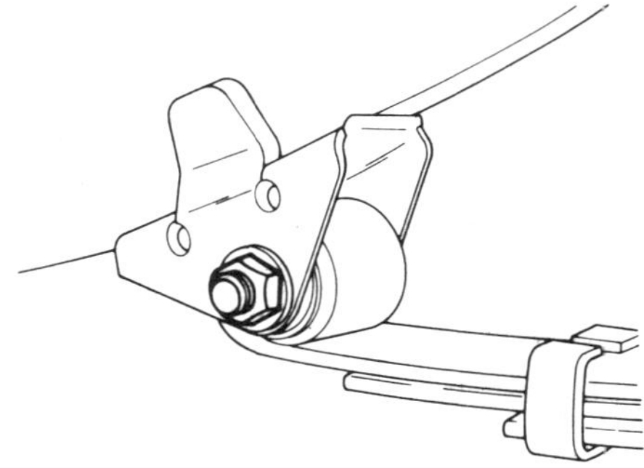

Spring pin

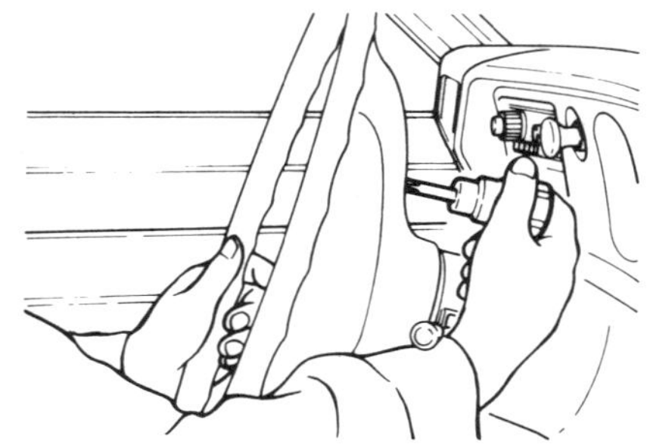

Removing the horn pad from the steering wheel

6. Install the spring in the reverse order of removal. The weight of the truck must be on the rear wheels before tightening the front pin, shackle, and shock absorber attaching nuts. Tighten the front pin and shackle nuts to 83–94 ft. lbs. (37–50 ft. lbs. for the spring shackle nuts, 1978–81), the U-bolt nuts to 53–72 ft. lbs., and the shock absorber lower end nut to 12–16 ft. lbs.

Shock Absorbers
INSPECTION AND TESTING

Inspect and test the rear shock absorbers in the same manner as outlined for the front shock absorbers.

REMOVAL AND INSTALLATION

The rear shock absorbers are removed simply by removing the upper and lower attaching nuts, and removing the component from the vehicle. They are installed in the reverse order. The weight of the vehicle must be on the rear wheels before tightening the shock absorber attaching nuts to 12–16 ft. lbs.

STEERING

Steering Wheel
REMOVAL AND INSTALLATION

1. Position the wheels in the straight-ahead position.
2. Disconnect the battery ground cable from the battery.
3. Remove the horn pad by unscrewing the two screws from the rear side of the steering wheel crossbar.
4. Punch mark the top of the steering column shaft and the steering wheel flange.
5. Remove the attaching nut and remove the steering wheel with a puller.

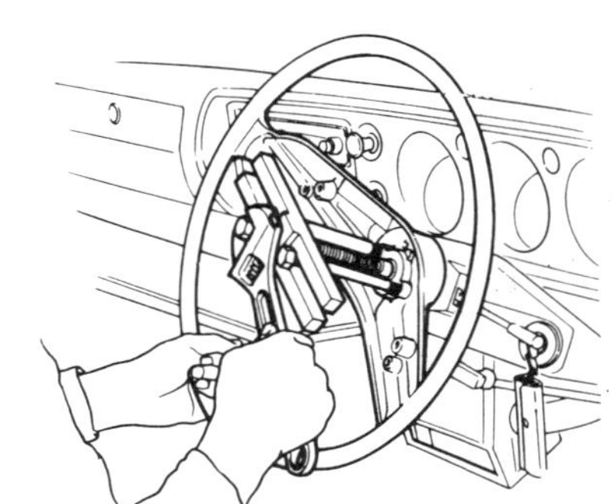

Removing the steering wheel with a puller

CAUTION: *Do not strike the shaft with a hammer, which may cause the column to collapse.*

6. Install the steering wheel in the reverse order of removal aligning the punch marks. Tighten the steering wheel attaching nut to 51–54 ft. lbs.

Turn Signal and Dimmer Switch
REMOVAL AND INSTALLATION

1. Disconnect the negative cable from the battery.
2. Remove the steering wheel.
3. Disconnect the wiring harness from the clip which retains it to the lower instrument panel.
4. Disconnect the multiple connector and lead wire from the instrument panel wiring harness.
5. Remove the steering column shell covers (upper and lower).
6. Loosen the two screws attaching the switch assembly to the steering column jacket and remove the switch assembly.
7. Install the turn signal and dimmer switch and the steering wheel in the reverse order of removal.

Ignition Switch
REMOVAL AND INSTALLATION

1. Disconnect the negative cable from the battery.

2. Unscrew and remove the escutcheon from the front of the ignition switch.

3. Remove the ignition switch and wiring harness with spacer from the steering shell cover.

4. Disconnect the wiring connector from the back of the ignition switch.

5. Install the ignition switch in the reverse order of removal.

NOTE: *On models with the optional steering lock cylinder, remove the switch by removing the two retaining screws from the back of the steering lock cylinder.*

Steering Lock
REMOVAL AND INSTALLATION

1. Remove the ignition switch.

2. Drill out the two shear screws.

3. Remove the two other normal type screws and dismount the steering lock from the steering jacket tube.

4. Install a new steering lock in the reverse order, being sure to tighten the two shear screws until they shear.

Steering Linkage
REMOVAL AND INSTALLATION

1. Jack up the front of the truck and support it with jackstands placed under the frame.

2. Remove the cotter pins and nuts securing the side rod ball studs to the steering knuckle/spindle arms.

3. Use a puller to disconnect the side rod ball studs from the steering knuckle arms. If a puller is not available, strike the side of the steering knuckle arm boss with a hammer, backing it up with a heavy hammer on the

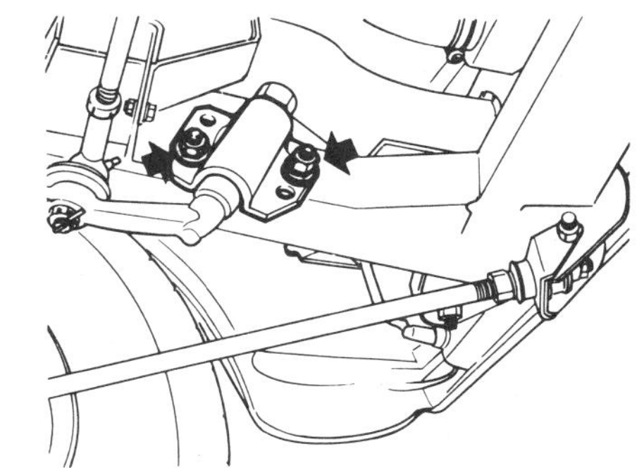

The idler arm attaching nuts

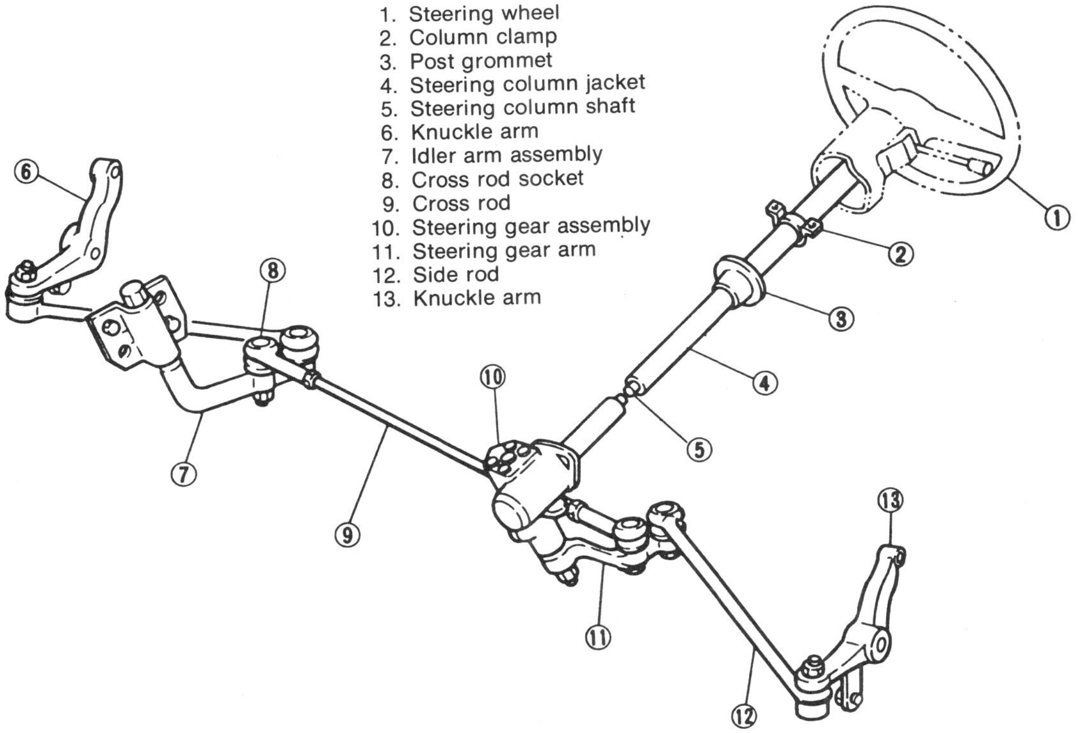

1. Steering wheel
2. Column clamp
3. Post grommet
4. Steering column jacket
5. Steering column shaft
6. Knuckle arm
7. Idler arm assembly
8. Cross rod socket
9. Cross rod
10. Steering gear assembly
11. Steering gear arm
12. Side rod
13. Knuckle arm

1970–79 Steering system

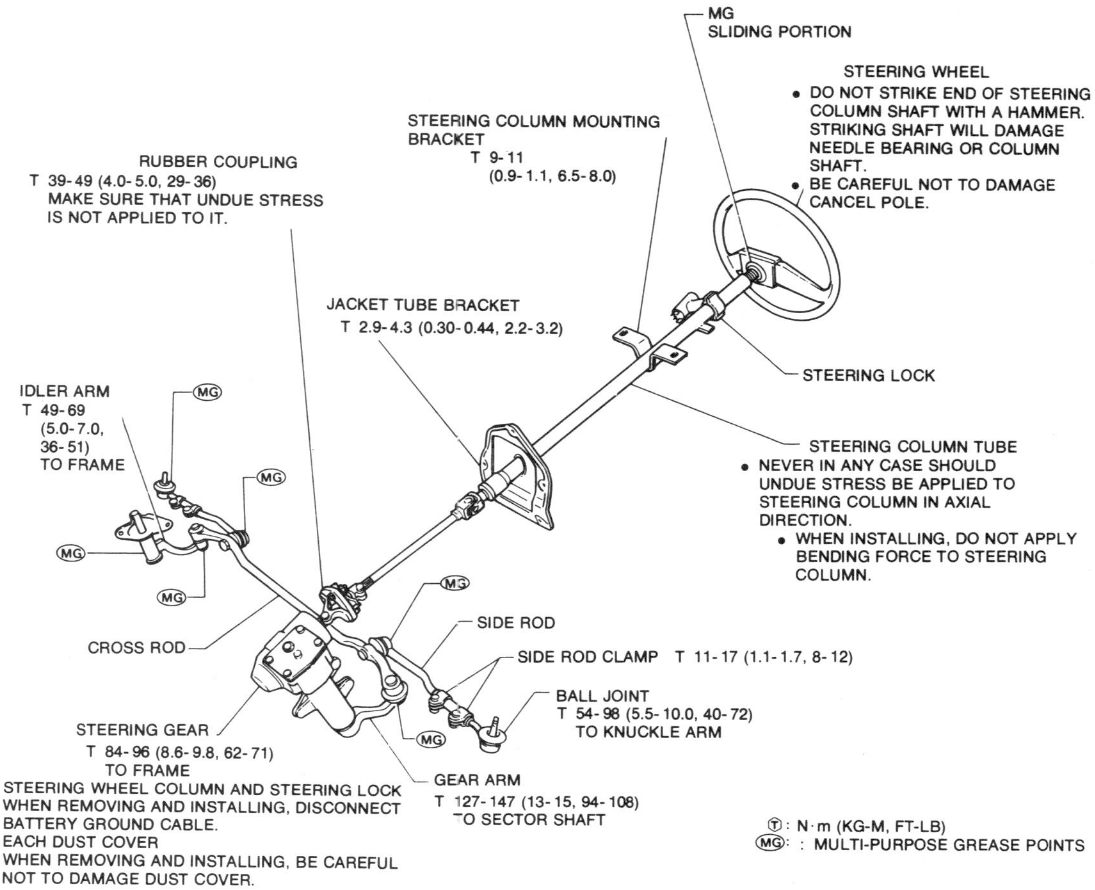

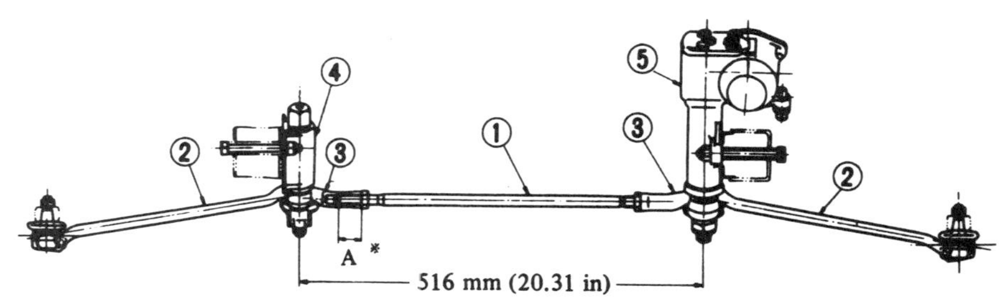

1980–81 steering system

*After adjustment of toe-in, be sure that dimension "A" at both ends
of cross rod is not less than 20 mm (0.79 in).

1. Cross rod (tie rod)
2. Side rod
3. Ball socket
4. Idler arm assembly
5. Steering gear assembly

Steering linkage installation dimension, all models

opposite side, and at the same time having an assistant pull the ball stud out of the steering knuckle arm.

NOTE: *Do not strike the ball stud head, the ball socket on the side rod, or the side rod with the hammer.*

4. Remove the nut securing the steering gear arm on the sector shaft and remove the gear arm with a puller. If a puller is not available, and the steering gear arm need not be removed, disconnect the side arm and tie-rod ball studs from the steering gear arm in the same manner as outlined in Step 3.

5. Remove the idler arm assembly from the frame by unscrewing the two attaching nuts.

6. Install the steering linkage in the reverse order of removal. Tighten the ball stud nuts to 40–55 lbs, idler arm assembly attaching nuts to 23–27 ft. lbs., and the tie-rod ad-

justment locknuts to 58–72 ft. lbs. Adjust the toe-in and steering angle.

Steering Gear
REMOVAL AND INSTALLATION

1. Raise and support the truck on jack stands.

2. Unbolt the wormshaft at the rubber coupling.

3. Matchmark the idler arm and sector shaft, and with the wheels in a straight ahead position, remove the idler arm-to-sector shaft nut and remove the idler arm with a puller.

4. Unbolt and remove the gear from the frame.

5. Installation is the reverse of removal. Torque the wormshaft coupling bolt to 29–36 ft. lb.; the idle arm nut to 94–108 ft. lb. and the gear-to-frame bolts to 62–71 ft. lb.

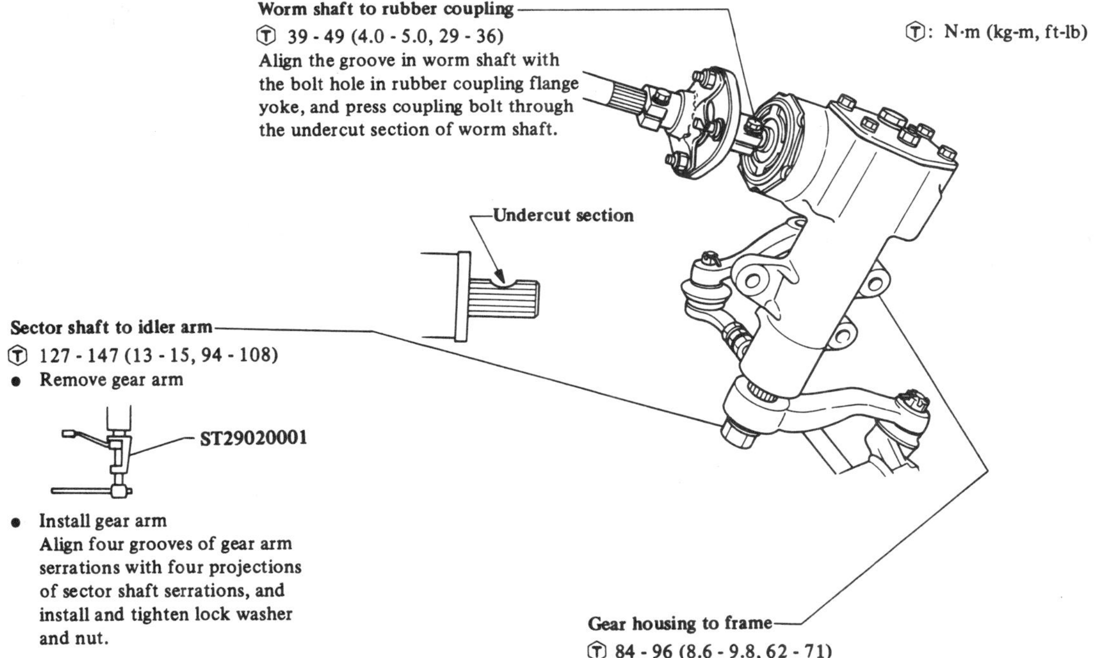

Steering gear removal and installation

Brakes

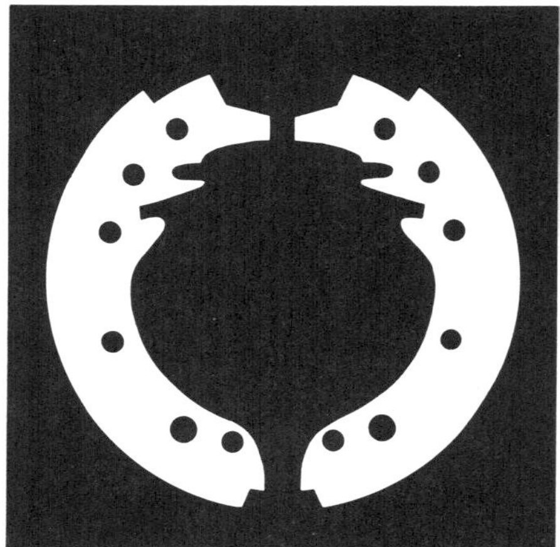

Datsun pick-ups are equipped with hydraulically-operated drum brakes through 1977, with single-servo brakes on the front wheels and duo-servo brakes on the rear wheels. 1978–81 models have single caliper, dual-piston disc brakes up front and duo-servo drums at the rear. A vacuum brake booster is standard equipment on all 1972 and later trucks. 1976 and later models have the Nissan Load Sensing Valve, a proportioning valve which senses changes in the vehicle load and brake fluid pressure, and regulates the amount of rear braking force accordingly, to reduce the possibility of rear brake lockup.

The single-servo front brake consists of two brake shoes, upper and lower return springs, an after shoe return spring, the adjuster assembly at the bottom, and the single piston wheel cylinder assembly at the top. The wheel cylinder has one hydraulically-operated piston which pushes out on the leading brake shoe. When the leading brake shoe contacts the rotating brake drum, servo action rotates the entire assembly slightly and pushes the trailing brake shoe against the brake drum.

The 1978–81 front disc brake caliper is surrounded by a yoke. When the brake is applied, hydraulic pressure forces the dual pistons apart. This presses the inner pad against the rotor, because it is directly actuated by the outer piston. The other (inner) piston slides the yoke inwards, which pulls the outer pad with it to contact the outside of the rotor. The disc brakes are self-adjusting.

The duo-servo rear brake consists of two brake shoes, the adjuster assembly at the bottom, the dual piston wheel cylinder at the top, upper and lower return springs, after shoe return spring, and the parking brake mechanism.

BRAKE SYSTEM

Adjustment

DISC BRAKES

The front disc brakes are inherently self-adjusting. No adjustments are either necessary or possible.

DRUM BRAKES

1. Jack up the wheel to be adjusted until it completely clears the ground.

2. Make sure that the parking brake is completely released if the rear brakes are being adjusted.

3. Remove the rubber boot from the rear of the brake backing plate.

4. Lightly tap the adjuster housing forward with a hammer and screwdriver.

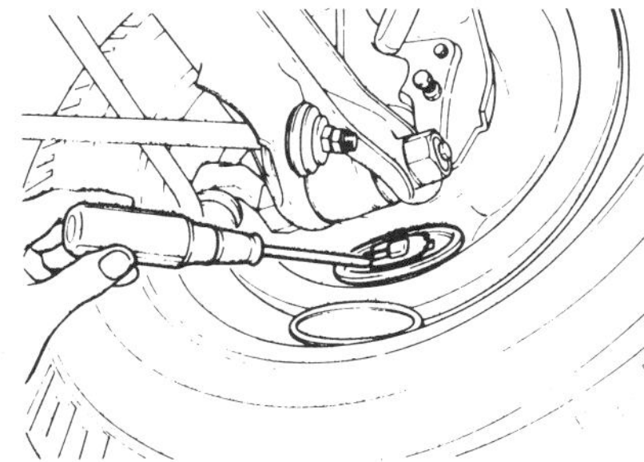

Adjusting the front brakes

5. Turn the adjuster wheel downward with a screwdriver to spread the brake shoes. Stop turning the adjuster wheel when the brake drum is locked and the wheel cannot be turned by hand.

6. Turn the adjuster wheel upward, backing off the shoes from the brake drum 12 notches, to obtain the correct clearance between the brake shoes and drum. Turn the wheel to make sure that the brake drum turns without dragging.

7. Install the rubber boot.

BRAKE PEDAL ADJUSTMENT

1970–71

These models have non-adjustable master cylinder pushrods. The free-play must be adjusted by the use of shims between the master cylinder and the firewall.

1. Loosen the pedal stopper so that it does not contact the pedal arm.

2. With the master cylinder pushrod completely extended, the pedal should be 5.45 inches from the toe board (rugs removed). The distance can be adjusted with shims. Use the same thickness of shim for both the upper and lower master cylinder mounting bolts.

3. Adjust the pedal height with the stopper, so that the front of the pedal pad is 5.33 inches from the toe board.

1972–79

These models have adjustable master cylinder pushrods for setting free-play.

1. Adjust the height of the stop light switch so that the top surface of the brake pedal is 5.5 in. off the surface of the floor board (without rugs) through 1975. The distance should be 5.8 in., 1976–77, 6.06 in., 1978–79, and 6.7 in. for 1980–81. Tighten the stop light switch locknut.

2. Adjust the length of the master cylinder pushrod (booster pushrod on vehicles so

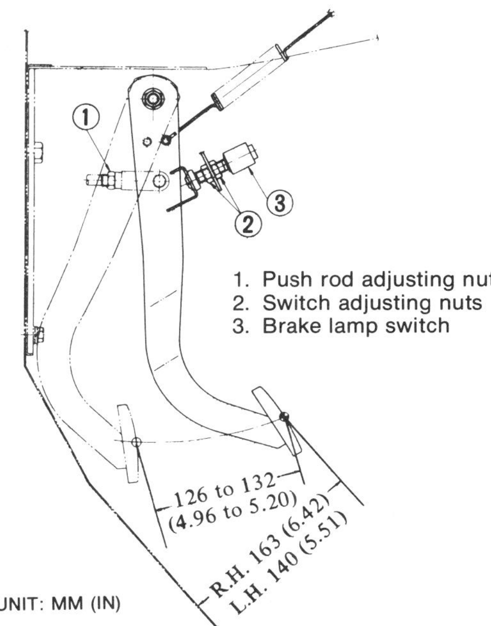

1. Push rod adjusting nut
2. Switch adjusting nuts
3. Brake lamp switch

Brake pedal adjustment—measurements shown are for 1972–75 models

equipped) clevis so that 0.04–0.12 in. of free-play exists between the brake pedal and the pushrod through 1975. Free-play should be 0.024–0.047 in., 1976–77, or 0.04–0.20 in., 1978–81.

3. Operate the brake pedal to make sure that it operates freely with no noise or interference.

HYDRAULIC SYSTEM

Master Cylinder

REMOVAL AND INSTALLATION

1. On a truck not equipped with power brakes, remove the cotter pin, pull out the clevis pin, and separate the brake pedal from the master cylinder pushrod.

2. Place a number of cloths or a container under the master cylinder to catch the brake fluid. Disconnect the brake tubes from the master cylinder; use a flare nut wrench if one is available.

NOTE: *Brake fluid eats paint; wipe up any spilled fluid immediately, then flush the area with clear water.*

3. Remove the master cylinder securing nuts and withdraw the master cylinder from the firewall or power brake booster.

4. Install the master cylinder in the reverse order of removal and bleed the brake hydraulic system.

OVERHAUL

This is a tedious, time-consuming job. You can save yourself a lot of trouble by buying a rebuilt master cylinder from your dealer or parts supply house. The small difference in price between a rebuilding kit and a rebuilt part usually makes it more economical, in terms of time and work, to buy the rebuilt part.

NOTE: *Datsun has two suppliers for brake parts: Nabco and Tokico. These parts are not interchangeable. Be certain to get the correct parts for the model installed on your truck. The manufacturer's name is clearly stamped on the part.*

Rebuilding procedures are the same for both the single piston master cylinders used 1971–71, and the double piston models used 1972 and later.

1. Remove the master cylinder from the vehicle.

2. Remove the reservoir caps and filters and drain the brake fluid.

3. Withdraw the pushrod and remove the rubber dust boot from the open end of the master cylinder.

4. Pry the piston stopper snap-ring from the open end of the master cylinder with a screwdriver.

5. Remove the stopper screw and washer from the bottom of the master cylinder and then remove the primary and secondary piston assemblies from the master cylinder bore.

6. Remove the caps on the underside of the master cylinder to gain access to the check valves for cleaning.

7. Discard all used rubber parts and gaskets. These parts should be replaced with new components which are usually contained in rebuilding kits.

NOTE: *Never remove the master cylinder reservoir tanks. If they are removed for any reason, they must be replaced with new ones.*

8. Clean all of the parts in clean brake fluid.

9. Check the cylinder bore and piston for wear, scoring, corrosion, or any other damage. The piston and cylinder bore can be dressed with crocus cloth soaked in brake fluid. Move the crocus cloth around the cylinder bore, not in and out. Do the same to

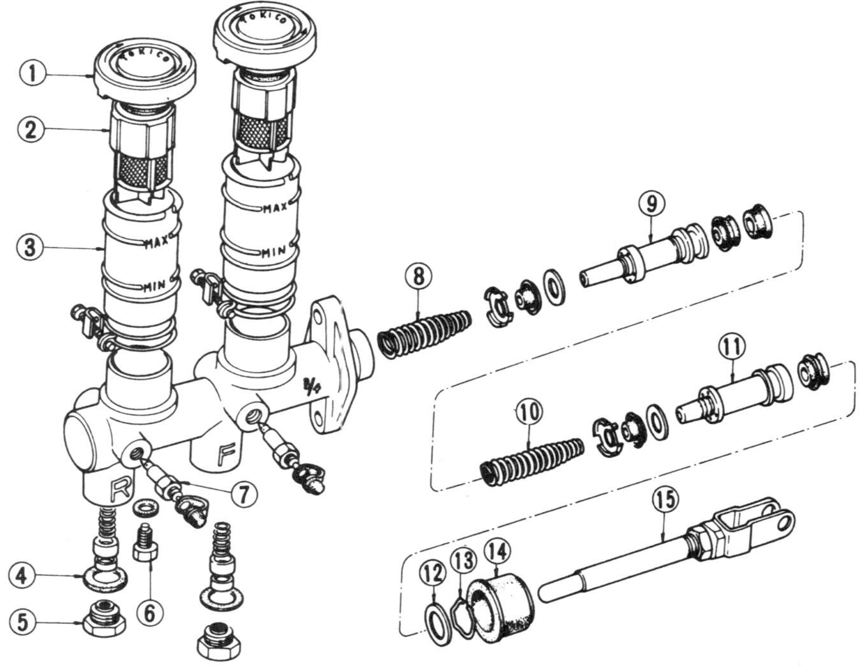

1. Reservoir cap	6. Secondary piston stopper	11. Primary piston
2. Oil filter	7. Bleeder screw	12. Piston stopper
3. Oil reservoir	8. Secondary return spring	13. Piston stopper ring
4. Packing	9. Secondary piston	14. Dust cover
5. Valve cap	10. Primary return spring	15. Pushrod

1972–79 master cylinder

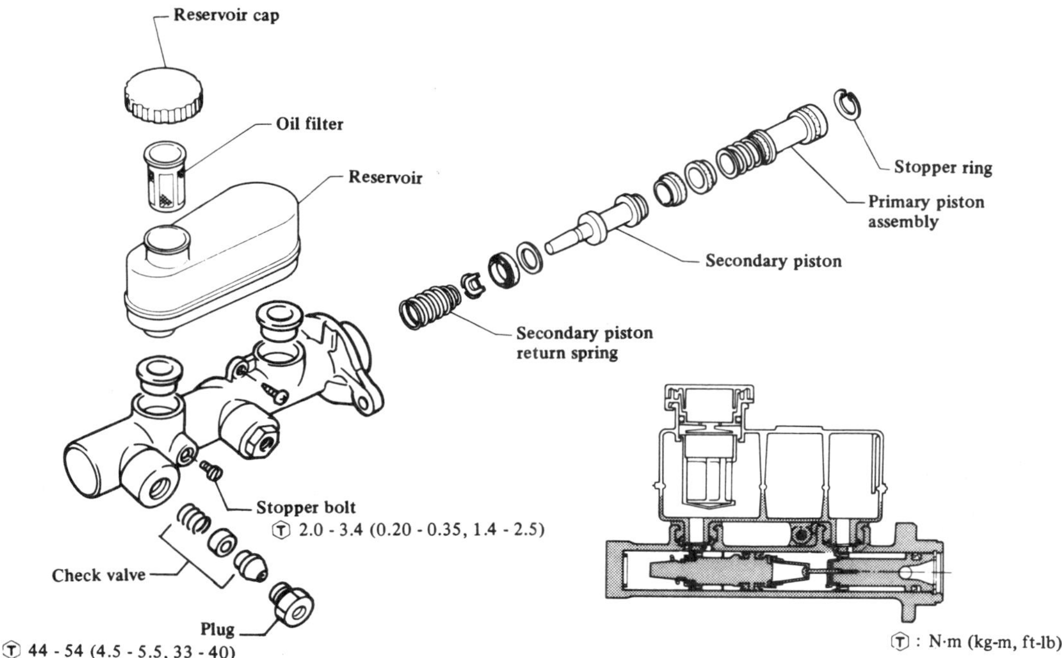

1980–81 master cylinder

the piston, if necessary. Wash both the cylinder bore and the piston in clean brake fluid after cleaning them with crocus cloth.

10. Check the piston-to-cylinder bore clearance; it should measure less than 0.15 mm (0.0059 in.) for all models. If greater clearance exists, replace the piston, or cylinder, or both. The cylinder bore diameter should be 19.05 mm (¾ in.) through 1977, 20.64 mm (¹³/₁₆ in.), 1978–79, or 22.23 mm (⁷/₈ in.) for 1980–81.

11. Assemble the master cylinder in the reverse order of removal. Soak all of the components in clean brake fluid before assembling them. Clamp the master cylinder in a vise. Fill the reservoir(s) with fresh fluid, and pump the piston with a screwdriver until fluid squirts from the outlet ports. Install the master cylinder and bleed the brakes.

NLSV

The Nissan Load Sensing Valve is a proportioning valve installed between the master cylinder and the brakes. It limits fluid pressure to the rear brakes to prevent rear wheel lock-up. The valve is not rebuildable, and must be replaced as an assembly if defective. It should be replaced if, in a stop on dry pavement from 30 mph, the rear wheels lock prior to the front wheels (although ideally, no wheel will lock).

Bleeding

The purpose of bleeding the brakes is to expel air trapped in the hydraulic system. The system must be bled whenever the pedal feels spongy, indicating that compressible air has entered the system. It must also be bled whenever the system has been opened or repaired. You will need a helper for this job.

CAUTION: *Never reuse brake fluid which has been bled from the brake system.*

1. The sequence for bleeding is as follows:
• 1970–71: Right rear, left rear, right front, left front;
• 1972–75: Master cylinder front, master cylinder rear, then the 1970–71 sequence;
• 1976–79: Master cylinder front, master cylinder rear, NLSV front, right front, left front, left rear, right rear, NLSV rear, NLSV center.

It is not necessary to run the engine on 1972–79 models with a vacuum booster.

2. Clean all the bleeder screws. You may want to give each one a shot of a penetrating lubricant to loosen it up; seizure is a common problem with bleeder screws, which then break off, sometimes requiring replacement of the part to which they are attached.

3. Fill the master cylinder with DOT 3 brake fluid.

NOTE: *Brake fluid picks up moisture from the air. Don't leave the master cylinder or the fluid container uncovered any longer than necessary. Be careful—brake fluid eats paint.*

Check the level of the fluid often when bleeding, and refill the reservoirs as necessary. Don't let them run dry, or you will have to repeat the process.

4. Attach a length of clear vinyl tubing to the bleeder screw on the wheel cylinder (or master cylinder). Insert the other end of the tube into a clear, clean jar half filled with brake fluid.

5. Have your helper slowly depress the brake pedal. As this is done, open the bleeder screw ⅓–½ of a turn, and allow the fluid to run through the tube. Then close the bleeder screw before the pedal reaches the end of its travel. Have your assistant slowly release the pedal. Repeat this process until no air bubbles appear in the expelled fluid.

NOTE: *If the brake pedal is depressed too fast, small air bubbles will form in the brake fluid.*

6. Repeat the procedure on the other three brakes, checking the level of fluid in the cylinder reservoirs often.

FRONT DISC BRAKES

Pads

INSPECTION

The pads should be removed so that the thickness of the remaining friction material can be measured. If the pads are less than 2 mm (0.08 in.) thick, they must be replaced. This measurement may disagree with your state inspection laws.

NOTE: *Always replace all pads on both wheels at the same time. The factory kit includes four pads, clips, pins, and springs; all parts should be used.*

REMOVAL AND INSTALLATION

1. Raise and support the front of the truck. Remove the wheels.

2. Remove the retaining clip from the outboard pad.

3. Remove the pad pins retaining the anti-squeal springs.

4. Remove the pads.

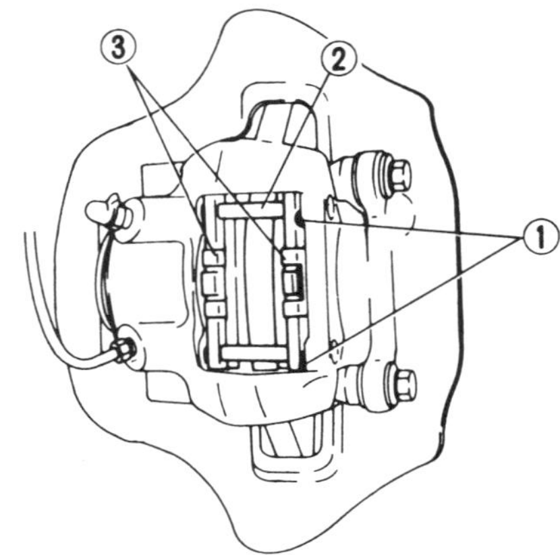

1. Clip
2. Pad pin
3. Anti-squeal spring

Removing the disc brake pads

NOTE: *When replacing the pads, always check the surface of the rotors for scoring or wear. The rotors should be removed for resurfacing if badly worn.*

5. To install, open the bleeder screw slightly and push the outer piston into the cylinder until the dust seal groove aligns with the end of the seal retaining ring, then close the bleeder screw. Be careful, because the piston can be pushed too far, requiring disassembly of the caliper to repair. Install the inner pad.

6. Pull the yoke to push the inner piston into place. Install the outer pad.

7. Lightly coat the areas where the pins touch the pads, and where the pads touch the caliper (at the top) with grease. Do not allow the grease to get on the pad friction surfaces.

8. Install the anti-squeal springs and pad pins. Install the clip.

9. Apply the brakes a few times to seat the pads. Check the master cylinder level; add fluid if necessary. Bleed the brakes if necessary.

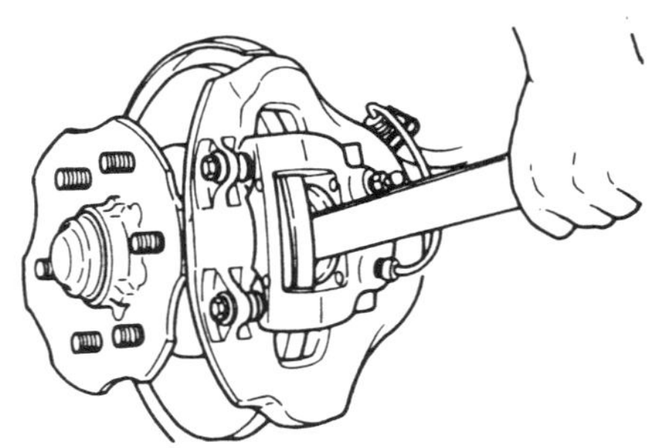

Pressing the piston into place

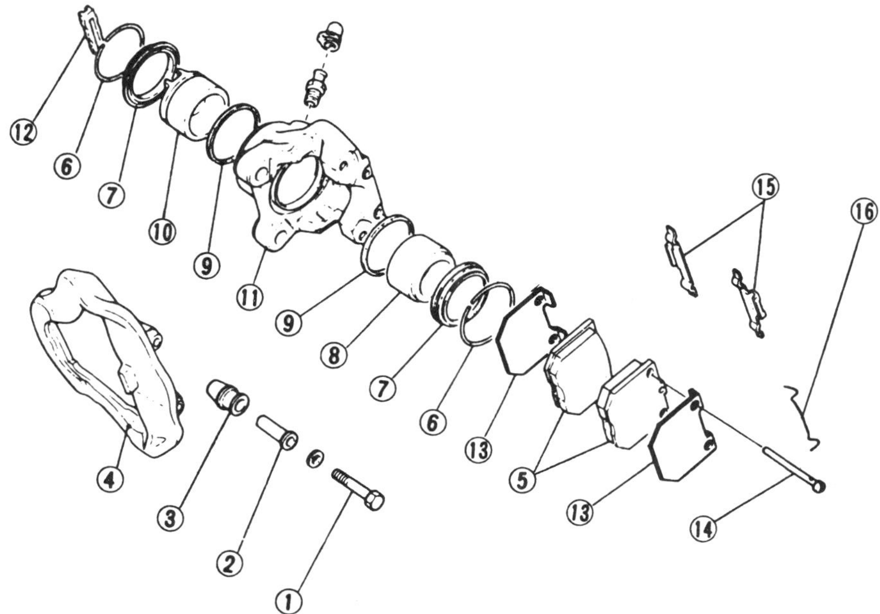

1. Fixing bolt
2. Collar
3. Gripper
4. Yoke
5. Pad
6. Retaining ring
7. Dust seal
8. Piston B
9. Piston seal
10. Piston A
11. Caliper body
12. Yoke holder
13. Shim
14. Pad pin
15. Anti-squeal spring
16. Clip

Exploded view of the front disc brake caliper

Caliper

REMOVAL AND INSTALLATION

1. Remove and plug the brake tube.
2. Unbolt and remove the caliper from the spindle.
3. Installation is the reverse. Tighten the mounting bolts to 53–72 ft. lbs.

Overhaul

1. Remove the caliper.
2. Remove the pads.
3. Remove the gripper pin attaching nuts and separate the yoke from the caliper body.
4. Remove the yoke holder from the piston and remove the retaining rings and dust seals from the ends of both pistons.
5. Apply air pressure gradually into the fluid chamber of the caliper, to force the pistons from the cylinders.
6. Remove the piston seals.
7. Inspect the parts for wear or damage. Check the inside surface of the cylinder for scoring or wear, and replace the caliper as necessary. Minor damage can be cleaned up with crocus cloth, but deep pitting or scoring warrants replacement of the caliper. The piston should be examined for wear, but do not polish it with crocus cloth; it has a plated sur-face which will be damaged by sanding. Replace the piston as necessary.
8. To assemble, coat the seals and pistons with clean brake fluid. Install the seals into the cylinder bore.
9. Slide the "A" piston into the cylinder, followed by the "B" piston, so that its yoke groove coincides with the yoke groove of the cylinder.
10. Install the dust seal and secure tightly with the retaining ring.
11. Install the yoke holder onto the "A" piston and install the gripper to the yoke. If you lightly coat the gripper pins with soapy water, they will be easier to install.
12. Support the end of the "B" piston, and

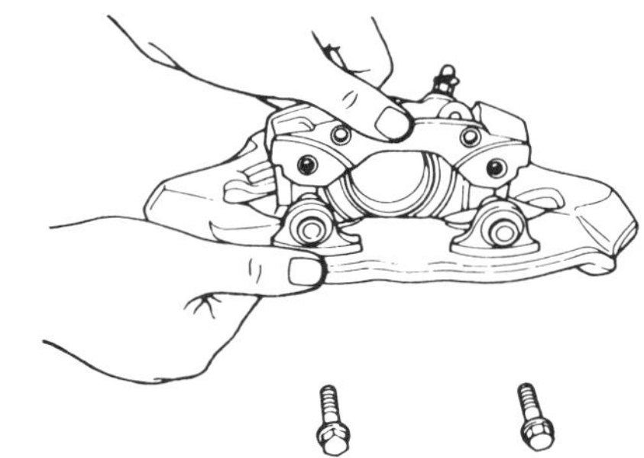

Separating the yoke from the caliper

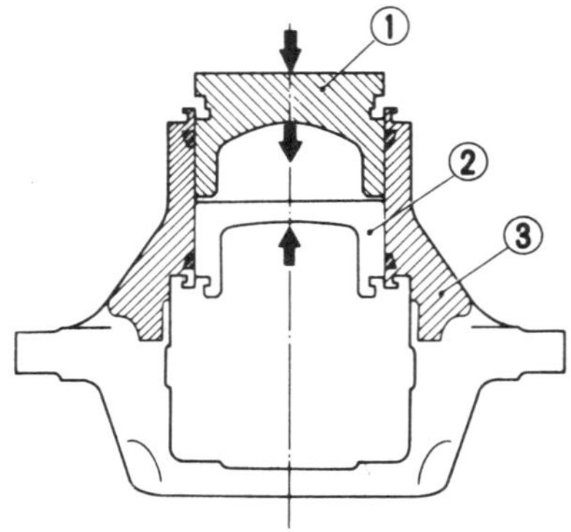

1. Piston A
2. Piston B
3. Caliper body

Installing the pistons in the caliper; the arrows show the direction of installation

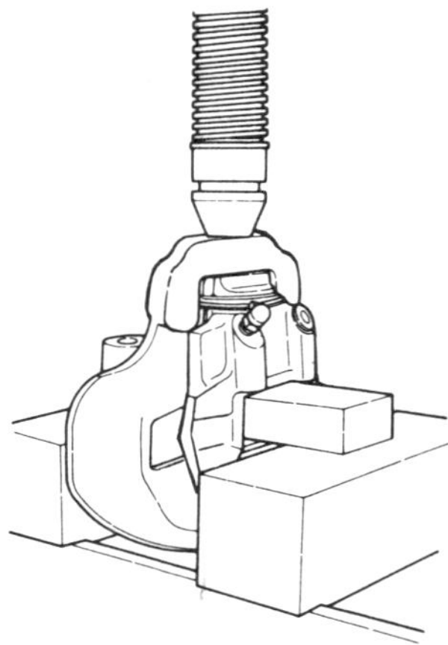

Installing the yoke

press the yoke into the yoke holder. This will require a good deal of force. Be careful to insert the yoke straight into the holder, to avoid cracking the yoke holder.

13. Install the pads, anti-squeal springs, and pad pins and retain with the clip.

14. Tighten the gripper pin attaching nuts to 12–15 ft. lbs., and install the caliper on the spindle.

Disc (Rotor)

REMOVAL AND INSTALLATION

1. Remove the claiper.

2. Remove the hub cap, cotter pin, and adjusting nut.

3. Remove the wheel bearing nut and remove the hub and rotor.

4. Hold the hub in a vise and loosen the bolts to remove the rotor from the hub.

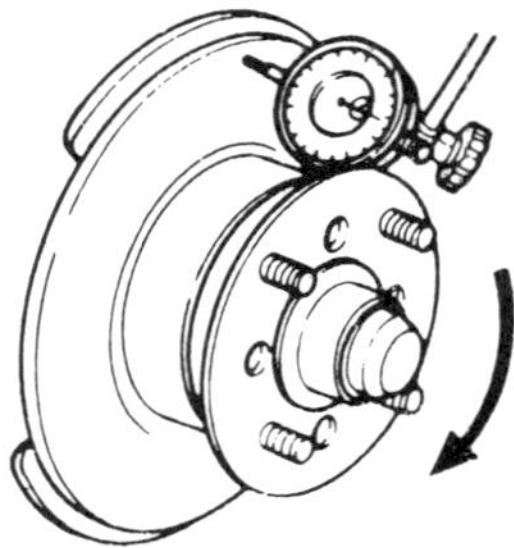

Measure the disc runout with a dial indicator

5. Installation is the reverse. Tighten the rotor-to-hub bolts to 28–38 ft. lbs. Adjust the wheel bearings after installation (see Chapter 1).

INSPECTION

1. Check the rotor for wear or scoring. Deep scoring, grooves, or rust pitting can be removed by refacing. Minimum disc thickness is 10.5 mm (0.413 in.). If the rotor will be thinner than this after refinishing, it must be replaced.

2. Check disc parallelism; it must be under 0.03 mm (0.0012 in.). If over this specification, the disc must be replaced.

3. With the disc and hub installed, and the wheel bearings adjusted to specification, measure the disc runout with a dial indicator. If runout exceeds 0.15 mm (0.0059 in.), the disc needs to be refinished or replaced.

FRONT DRUM BRAKES

Brake Drums

REMOVAL AND INSTALLATION

1. Jack up the front of the vehicle so that the wheel which is to be serviced is off the ground. Be sure to loosen the lug nuts before the wheel comes off the ground.

2. Remove the wheel and tire assembly.

3. Pull the brake drum off of the hub. If the drum cannot be easily removed, back off on the brake adjustment.

NOTE: *Never depress the brake pedal while the brake drum is removed.*

4. Install the brake drum in the reverse order of removal and adjust the brakes.

INSPECTION

After removing the brake drum, wipe out the accumulated dust with a damp cloth.

CAUTION: *Do not blow the brake dust out of the drums with compressed air or*

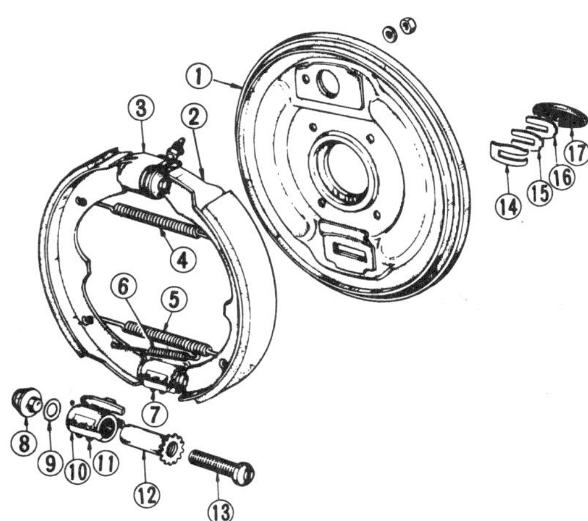

1. Brake backing plate
2. Brake shoe assembly
3. Wheel cylinder assembly
4. Brake shoe upper return spring
5. Brake shoe lower return spring
6. After shoe return spring
7. Adjuster assembly
8. Adjuster head
9. Adjuster head shim
10. Lock spring
11. Adjuster housing
12. Adjuster wheel
13. Adjuster screw
14. Retaining spring
15. Lockplate
16. Adjuster shim
17. Rubber boot

Exploded view of the front drum brake assembly

lung power. Brake linings contain asbestos, a known cancer causing substance. Dispose of the cloth after use.

Inspect the drum for cracks, deep grooves, roughness, scoring, or out-of-roundness. Replace any brake drum which is cracked.

Smooth any slight scores by polishing the friction surface with fine emery cloth. Heavy or extesnive scoring will cause excessive brake lining wear and should be removed from the brake drum through resurfacing. The maximum finished diameter of the brake drums must not exceed 10.059 in. The brake drum must be replaced if the diameter is 10.079 in. or greater.

Brake Shoes

INSPECTION

After removing the brake drum, inspect the brake shoes. If the lining is worn down so that the thickness is less than 0.0394 in., the brake shoes must be replaced.

NOTE: *This figure may disagree with your state inspection laws.*

If the brake lining is soaked with brake fluid, it must be replaced. Also the brake drum should be sanded with crocus cloth to remove all traces of brake fluid and the wheel cylinders rebuilt. Clean all grit off the brake drum before installing it.

If the brake lining is chipped, cracked, or otherwise damaged, it must be replaced with new lining.

NOTE: *Always replace the brake linings (shoes) in sets of two on both ends of the axle. Never replace just one shoe or both shoes on just one side.*

Check the condition of the shoes, retracting springs and hold-down springs for signs of overheating. If the shoes or springs have a slight blue color, this indicates overheating and replacement of the springs and shoes is recommended. The wheel cylinders should be rebuilt as a precaution against future problems.

REMOVAL AND INSTALLATION

1. Jack up the vehicle until the wheel which is to be serviced is off the ground. Remove the wheel and brake drum.

NOTE: *It is not absolutely essential to remove the hub assembly from the spindle, but it makes the job a great deal easier. If you can work with the hub in place, skip down to Step 7.*

2. Remove the hub dust cap.

3. Straighten the cotter pin and remove it from the spindle.

4. Unscrew the spindle nut and remove the adjusting cap, spindle nut, and spindle washer.

5. Wiggle the hub assembly until the outer bearing comes unseated and can be removed from the hub. Remove the outer bearing.

6. Pull the hub assembly off the spindle.

7. Unhook the upper, lower, and after shoe return springs, and remove them from the brake assembly.

8. Remove the brake shoes from the wheel cylinder at the top and the adjuster assembly at the bottom.

9. Clean the backing plate and adjuster assembly free of all dust and dirt. To remove the adjuster assembly for cleaning, remove the rubber boot at the back of the adjuster assembly and slide the adjuster shim, lockplate and retaining spring off the rear of the adjuster assembly.

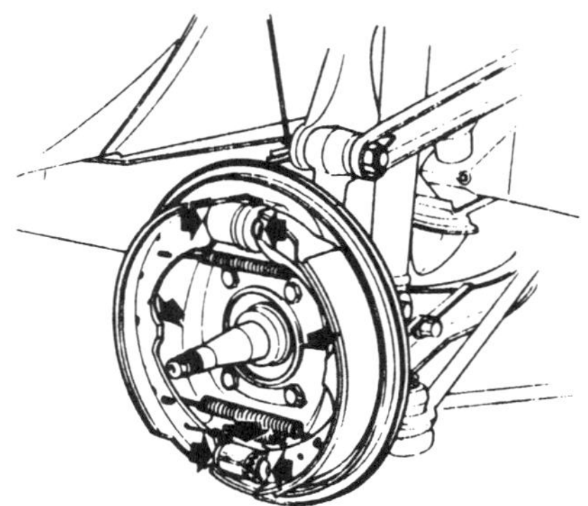

Greasing points on the front brakes indicated by the arrows

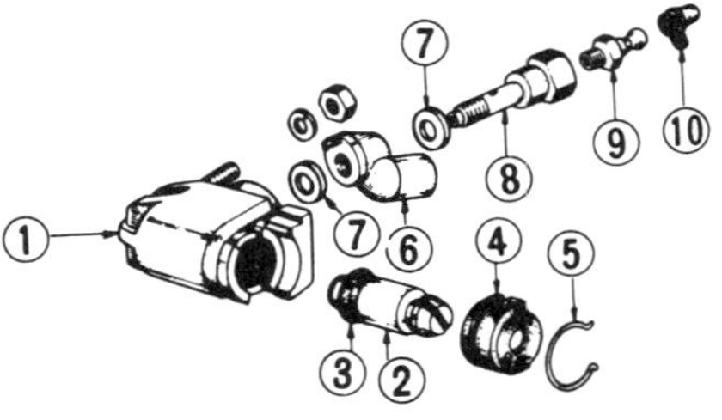

1. Wheel cylinder	6. Connector
housing	7. Packings
2. Piston	8. Connector bolt
3. Piston cup	9. Bleeder screw
4. Dust cover	10. Bleeder cap
5. Snap-ring	

Exploded view of the front wheel cylinder

10. Check the wheel cylinders by carefully pulling the lower edges of the wheel cylinder boots away from the cylinders. If there is leakage, the inside of the cylinder will be wet with fluid. If leakage exists, a wheel cylinder overhaul is in order. Do not delay, because brake failure could result.

NOTE: *A trace of fluid will be present, which acts as a lubricant for the wheel cylinder pistons.*

11. Apply brake grease to the adjuster assembly housing bore, adjuster wheel, and adjuster screw. Assemble the adjuster assembly with the adjuster screw turned all the way *in.* Apply brake grease to the sliding surfaces of the adjuster assembly, brake backing plate, and the retaining spring. Install the adjuster assembly to the backing plate in the reverse order of removal.

12. Before installing the brake shoes, apply brake grease to the notches into which the brake shoes fit on the wheel cylinder and adjuster mechanism, and brake shoe-to-backing plate contact surfaces.

13. Install the brake shoes in the reverse order of removal.

14. Install the hub, brake drum, and wheel in the reverse order of removal and adjust the wheel bearings and the brake shoes. Bleed the brakes. Lower the vehicle and roadtest it.

Wheel Cylinders

REMOVAL AND INSTALLATION

1. Jack up the wheel to be serviced.
2. Remove the wheel, brake drum, hub assembly, and brake shoes.
3. Disconnect the brake hose from the wheel cylinder.
4. Unscrew the wheel cylinder securing nut and remove the wheel cylinder from the brake backing plate.

5. Install the wheel cylinder in the reverse order of removal, assemble the remaining components, and bleed the brake hydraulic system.

OVERHAUL

NOTE: *This is one of those jobs where it is usually easier just to replace the part rather than rebuild it. If you decide to rebuild the wheel cylinders, be sure you get the correct parts for your truck. Datsun obtains parts from two manufacturers: Nabco and Tokico. Parts are not interchangeable. The name of the manufacturer is on the part.*

1. Remove the wheel cylinder from the backing plate.
2. Remove the snap-ring from the piston bore.
3. Remove the dust boot and take out the piston. Discard the piston cup. The dust boot can be reused, if necessary, but it is better to replace it.
4. Wash all of the components in clean brake fluid.
5. Inspect the piston and piston bore. Replace any components which are severely corroded, scored, or worn. The piston and piston bore can be polished lightly with crocus cloth. Move the crocus cloth around the piston bore; *not* in and out of the piston bore.
6. Wash the wheel cylinder and piston thoroughly in clean brake fluid, allowing them to remain lubricated for assembly.
7. Coat all of the new components to be installed in the wheel cylinder with clean brake fluid prior to assembly.
8. Assemble the wheel cylinder and install it in the reverse order of removal. Assemble

the remaining components and bleed the brake hydraulic system.

Wheel Bearings

See Chapter 1 for all maintenance and adjustment procedures for the front wheel bearings on 2 wheel drive models, and Chapter 7 for 4 wheel drive models.

REAR DRUM BRAKES

Brake Drums

INSPECTION

The inspection procedures for the rear brake drums are the same as those previously outlined for the front brake drums.

REMOVAL AND INSTALLATION

Remove the rear brake drum in the same manner as outlined for the front brake drums.

Brake Shoes

REMOVAL AND INSTALLATION

1. Jack up the vehicle until the wheel to be serviced is off the ground and remove the wheel and brake drum.

2. With a pair of pliers, remove the brake shoe hold-down anti-rattle spring retainers. Depress the retainer while rotating it 90° to align the slot in the retainer with the flanged end of the pin. Remove the retainers, springs, spring seats, and pins.

3. Open the brake shoes outward against

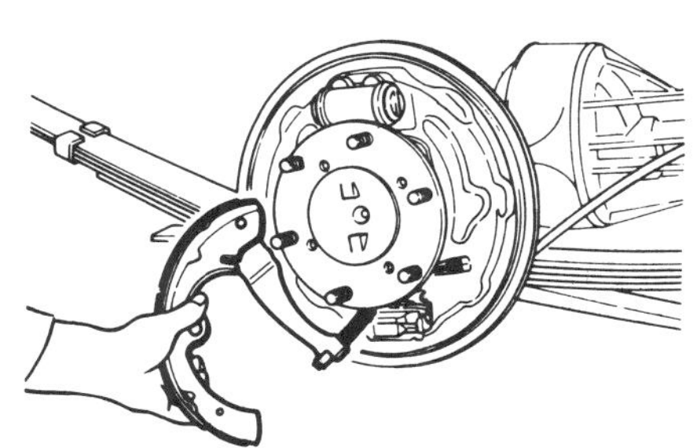

Remove the parking brake toggle lever from the secondary shoe

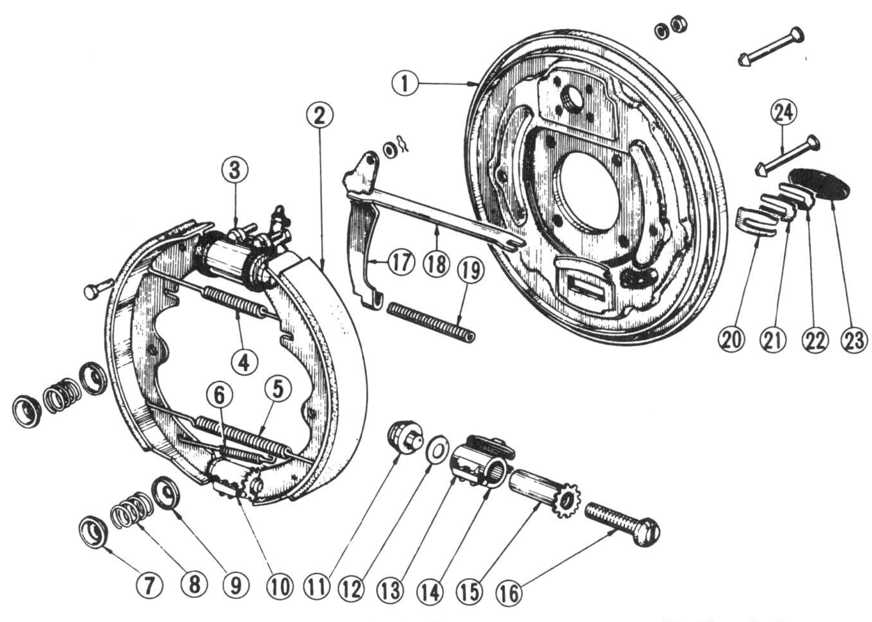

1. Brake backing plate	9. Spring seat	17. Toggle lever
2. Brake shoe	10. Adjuster assembly	18. Extension link
3. Wheel cylinder	11. Adjuster head	19. Return spring
4. Return upper spring	12. Adjuster head shim	20. Adjuster spring
5. Return lower spring	13. Lock-spring	21. Lockplate
6. After shoe return spring	14. Adjuster housing	22. Adjuster shim
7. Retainer	15. Adjuster wheel	23. Rubber boot
8. Anti-rattle spring	16. Adjuster screw	24. Anti-rattle pin

Exploded view of the rear brake assembly

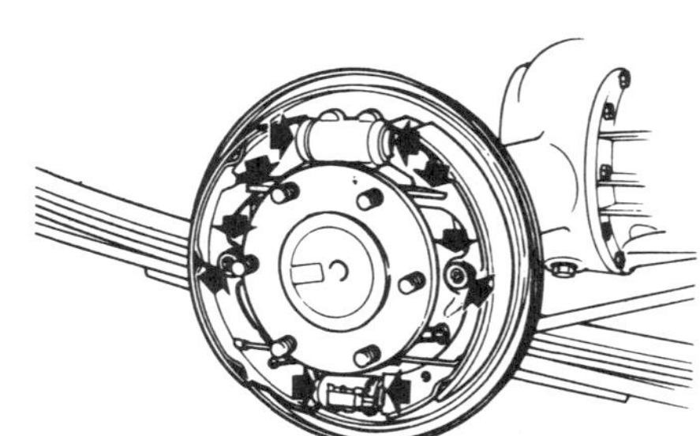

Greasing points on the rear brakes indicated by arrows

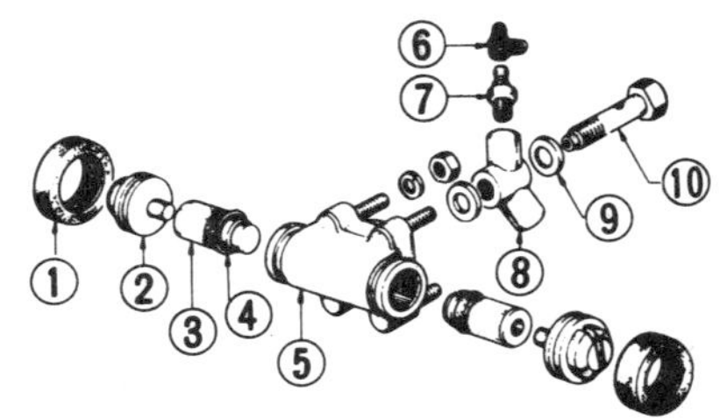

1. Dust cover
2. Piston head
3. Piston
4. Piston cup
5. Wheel cylinder housing
6. Bleeder cap
7. Bleeder screw
8. Connector
9. Washer
10. Connector bolt

Exploded view of the rear wheel cylinder

the return springs and remove the parking brake extension link.

4. Disconnect the brake shoe return springs.

5. Remove the brake shoes from the backing plate. The secondary (after) brake shoe must be disconnected from the parking brake toggle lever after withdrawing the toggle lever clevis pin.

6. Remove the rubber boot from behind the brake backing plate and slide the adjuster shim, lockplate, and adjustdr springs off the back of the adjuster assembly. Remove the adjuster assembly from the backing plate.

7. Clean the backing plate and adjuster assembly free of all dust and dirt.

8. Check the wheel cylinders as outlined in Step 10 of the front brake procedure.

9. Apply brake grease to the adjuster assembly housing bore, adjuster wheel, and adjuster screw. Assemble the adjuster assembly with the adjuster screw turned all the way *in*. Apply brake grease to the sliding surfaces of the adjuster assembly, brake backing plate, and the retaining spring. Install the adjuster assembly to the backing plate in the reverse order of removal.

10. Assemble the secondary (after) brake shoe to the parking brake toggle lever and adjust the clearance between the toggle lever and the brake shoe. Use the proper thickness toggle pin washer to adjust the clearance which should be 0.012 in. Toggle pin washers are available in the following thicknesses: 0.079, 0.091, 0.102, 0.114, and 0.126 in.

11. Before assembling the brake shoes to the backing plate apply brake grease to the following areas: the brake shoe grooves in the parking brake extension link, the inside surfaces of the anti-rattle (retaining) spring seats, and the contact surfaces between the brake backing plate and the brake shoes.

12. Assemble the brake shoes to the backing plate in the reverse order of removal.

13. Install the brake drum and the wheel.

14. Adjust the brakes. Bleed the brakes.

Wheel Cylinders
REMOVAL AND INSTALLATION

1. Jack up the vehicle, remove the wheel, brake drum, and brake shoes.

2. Disconnect the brake tube from the rear of the wheel cylinder.

3. Remove the wheel cylinder retaining nuts and remove the wheel cylinder from the backing plate.

4. Install the wheel cylinder and assemble the brake in the reverse order of removal and disassembly.

5. Bleed the brake hydraulic system.

OVERHAUL

The rear wheel cylinders are overhauled in the same manner as is outlined for the front wheel cylinders. The only difference is that the rear wheel cylinder bores are open at both ends and contain two pistons, piston heads, and dust boots.

PARKING BRAKE

Cable
ADJUSTMENT

1. Jack up the rear of the vehicle until the rear wheels clear the ground.

2. Adjust the rear brakes as outlined under "Brake System Adjustment."

3. Loosen the locknut at the parking cable lever assembly mounted on the driveshaft center bearing crossmember.

4. Turn the adjusting nut until the parking brake control lever operating stroke is between 3 in. and 4 in.

Brake Specifications
(All measurements are given in in.)

| Year | Master Cylinder Bore | Wheel Cylinder or Caliper Bore | | Piston-to-Bore Clearance | Brake Drum Diameter | | Minimum Lining Thickness | Brake Disc | |
		Front	Rear		Front	Rear		Minimum Thickness	Maximum Run-out
1970–71	0.750	0.813	0.813	0.006	10.00	10.00	0.0394	—	—
1972–75	0.750	0.750	0.688	0.006	10.00	10.00	0.0394	—	—
1976–77	0.750	0.750	0.750	0.006	10.00	10.00	0.0394	—	—
1978–79	0.813	2.125	0.625	0.006	—	10.00	0.08 disc) 0.06 (drum)	0.413	0.0059
1980–81	0.875	2.125	0.625	0.006	—	10.00	0.08 (disc) 0.06 (drum)	0.413	0.0059

5. Release the parking brake and make sure that the rear wheels turn freely with no drag.

6. Lower the vehicle.

REMOVAL AND INSTALLATION

1. Fully release the parking brake control lever.

2. Loosen the adjusting nut at the cable lever mounted to the frame crossmember.

3. Disconnect the cable from the control lever.

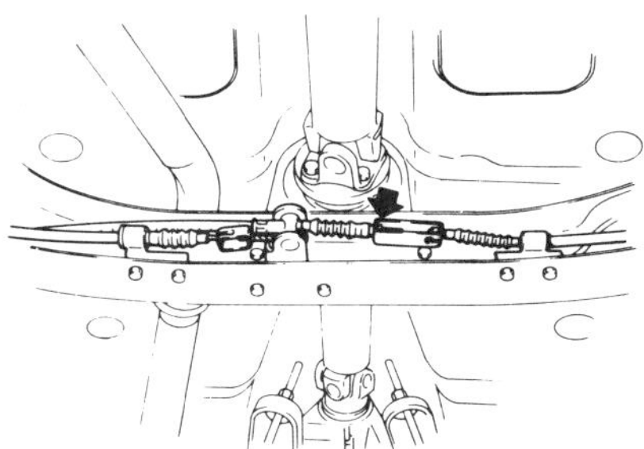

Parking brake adjusting nut, 1975 and later

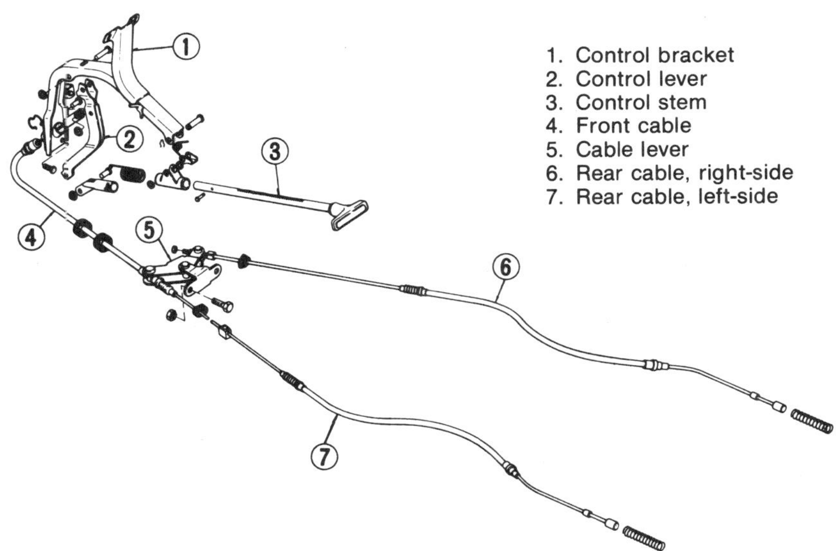

The parking brake system through 1974; later years similar

4. Remove the rear brake drums, and disconnect the parking brake cables from the parking brake toggle levers of the rear service brake assemblies.

5. Remove the lockplate, spring and clip, and pull the parking brake cable out toward the cable lever.

6. Remove the cotter pin at the cable lever and disconnect the cable.

7. Install the cables in the reverse order of removal. Apply a light coat of grease to the cables to make sure that they slide properly. Adjust the parking brake cables.

Body

You can repair most minor auto body damage yourself. Minor damage usually falls into one of several categories: (1) small scratches and dings in the paint that can be repaired without the use of body filler, (2) deep scratches and dents that require body filler, but do not require pulling, or hammering metal back into shape and (3) rust-out repairs. The repair sequences illustrated in this chapter are typical of these types of repairs. If you want to get involved in more complicated repairs including pulling or hammering sheet metal back into shape, you will probably need more detailed instructions. Chilton's *Minor Auto Body Repair, 2nd Edition* is a comprehensive guide to repairing auto body damage yourself.

TOOLS AND SUPPLIES

The list of tools and equipment you may need to fix minor body damage ranges from very basic hand tools to a wide assortment of specialized body tools. Most minor scratches, dings and rust holes can be fixed using an electric drill, wire wheel or grinder attachment, half-round plastic file, sanding block, various grades of sandpaper (#36, which is coarse through #600, which is fine) in both wet and dry types, auto body plastic, primer, touch-up paint, spreaders, newspaper and masking tape.

Most manufacturers of auto body repair products began supplying materials to professionals. Their knowledge of the best, most-used products has been translated into body repair kits for the do-it-yourselfer. Kits are available from a number of manufacturers and contain the necessary materials in the required amounts for the repair identified on the package.

Kits are available for a wide variety of uses, including:
- Rusted out metal
- All purpose kit for dents and holes
- Dents and deep scratches
- Fiberglass repair kit
- Epoxy kit for restyling.

Kits offer the advantage of buying what you need for the job. There is little waste and little chance of materials going bad from not being used. The same manufacturers also merchandise all of the individual products used—spreaders, dent pullers, fiberglass cloth, polyester resin, cream hardener, body filler, body files, sandpaper, sanding discs and holders, primer, spray paint, etc.

CAUTION: *Most of the products you will be using contain harmful chemicals, so be extremely careful. Always read the complete label before opening the containers. When* *you put them away for future use, be sure they are out of children's reach!*

Most auto body repair kits contain all the materials you need to do the job right in the kit. So, if you have a small rust spot or dent you want to fix, check the contents of the kit before you run out and buy any additional tools.

ALIGNING BODY PANELS

Doors

There are several methods of adjusting doors. Your vehicle will probably use one of those illustrated.

Whenever a door is removed and is to be reinstalled, you should matchmark the position of the hinges on the door pillars. The holes of the hinges and/or the hinge attaching points are usually oversize to permit alignment of doors. The striker plate is also moveable, through oversize holes, permitting up-and-down, in-and-out and fore-and-aft movement. Fore-and-aft movement is made by adding or subtracting shims from behind the striker and pillar post. The striker should be adjusted so that the door closes fully and remains closed, yet enters the lock freely.

DOOR HINGES

Don't try to cover up poor door adjustment with a striker plate adjustment. The gap on each side of the door should be equal and uniform and there should be no metal-to-metal contact as the door is opened or closed.

1. Determine which hinge bolts must be loosened to move the door in the desired direction.

2. Loosen the hinge bolt(s) just enough to allow the door to be moved with a padded pry bar.

3. Move the door a small amount and check the fit, after tightening the bolts. Be sure that there is no bind or interference with adjacent panels.

4. Repeat this until the door is properly positioned, and tighten all the bolts securely.

Hood, Trunk or Tailgate

As with doors, the outline of hinges should be scribed before removal. The hood and trunk can be aligned by loosening the hinge bolts in their slotted mounting holes and moving the hood or trunk lid as necessary.

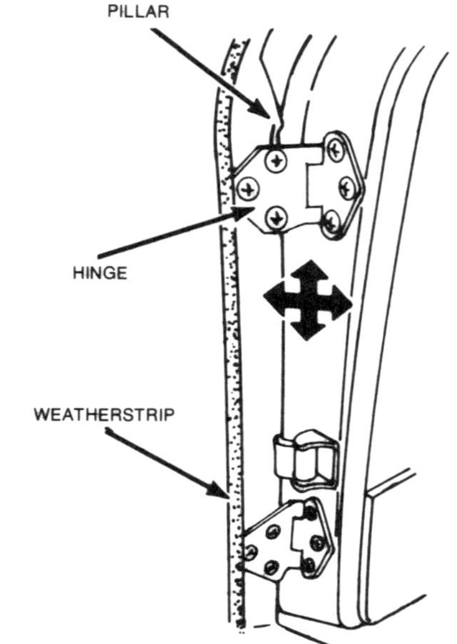

Door hinge adjustment

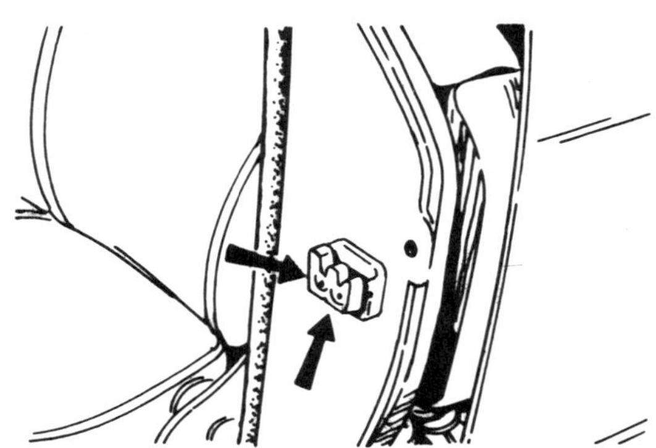

Move the door striker as indicated by arrows

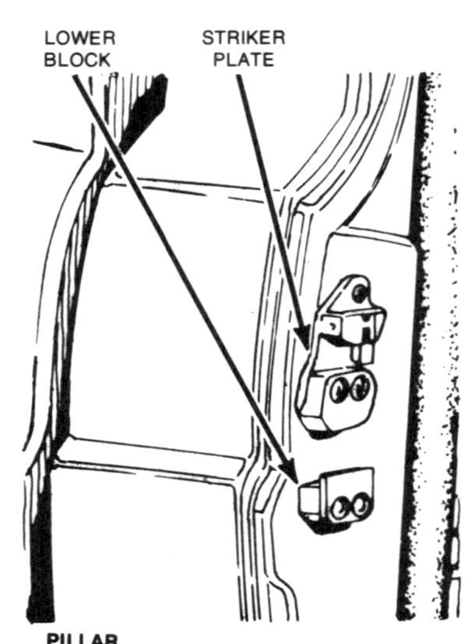

Striker plate and lower block

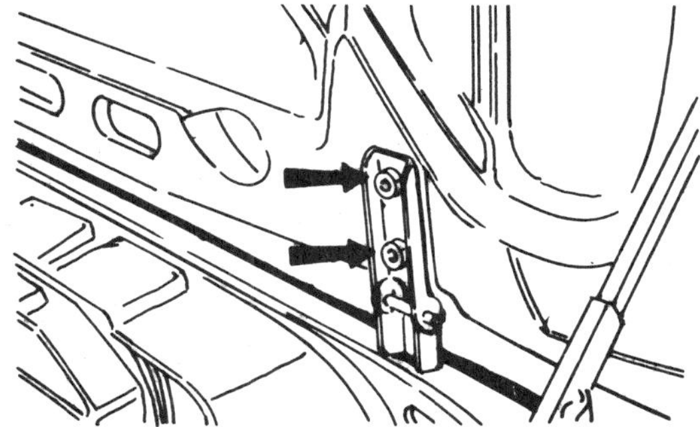

Loosen the hinge boots to permit fore-and-aft and horizontal adjustment

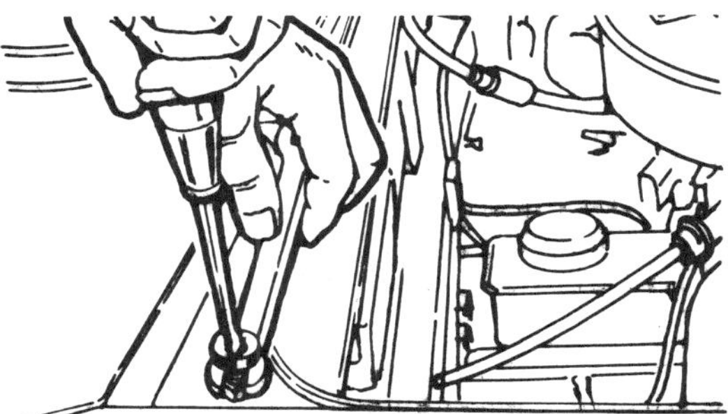

The hood is adjusted vertically by stop-screws at the front and/or rear

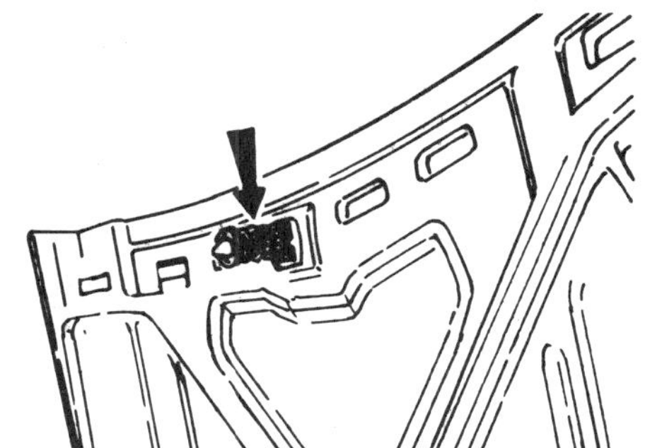

The hood pin can be adjusted for proper lock engagement

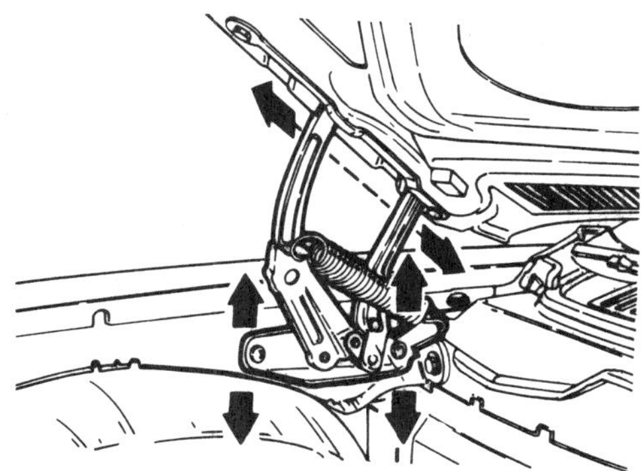

The height of the hood at the rear is adjusted by loosening the bolts that attach the hinge to the body and moving the hood up or down

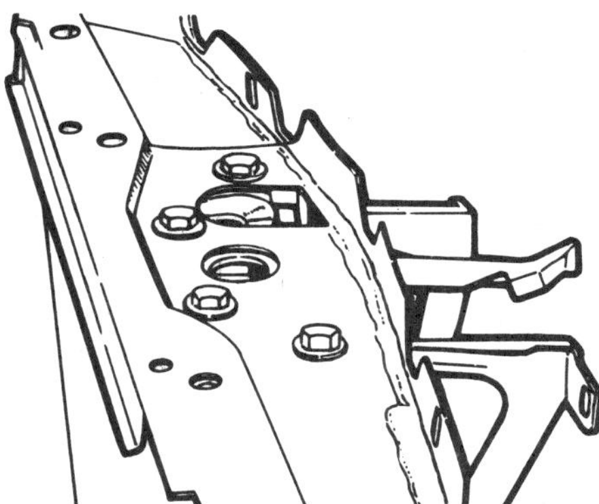

The base of the hood lock can also be re-positioned slightly to give more positive lock engagement

The hood and trunk have adjustable catch locations to regulate lock engagement. Bumpers at the front and/or rear of the hood provide a vertical adjustment and the hood lockpin can be adjusted for proper engagement.

The tailgate on the station wagon can be adjusted by loosening the hinge bolts in their slotted mounting holes and moving the tailgate on its hinges. The latchplate and latch striker at the bottom of the tailgate opening can be adjusted to stop rattle. An adjustable bumper is located on each side.

RUST, UNDERCOATING, AND RUSTPROOFING

Rust

Rust is an electrochemical process. It works on ferrous metals (iron and steel) from the inside out due to exposure of unprotected surfaces to air and moisture. The possibility of rust exists practically nationwide—anywhere humidity, industrial pollution or chemical salts are present, rust can form. In coastal areas, the problem is high humidity and salt air; in snowy areas, the problem is chemical salt (de-icer) used to keep the roads clear, and in industrial areas, sulphur dioxide is present in the air from industrial pollution and is changed to sulphuric acid when it rains. The rusting process is accelerated by high temperatures, especially in snowy areas, when vehicles are driven over slushy roads and then left overnight in a heated garage.

Automotive styling also can be a contributor to rust formation. Spot welding of panels creates small pockets that trap moisture and form an environment for rust formation. Fortunately, auto manufacturers have been working hard to increase the corrosion protection of their products. Galvanized sheet metal enjoys much wider use, along with the increased use of plastic and various rust retardant coatings. Manufacturers are also designing out areas in the body where rust-forming moisture can collect.

To prevent rust, you must stop it before it gets started. On new vehicles, there are two ways to accomplish this.

First, the car or truck should be treated with a commercial rustproofing compound. There are many different brands of franchised rustproofers, but most processes involve spraying a waxy "self-healing" compound under the chassis, inside rocker panels, inside doors and fender liners and similar places where rust is likely to form. Prices for a quality rustproofing job range from $100–$250, depending on the area, the brand name and the size of the vehicle.

Ideally, the vehicle should be rustproofed as soon as possible following the purchase. The surfaces of the car or truck have begun to oxidize and deteriorate during shipping. In addition, the car may have sat on a dealer's lot or on a lot at the factory, and once the rust has progressed past the stage of light, powdery surface oxidation rustproofing is not likely to be worthwhile. Professional rustproofers feel that once rust has formed, rustproofing will simply seal in moisture already present. Most franchised rustproofing operations offer a 3–5 year warranty against rust-through, but will not support that warranty if the rustproofing is not applied within three months of the date of manufacture.

Undercoating should not be mistaken for rustproofing. Undercoating is a black, tar-like substance that is applied to the underside of a vehicle. Its basic function is to deaden noises that are transmitted from under the car. It simply cannot get into the crevices and seams where moisture tends to collect. In fact, it may clog up drainage holes and ventilation passages. Some undercoatings also tend to crack or peel with age and only create more moisture and corrosion attracting pockets.

The second thing you should do immediately after purchasing the car is apply a paint sealant. A sealant is a petroleum based product marketed under a wide variety of brand names. It has the same protective properties as a good wax, but bonds to the paint with a chemically inert layer that seals it from the air. If air can't get at the surface, oxidation cannot start.

The paint sealant kit consists of a base coat and a conditioning coat that should be applied every 6–8 months, depending on the manufacturer. The base coat must be applied before waxing, or the wax must first be removed.

Third, keep a garden hose handy for your car in winter. Use it a few times on nice days during the winter for underneath areas, and it will pay big dividends when spring arrives. Spraying under the fenders and other areas which even car washes don't reach will help remove road salt, dirt and other build-ups which help breed rust. Adjust the nozzle to a high-force spray. An old brush will help break up residue, permitting it to be washed away more easily.

It's a somewhat messy job, but worth it in the long run because rust often starts in those hidden areas.

At the same time, wash grime off the door sills and, more importantly, the under portions of the doors, plus the tailgate if you have a station wagon or truck. Applying a coat of wax to those areas at least once before and once during winter will help fend off rust.

When applying the wax to the under parts of the doors, you will note small drain holes. These holes often are plugged with undercoating or dirt. Make sure they are cleaned out to prevent water build-up inside the doors. A small punch or penknife will do the job.

Water from the high-pressure sprays in car washes sometimes can get into the housings for parking and taillights, so take a close look. If they contain water merely loosen the retaining screws and the water should run out.

Repairing Scratches and Small Dents

Step 1. This dent (arrow) is typical of a deep scratch or minor dent. If deep enough, the dent or scratch can be pulled out or hammered out from behind. In this case no straightening is necessary

Step 2. Using an 80-grit grinding disc on an electric drill grind the paint from the surrounding area down to bare metal. This will provide a rough surface for the body filler to grab

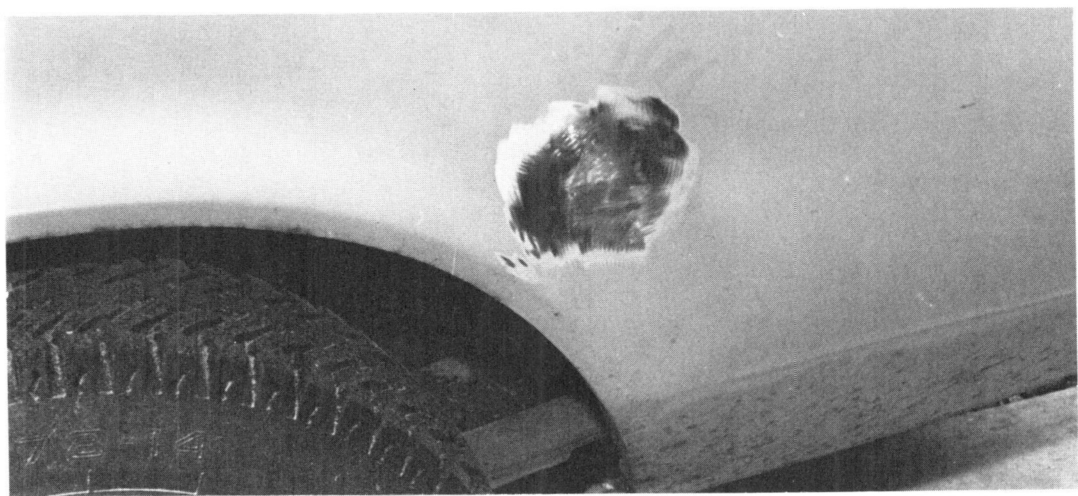

Step 3. The area should look like this when you're finished grinding

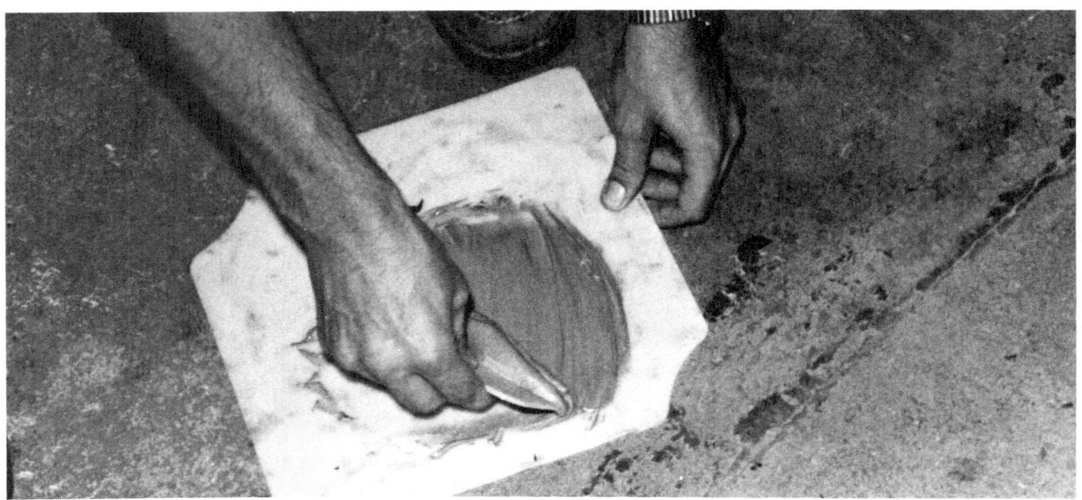

Step 4. Mix the body filler and cream hardener according to the directions

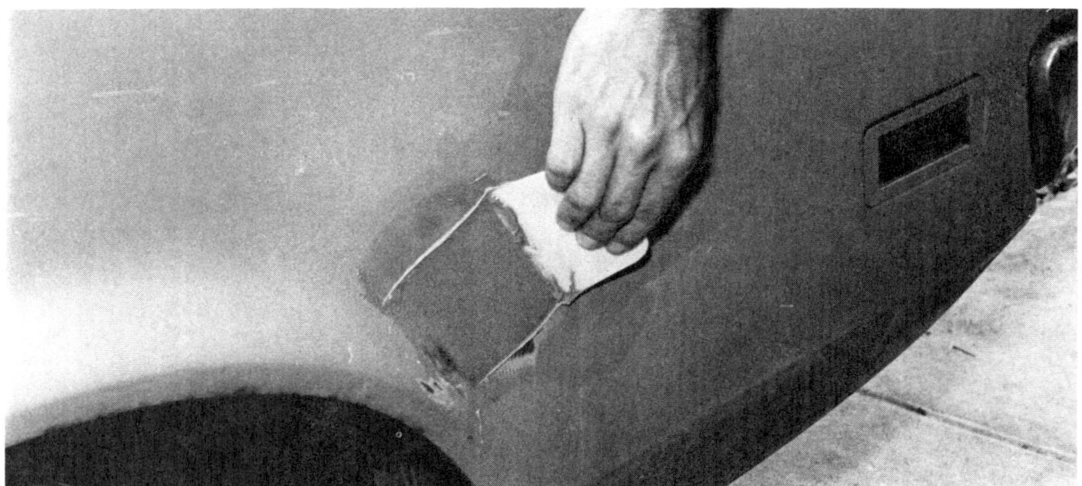

Step 5. Spread the body filler evenly over the entire area. Be sure to cover the area completely

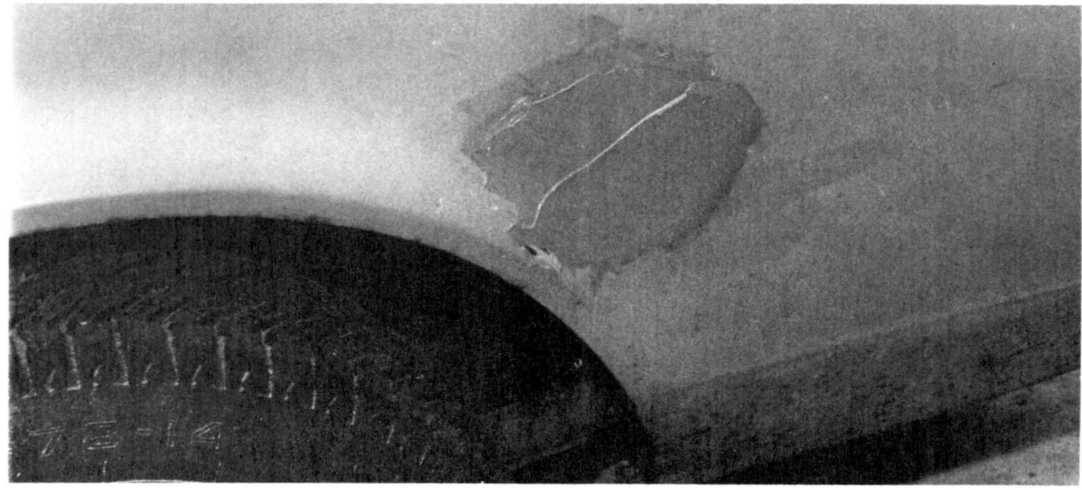

Step 6. Let the body filler dry until the surface can just be scratched with your fingernail

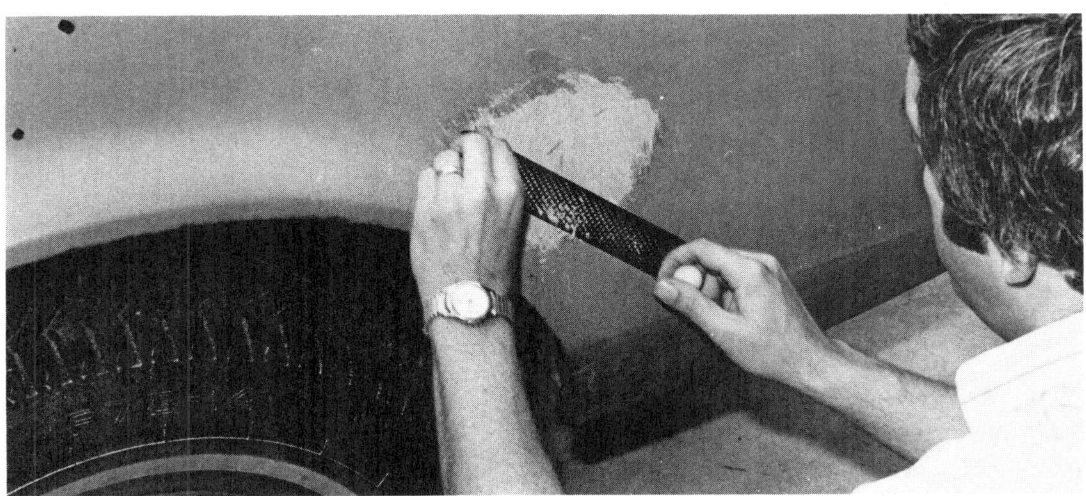

Step 7. Knock the high spots from the body filler with a body file

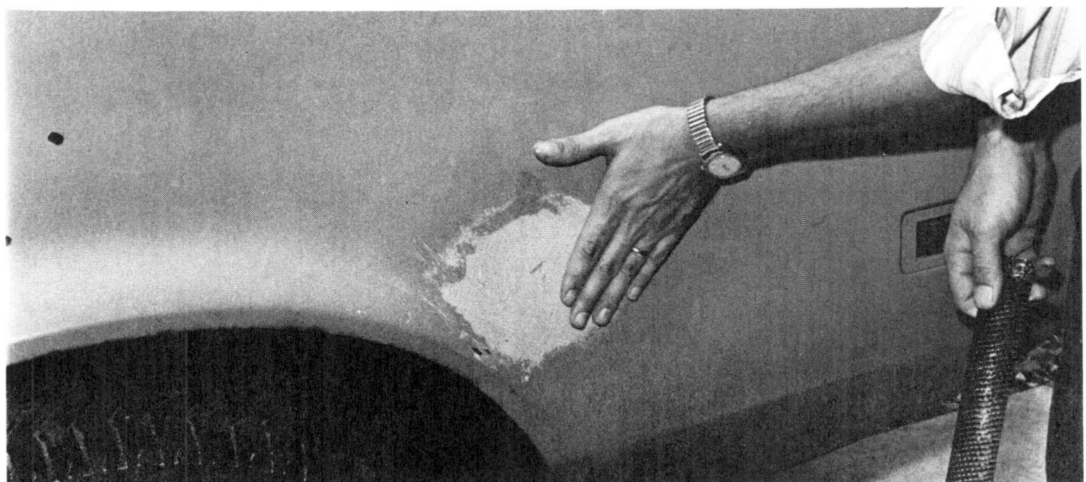

Step 8. Check frequently with the palm of your hand for high and low spots. If you wind up with low spots, you may have to apply another layer of filler

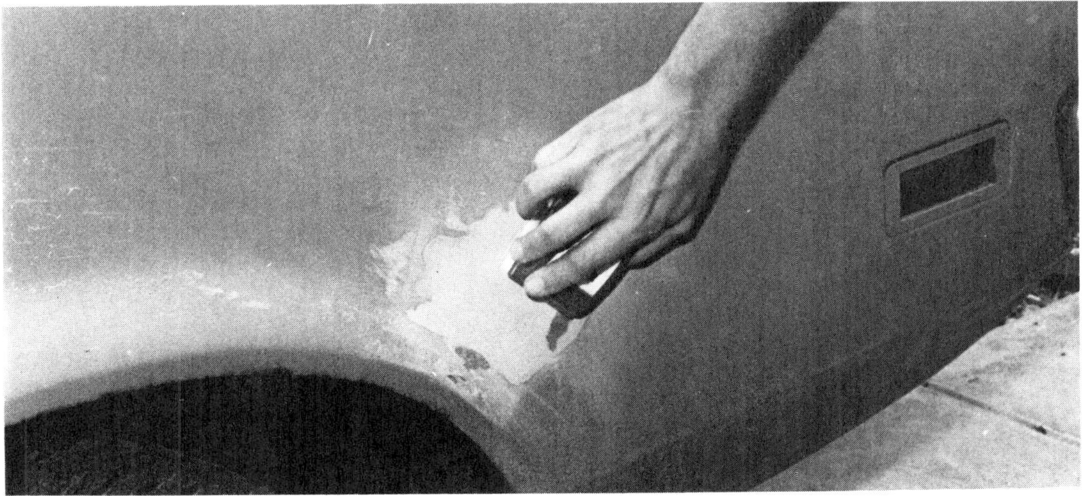

Step 9. Block sand the entire area with 320 grit paper

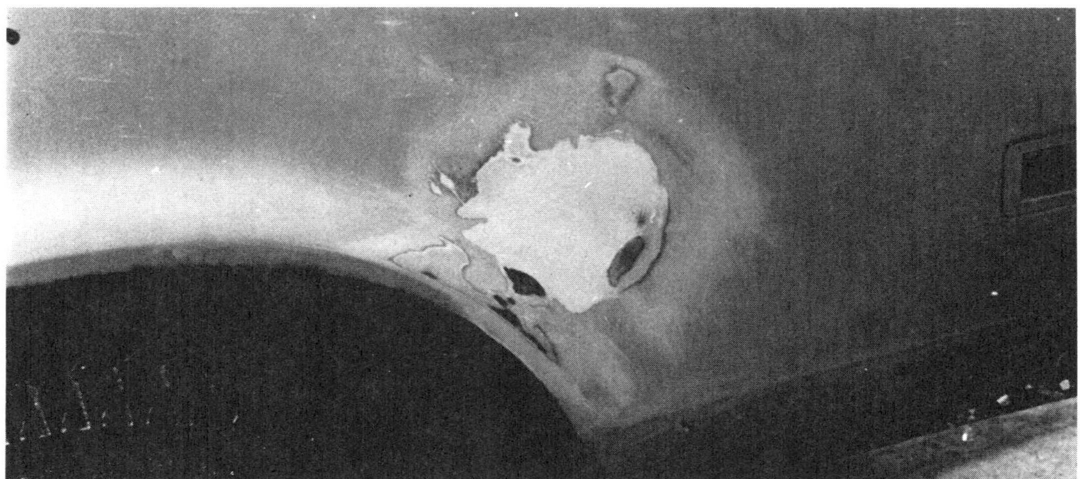

Step 10. When you're finished, the repair should look like this. Note the sand marks extending 2—3 inches out from the repaired area

Step 11. Prime the entire area with automotive primer

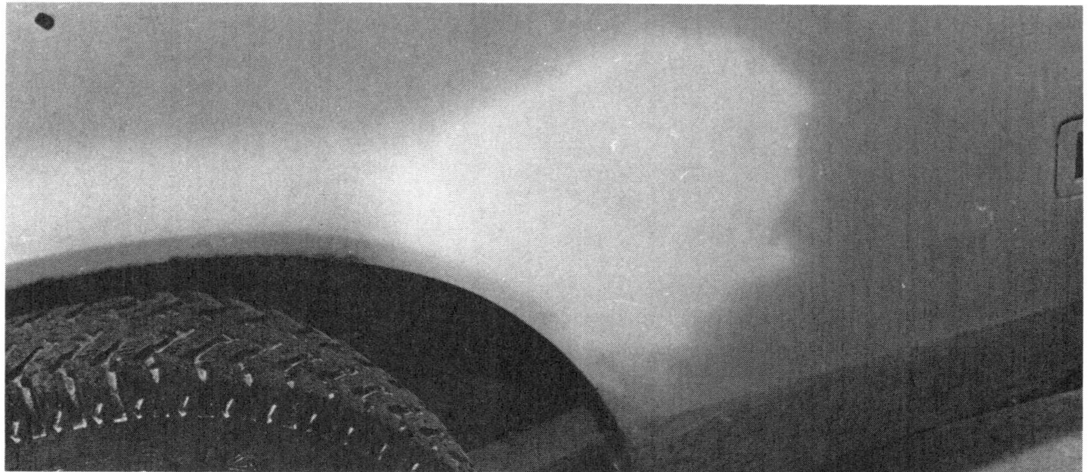

Step 12. The finished repair ready for the final paint coat. Note that the primer has covered the sanding marks (see Step 10). A repair of this size should be able to be spotpainted with good results

REPAIRING RUST HOLES

One thing you have to remember about rust: even if you grind away all the rusted metal in a panel, and repair the area with any of the kits available, *eventually* the rust will return. There are two reasons for this. One, rust is a chemical reaction that causes pressure under the repair from the inside out. That's how the blisters form. Two, the back side of the panel (and the repair) is wide open to moisture, and unpainted body filler acts like a sponge. That's why the best solution to rust problems is to remove the rusted panel and install a new one or have the rusted area cut out and a new piece of sheet metal welded in its place. The trouble with welding is the expense; sometimes it will cost more than the car or truck is worth.

One of the better solutions to do-it-yourself rust repair is the process using a fiberglass cloth repair kit (shown here). This will give a strong repair that resists cracking and moisture and is relatively easy to use. It can be used on large or small holes and also can be applied over contoured surfaces.

Step 1. Rust areas such as this are common and are easily fixed

Step 2. Grind away all traces of rust with a 24-grit grinding disc. Be sure to grind back 3—4 inches from the edge of the hole down to bare metal and be sure all traces of rust are removed

Step 3. Be sure all rust is removed from the edges of the metal. The edges must be ground back to un-rusted metal

Step 4. If you are going to use release film, cut a piece about 2″ larger than the area you have sanded. Place the film over the repair and mark the sanded area on the film. Avoid any unnecessary wrinkling of the film

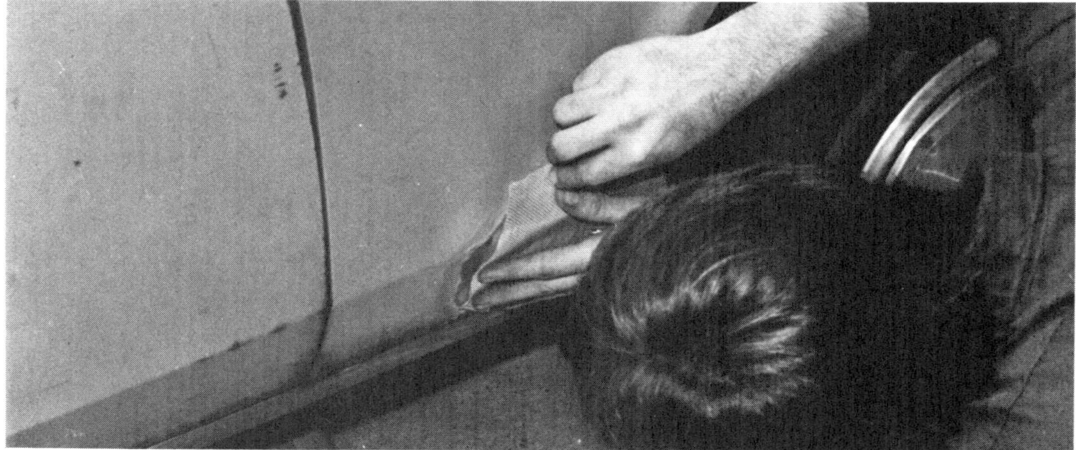

Step 5. Cut 2 pieces of fiberglass matte. One piece should be about 1″ smaller than the sanded area and the second piece should be 1″ smaller than the first. Use sharp scissors to avoid loose ends

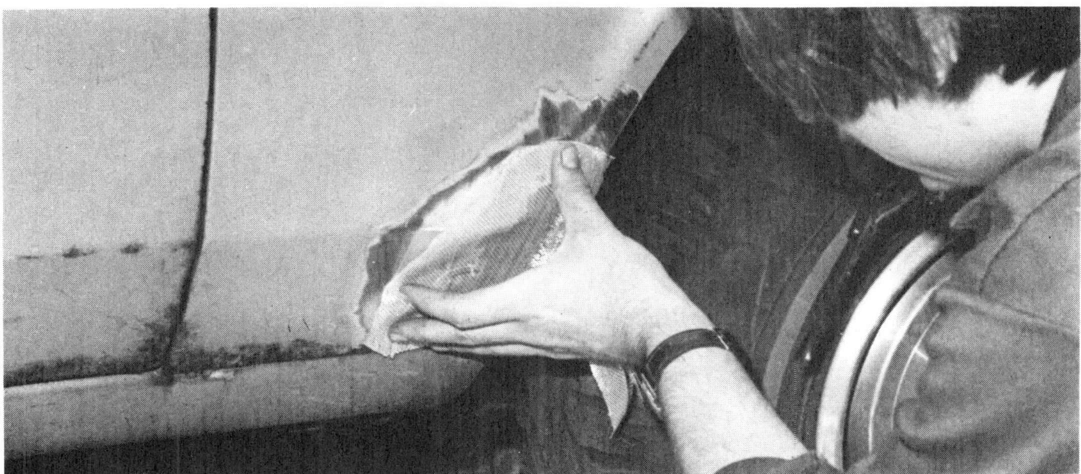

Step 6. Check the dimensions of the release film and cloth by holding them up to the repair area

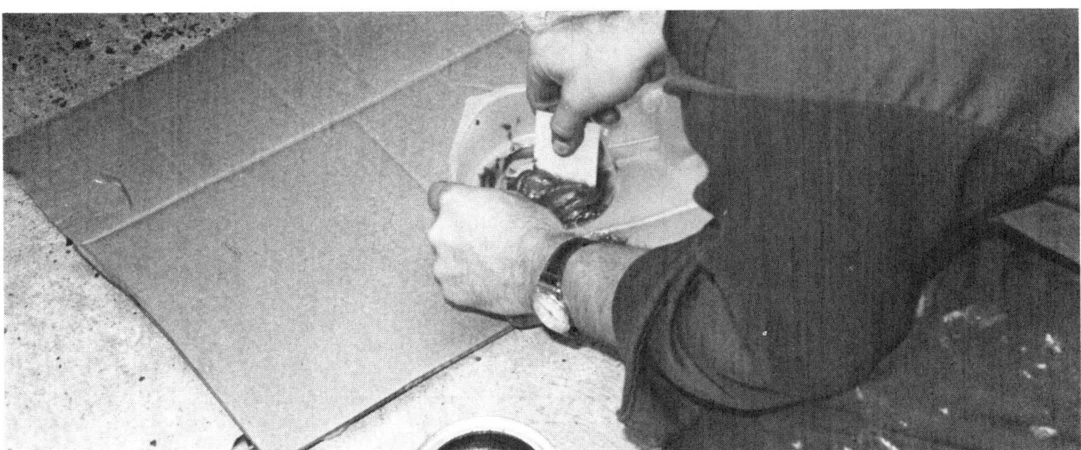

Step 7. Mix enough repair jelly and cream hardener in the mixing tray to saturate the fiberglass material or fill the repair area. Follow the directions on the container

Step 8. Lay the release sheet on a flat surface and spread an even layer of filler, large enough to cover the repair. Lay the smaller piece of fiberglass cloth in the center of the sheet and spread another layer of repair jelly over the fiberglass cloth. Repeat the operation for the larger piece of cloth. If the fiberglass cloth is not used, spread the repair jelly on the release film, concentrated in the middle of the repair

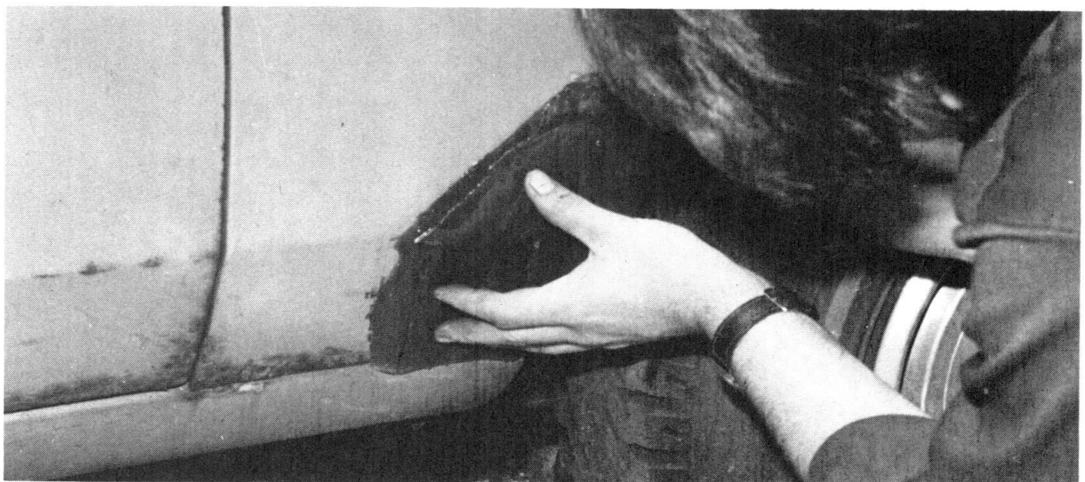

Step 9. Place the repair material over the repair area, with the release film facing outward

Step 10. Use a spreader and work from the center outward to smooth the material, following the body contours. Be sure to remove all air bubbles

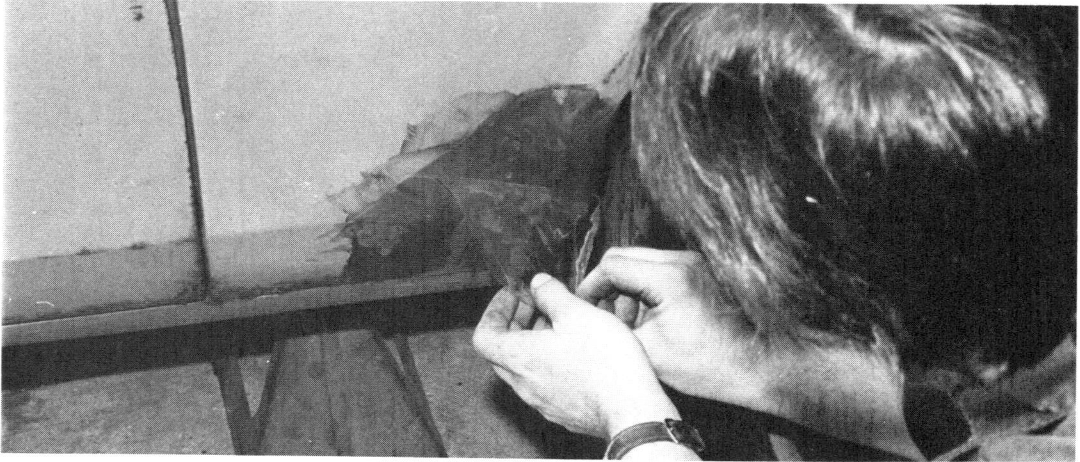

Step 11. Wait until the repair has dried tack-free and peel off the release sheet. The ideal working temperature is 65—90° F. Cooler or warmer temperatures or high humidity may require additional curing time

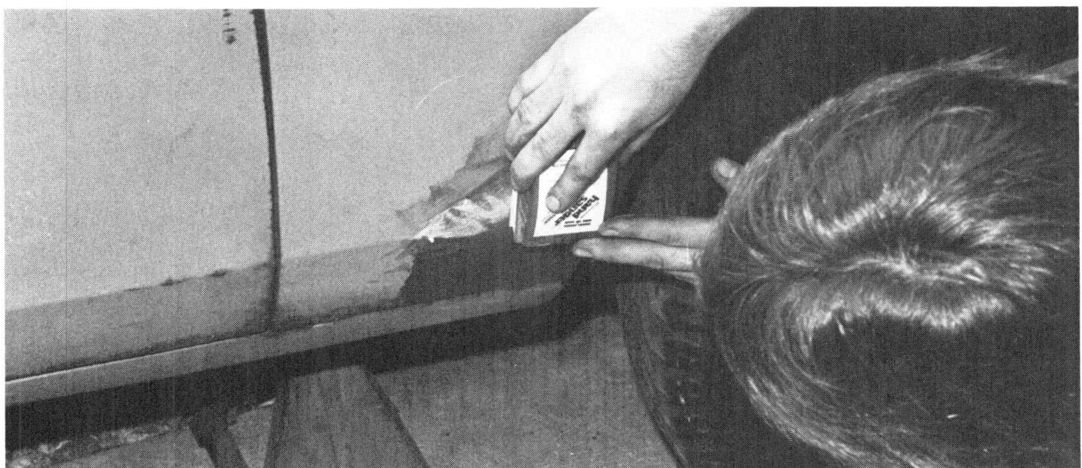

Step 12. Sand and feather-edge the entire area. The initial sanding can be done with a sanding disc on an electric drill if care is used. Finish the sanding with a block sander

Step 13. When the area is sanded smooth, mix some topcoat and hardener and apply it directly with a spreader. This will give a smooth finish and prevent the glass matte from showing through the paint

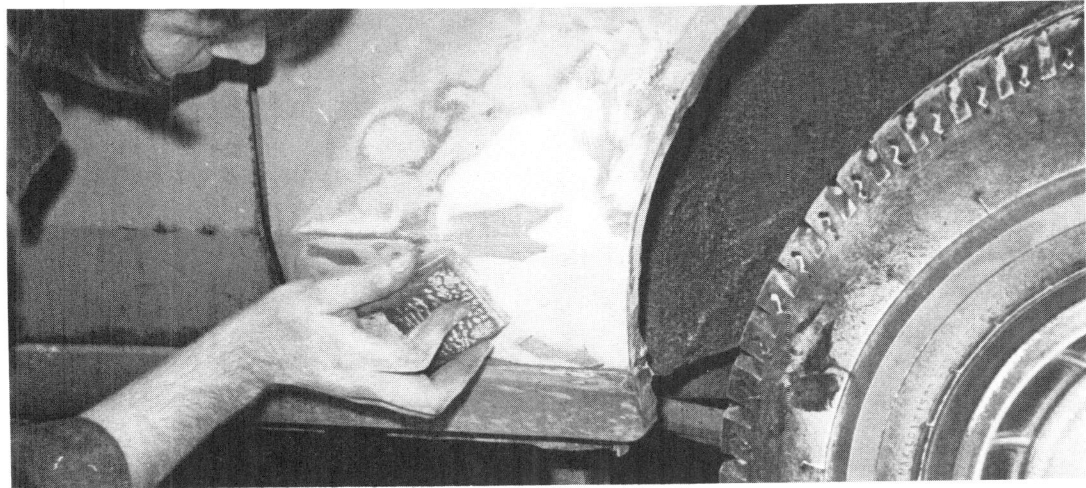

Step 14. Block sand the topcoat with finishing sandpaper

Step 15. To finish this repair, grind out the surface rust along the top edge of the rocker panel

Step 16. Mix some more repair jelly and cream hardener and apply it directly over the surface

Step 17. When it dries tack-free, block sand the surface smooth

Step 18. If necessary, mask off adjacent panels and spray the entire repair with primer. You are now ready for a color coat

AUTO BODY CARE

There are hundreds—maybe thousands—of products on the market, all designed to protect or aid your car's finish in some manner. There are as many different products as there are ways to use them, but they all have one thing in common—the surface must be clean.

Washing

The primary ingredient for washing your car is water, preferably "soft" water. In many areas of the country, the local water supply is "hard" containing many minerals. The little rings or film that is left on your car's surface after it has dried is the result of "hard" water.

Since you usually can't change the local water supply, the next best thing is to dry the surface before it has a chance to dry itself.

Into the water you usually add soap. Don't use detergents or common, coarse soaps. Your car's paint never truly dries out, but is always evaporating residual oils into the air. Harsh detergents will remove these oils, causing the paint to dry faster than normal. Instead use warm water and a non-detergent soap made especially for waxed surfaces or a liquid soap made for waxed surfaces or a liquid soap made for washing dishes by hand.

Other products that can be used on painted surfaces include baking soda or plain soda water for stubborn dirt.

Wash the car completely, starting at the top, and rinse it completely clean. Abrasive grit should be loaded off under water pressure; scrubbing grit off will scratch the finish. The best washing tool is a sponge, cleaning mitt or soft towel. Whichever you choose, replace it often as each tends to absorb grease and dirt.

Other ways to get a better wash include:

• Don't wash your car in the sun or when the finish is hot.

• Use water pressure to remove caked-on dirt.

• Remove tree-sap and bird effluence immediately. Such substances will eat through wax, polish and paint.

One of the best implements to dry your car is a turkish towel or an old, soft bath towel. Anything with a deep nap will hold any dirt in suspension and not grind it into the paint.

Harder cloths will only grind the grit into the paint making more scratches. Always start drying at the top, followed by the hood and trunk and sides. You'll find there's always more dirt near the rocker panels and wheelwells which will wind up on the rest of the car if you dry these areas first.

Cleaners, Waxes and Polishes

Before going any farther you should know the function of various products.

Cleaners—remove the top layer of dead pigment or paint.

Rubbing or polishing compounds—used to remove stubborn dirt, get rid of minor scratches, smooth away imperfections and partially restore badly weathered paint.

Polishes—contain no abrasives or waxes; they shine the paint by adding oils to the paint.

Waxes—are a protective coating for the polish.

CLEANERS AND COMPOUNDS

Before you apply any wax, you'll have to remove oxidation, road film and other types of pollutants that washing alone will not remove.

The paint on your car never dries completely. There are always residual oils evaporating from the paint into the air. When enough oils are present in the paint, it has a healthy shine (gloss). When too many oils evaporate the paint takes on a whitish cast known as oxidation. The idea of polishing and waxing is to keep enough oil present in the painted surface to prevent oxidation; but when it occurs, the only recourse is to remove the top layer of "dead" paint, exposing the healthy paint underneath.

Products to remove oxidation and road film are sold under a variety of generic names—polishes, cleaner, rubbing compound, cleaner/polish, polish/cleaner, self-polishing wax, pre-wax cleaner, finish restorer and many more. Regardless of name there are two types of cleaners—abrasive cleaners (sometimes called polishing or rubbing compounds) that remove oxidation by grinding away the top layer of "dead" paint, or chemical cleaners that dissolve the "dead" pigment, allowing it to be wiped away.

Abrasive cleaners, by their nature, leave thousands of minute scratches in the finish, which must be polished out later. These should only be used in extreme cases, but are usually the only thing to use on badly oxidized paint finishes. Chemical cleaners are much milder but are not strong enough for severe cases of oxidation or weathered paint.

The most popular cleaners are liquid or paste abrasive polishing and rubbing compounds. Polishing compounds have a finer abrasive grit for medium duty work. Rubbing compounds are a coarser abrasive and for heavy duty work. Unless you are familiar with how to use compounds, be very careful. Excessive rubbing with any type of compound or cleaner can grind right through the paint to primer or bare metal. Follow the directions on the container—depending on type, the cleaner may or may not be OK for your paint. For example, some cleaners are not formulated for acrylic lacquer finishes.

When a small area needs compounding or heavy polishing, it's best to do the job by hand. Some people prefer a powered buffer for large areas. Avoid cutting through the paint along styling edges on the body. Small, hand operations where the compound is applied and rubbed using cloth folded into a thick ball allow you to work in straight lines along such edges.

To avoid cutting through on the edges when using a power buffer, try masking tape. Just cover the edge with tape while using power. Then finish the job by hand with the tape removed. Even then work carefully. The paint tends to be a lot thinner along the sharp ridges stamped into the panels.

Whether compounding by machine or by hand, only work on a small area and apply the compound sparingly. If the materials are spread too thin, or allowed to sit too long, they dry out. Once dry they lose the ability to deliver a smooth, clean finish. Also, dried out polish tends to cause the buffer to stick in one spot. This in turn can burn or cut through the finish.

WAXES AND POLISHES

Your car's finish can be protected in a number of ways. A cleaner/wax or polish/cleaner followed by wax or variations of each all provide good results. The two-step approach (polish followed by wax) is probably slightly better but consumes more time and effort. Properly fed with oils, your paint should never need cleaning, but despite the best polishing job, it won't last unless it's protected with wax. Without wax, polish must be renewed at least once a month to prevent oxidation. Years ago (some still swear by it today), the best wax was made from the Brazilian palm, the Carnuba, favored for its vegetable base and high melting point. However, modern synthetic waxes are harder, which means they protect against moisture better, and chemically inert silicone is used for a long lasting protection. The only problem with silicone wax is that it penetrates all layers of paint. To repaint or touch up a panel or car protected by silicone wax, you have to completely strip the finish to avoid "fish-eyes."

Under normal conditions, silicone waxes will last 4–6 months, but you have to be careful of wax build-up from too much waxing. Too thick a coat of wax is just as bad as no wax at all; it stops the paint from breathing.

Combination cleaners/waxes have become popular lately because they remove the old layer of wax plus light oxidation, while putting on a fresh coat of wax at the same time. Some cleaners/waxes contain abrasive cleaners which require caution, although many cleaner/waxes use a chemical cleaner.

Applying Wax or Polish

You may view polishing and waxing your car as a pleasant way to spend an afternoon, or as a boring chore, but it has to be done to keep the paint on your car. Caring for the paint doesn't require special tools, but you should follow a few rules.

1. Use a good quality wax.

2. Before applying any wax or polish, be sure the surface is completely clean. Just because the car looks clean, doesn't mean it's ready for polish or wax.

3. If the finish on your car is weathered, dull, or oxidized, it will probably have to be compounded to remove the old or oxidized paint. If the paint is simply dulled from lack of care, one of the non-abrasive cleaners known as polishing compounds will do the trick. If the paint is severely scratched or really dull, you'll probably have to use a rubbing compound to prepare the finish for waxing. If you're not sure which one to use, use the polishing compound, since you can easily ruin the finish by using too strong a compound.

4. Don't apply wax, polish or compound in direct sunlight, even if the directions on the can say you can. Most waxes will not cure properly in bright sunlight and you'll probably end up with a blotchy looking finish.

5. Don't rub the wax off too soon. The result will be a wet, dull looking finish. Let the wax dry thoroughly before buffing it off.

6. A constant debate among car enthusiasts is how wax should be applied. Some maintain pastes or liquids should be applied in a circular motion, but body shop experts have long thought that this approach results in barely detectable circular abrasions, especially on cars that are waxed frequently. They advise rubbing in straight lines, especially if any kind of cleaner is involved.

7. If an applicator is not supplied with the wax, use a piece of soft cheesecloth or very soft lint-free material. The same applies to buffing the surface.

SPECIAL SURFACES

One-step combination cleaner and wax formulas shouldn't be used on many of the special surfaces which abound on cars. The one-step materials contain abrasives to achieve a clean surface under the wax top coat. The abrasives are so mild that you could clean a car every week for a couple of years without fear of rubbing through the paint. But this same level of abrasiveness might, through repeated use, damage decals used for special trim effects. This includes wide stripes, wood-grain trim and other appliques.

Painted plastics must be cleaned with care. If a cleaner is too aggressive it will cut through the paint and expose the primer. If bright trim such as polished aluminum or chrome is painted, cleaning must be performed with even greater care. If rubbing compound is being used, it will cut faster than polish.

Abrasive cleaners will dull an acrylic finish. The best way to clean these newer finishes is with a non-abrasive liquid polish. Only dirt and oxidation, not paint, will be removed.

Taking a few minutes to read the instructions on the can of polish or wax will help prevent making serious mistakes. Not all preparations will work on all surfaces. And some are intended for power application while others will only work when applied by hand.

Don't get the idea that just pouring on some polish and then hitting it with a buffer will suffice. Power equipment speeds the operation. But it also adds a measure of risk. It's very easy to damage the finish if you use the wrong methods or materials.

Caring for Chrome

Read the label on the container. Many products are formulated specifically for chrome, but others contain abrasives that will scratch the chrome finish. If it isn't recommended for chrome, don't use it.

Never use steel wool or kitchen soap pads to clean chrome. Be careful not to get chrome cleaner on paint or interior vinyl surfaces. If you do, get it off immediately.

Troubleshooting

This section is designed to aid in the quick, accurate diagnosis of automotive problems. While automotive repairs can be made by many people, accurate troubleshooting is a rare skill for the amateur and professional alike.

In its simplest state, troubleshooting is an exercise in logic. It is essential to realize that an automobile is really composed of a series of systems. Some of these systems are interrelated; others are not. Automobiles operate within a framework of logical rules and physical laws, and the key to troubleshooting is a good understanding of all the automotive systems.

This section breaks the car or truck down into its component systems, allowing the problem to be isolated. The charts and diagnostic road maps list the most common problems and the most probable causes of trouble. Obviously it would be impossible to list every possible problem that could happen along with every possible cause, but it will locate MOST problems and eliminate a lot of unnecessary guesswork. The systematic format will locate problems within a given system, but, because many automotive systems are interrelated, the solution to your particular problem may be found in a number of systems on the car or truck.

USING THE TROUBLESHOOTING CHARTS

This book contains all of the specific information that the average do-it-yourself mechanic needs to repair and maintain his or her car or truck. The troubleshooting charts are designed to be used in conjunction with the specific procedures and information in the text. For instance, troubleshooting a point-type ignition system is fairly standard for all models, but you may be directed to the text to find procedures for troubleshooting an individual type of electronic ignition. You will also have to refer to the specification charts throughout the book for specifications applicable to your car or truck.

TOOLS AND EQUIPMENT

The tools illustrated in Chapter 1 (plus two more diagnostic pieces) will be adequate to troubleshoot most problems. The two other tools needed are a voltmeter and an ohmmeter. These can be purchased separately or in combination, known as a VOM meter.

In the event that other tools are required, they will be noted in the procedures.

Troubleshooting Diesel Engine Problems
See chapters 2, 3, and 4 for more information and service procedures

NOTE: *The following troubleshooting procedures cover problems usually associated with diesel engines. These problems common to both gasoline and diesel engines are covered in the gasoline engine troubleshooting procedures later in this chapter.*

Index to Problems

Problem: Symptom	Begin at Specific Diagnosis, Number
Fuel System:	Section 1
Engine Starting Difficulty	1.1
Feed pump does not feed fuel	1.2
Injection pump does not feed fuel	1.3
Incorrect injection timing	1.4
Defective injection nozzles	
Engine Operating Instability:	
Engine shuts off immediately after starting	1.5
Uneven idling	1.6
Engine will not reach maximum rated speed	1.7
Engine exceeds maximum rated speed	1.8
Loss of power	1.9
Engine Knock:	
Associated with exhaust gas problems	1.10
Not associated with exhaust gas problems	1.11
Engine Mechanical:	Section 2
Engine Starting Difficulty	2.1
Unusual Noises	2.2
Engine Operating Instability	2.3
Loss of Power	2.4
Exhaust gas Problem	2.5
Engine Shut-Off	2.6
Loss of Oil Pressure	2.7
Oil Leakage	2.8
Compression Pressure Leakage	2.9

Section 1—Fuel System

Test and Procedure	Results and Indication	Proceed To
1.1a —Check for pressure at the outlet of the feed pump	If pressure exists, there is a clog in the supply line. Clean or replace it. If there is little or no pressure at the outlet, the filter is clogged. Clean or replace the filter If the filter is clear, the feed pump piston is inoperative. Replace it.	**1.1b**
1.1b —Check the feed pump valves	If the inlet and outlet valves do not operate, the check valve or spring is broken. Replace it.	**1.2a**
1.2a —Check for fuel leakage at the overflow or return line	A clogged filter can result in high pressure causing leakage. Replace the filter.	**1.2b**
1.2b —Check for fuel in the filter leaking at the overflow valve	If leakage is found, the overflow valve is damaged. Replace it.	**1.2c**
1.2c —Check for leakage at the injection pump overflow valve	If leakage is found, it is caused by: damaged overflow valve, sticking plunger, or sticking delivery valve. Replace the defective part(s).	**1.2d**
1.2d —Check the injection pump plunger feed pressures	If pressure at the plungers is low, replace the plunger(s).	**1.2e**

Test and Procedure	Results and Indication	Proceed To
1.2e —Check to make sure the injection pump is operating	An inoperative pump is caused by: a damaged or missing shaft key, or a damaged drive gear train.	**1.3a**
1.3a —Check that the pump timing marks are correctly aligned in the gear train	Incorrect timing marks alignment must be corrected.	**1.3b**
1.3b —Check that the injection pump is properly mounted	Remove and install the pump correctly	**1.4a**
1.4a —Install an injection nozzle on a tester and make sure that fuel is continuously ejected	A broken or intermittent stream is caused by a damaged spring or a sticking nozzle needle	**1.4b**
1.4b —With the nozzle on the tester as in 1.4a, check that shutoff is clean with no dribble or afterdrip	Dribble is caused by a defective nozzle valve seat. Replace the nozzle.	**1.4c**
1.4c —Using a tester, check injection pressure	Low pressure is a result of a weak spring. Replace the spring or adjust the initial injection pressure.	**1.5a**
1.5a —see 1.2a	Proceed as in 1.2a	**1.5b**
1.5b —Check for water in the fuel	Drain and clean the tank	**1.5c**
1.5c —Check for air in the fuel lines	Air can be introduced through a damaged fuel inlet line, a loose inlet line connector or a damaged gasket	**1.5d**
1.5d —Check for insufficient fuel feed	Insufficient fuel feed is caused by: a damaged feed pump, a clogged tank vent, or a clogged filter. Replace or repair as necessary.	**1.6a**
1.6a —Check the control rack action for smooth operation	Uneven control rack operation is caused by: a sticking plunger, improper meshing of the rack and pinion, poor seating of the plunger spring, insufficient clearance between the plunger and lower spring seat, or an overly tight delivery valve holder. Replace or adjust as necessary.	**1.6b**
1.6b —Check that the injection pump discharge is uniform	If the output is uneven, adjust as necessary	**1.6c**
1.6c —Check that the injection pump discharge volume is adequate	An inadequate discharge volume is caused by a worn plunger or a broken plunger spring	**1.6d**
1.6d —Check for even low speed engine performance	If the engine performs unevenly or erratically at low speed only, a worn feed pump piston or defective feed pump valve is the cause.	**1.6e**
1.6e —Check for smooth engine operation throughout the operating range	This problem is usually caused by mechanical governor defects such as: a defective low speed spring, defective damper spring, or excessive friction among moving parts. Replace the defective parts.	**1.6f**
1.6f —Check the injectors on a tester	Improper nozzle operation should be corrected accordingly	**1.7a**

Test and Procedure	Results and Indications	Proceed To
1.7a —Check the operating governor	A broken or weak spring in the governor will prevent full speed operation.	1.7b
1.7b —Check the injectors on a tester for a drop in injector output	A drop in output is caused by a sticking needle or a dirty nozzle. Replace or clean as necessary.	1.8a
1.8a —Check the injection pump for proper rack and pinion action	A catching or dirty rack and pinion will cause overspeeding.	1.8b
1.8b —Check the governor adjustment	An improperly adjusted governor will cause overspeeding. Adjust.	1.9a
1.9a —Check the injection pump output	Low output can be caused by: Incorrect adjustment—Adjust Loose delivery valve—Tighten Broken delivery valve seal—Replace Poor valve seat contact—Replace Broken/weak delivery valve spring —Replace	1.9b
1.9b —Check for unusual noise at the injection pump	A noisy pump is an indication of a broken plunger spring	1.9c
1.9c —Check plunger operation	A sticking injection pump plunger will cause power loss. Replace	1.9d
1.9d —Check the injection timer	A lag in injection timing is caused by large clearances in the timer due to wear. Replace.	1.9e
1.9e —Check for air or water in the fuel	Bleed the air or drain the fuel and clean the tank and lines	1.9f
1.9f —Check the injection timing	Readjust timing if necessary	1.10a
1.10a—Check the initial injection timing	Adjust if necessary	1.10b
1.10b—Check the injection pressure	High pressure will cause knock. Adjust as necessary	1.10c
1.10c—Check the injector nozzle	A clogged nozzle causes knock. Clean or replace the nozzle	1.11a
1.11a—Check the injection pump output and timing	Excessive output, coupled with incorrect timing causes knock. Adjust as necessary	1.11b
1.11b—Check the delivery valve seat	Replace a defective seat	1.11c
1.11c—Check the pump plungers	Replace badly worn plungers	1.11d
1.11d—Check the injector opening pressure on a tester	Adjust as necessary	1.11e
1.11e—Check the injector	Replace a broken nozzle spring or sticking needle.	

Section 2—Engine Mechanical

Test and Procedure	Results and Indications	Proceed To
2.1a—Check for piston seizing	Seized pistons are caused by low oil pressure, oil breakdown, or overheating. Replace the pistons and lin-	

Section 2—Engine Mechanical (Cont.)

Test and Procedure	Results and Indications	Proceed To
2.1b—Check for a damaged flywheel ring gear	A damaged ring gear will cause poor meshing with the starter. Replace the ring gear.	**2.1c**
2.1c—Make a compression check	Low compression can be caused by: sticking rings, worn rings, worn liners. Replace the rings or liners.	**2.2a**
2.2a—A knocking noise at idle or during acceleration can be caused by a variety of wear problems.	Use a stethoscope or similar listening device to try to pinpoint the source of the noise. Among other reasons for knocking are: piston pins, rod bearings, loose rod caps, crankshaft journals and/or bearings, crankshaft thrust washer. Replace any worn parts.	**2.2b**
2.2b—An infrequently encountered noise is a continuous growl during acceleration	This problem is usually caused by problems in the engine timing gears. Poor contact, excessive backlash or loose gears are usually at fault.	**2.2c**
2.2c—Intermittant noises are the hardest to find. They are usually caused by broken moving parts	Check the gear train for a chipped or cracked gear; the oil pan for broken parts or foreign objects or the cylinder head for a broken valve or valve spring.	**2.3**
2.3 —Check for oil in the combustion chambers	Oil entering the combustion chambers will cause the engine to overspeed if the amount of oil is too great, or run unevenly if it is great. Check for broken or sticking rings, bad head gasket(s) or worn valve guides.	**2.4**
2.4 —Check the compression Caution: Use a screw-in type compression tester only. Very high compression pressures will be recorded.	Low compression is the main cause of power loss. The main causes for low compression are: worn rings or liners, cracked valves, warped head or block, and bad head gasket.	**2.5**
2.5 —A large amount of black exhaust is caused by low compression	See 2.4 above	**2.6**
2.6 —If the engine stops suddenly during operation, the cause is usually sudden damage	Check the pistons, main bearings or rod bearings for lack of lubrication. A seized camshaft is also a result of low or no lubrication. Check the timing gears for damage. Replace as necessary.	**2.7**
2.7 —Check for excessive clearance between the bearings and journals on both the mains and rod bearings. Check the oil pressure.	Replace the pump as necessary.	**2.8**
2.8 —Aside from the usual leaking gasket problems, check the condition of the combustion chamber O-rings.	Replace as necessary	**2.9**
2.9 —Compression leakage is usually caused by a seal defect between the head and the block	Check the head gasket; check for loose head bolts; check for head or block warpage. Replace or repair as necessary.	

Troubleshooting Gasoline Engine Problems
See Chapters 2, 3, 4 for more information and service procedures.

Index to Systems

System	To Test	Group
Battery	Engine need not be running	1
Starting system	Engine need not be running	2
Primary electrical system	Engine need not be running	3
Secondary electrical system	Engine need not be running	4
Fuel system	Engine need not be running	5
Engine compression	Engine need not be running	6
Engine vacuum	Engine must be running	7
Secondary electrical system	Engine must be running	8
Valve train	Engine must be running	9
Exhaust system	Engine must be running	10
Cooling system	Engine must be running	11
Engine lubrication	Engine must be running	12

Index to Problems

Problem: Symptom	Begin at Specific Diagnosis, Number ____
Engine Won't Start:	
Starter doesn't turn	1.1, 2.1
Starter turns, engine doesn't	2.1
Starter turns engine very slowly	1.1, 2.4
Starter turns engine normally	3.1, 4.1
Starter turns engine very quickly	6.1
Engine fires intermittently	4.1
Engine fires consistently	5.1, 6.1
Engine Runs Poorly:	
Hard starting	3.1, 4.1, 5.1, 8.1
Rough idle	4.1, 5.1, 8.1
Stalling	3.1, 4.1, 5.1, 8.1
Engine dies at high speeds	4.1, 5.1
Hesitation (on acceleration from standing stop)	5.1, 8.1
Poor pickup	4.1, 5.1, 8.1
Lack of power	3.1, 4.1, 5.1, 8.1
Backfire through the carburetor	4.1, 8.1, 9.1
Backfire through the exhaust	4.1, 8.1, 9.1
Blue exhaust gases	6.1, 7.1
Black exhaust gases	5.1
Running on (after the ignition is shut off)	3.1, 8.1
Susceptible to moisture	4.1
Engine misfires under load	4.1, 7.1, 8.4, 9.1
Engine misfires at speed	4.1, 8.4
Engine misfires at idle	3.1, 4.1, 5.1, 7.1, 8.4

Sample Section

Test and Procedure	Results and Indications	Proceed to
4.1—Check for spark: Hold each spark plug wire approximately ¼″ from ground with gloves or a heavy, dry rag. Crank the engine and observe the spark.	→ If no spark is evident:	→ 4.2
	→ If spark is good in some cases:	→ 4.3
	→ If spark is good in all cases:	→ 4.6

Specific Diagnosis

This section is arranged so that following each test, instructions are given to proceed to another, until a problem is diagnosed.

Section 1—Battery

Test and Procedure	Results and Indications	Proceed to
1.1—Inspect the battery visually for case condition (corrosion, cracks) and water level.	If case is cracked, replace battery:	**1.4**
	If the case is intact, remove corrosion with a solution of baking soda and water (**CAUTION:** *do not get the solution into the battery*), and fill with water:	**1.2**

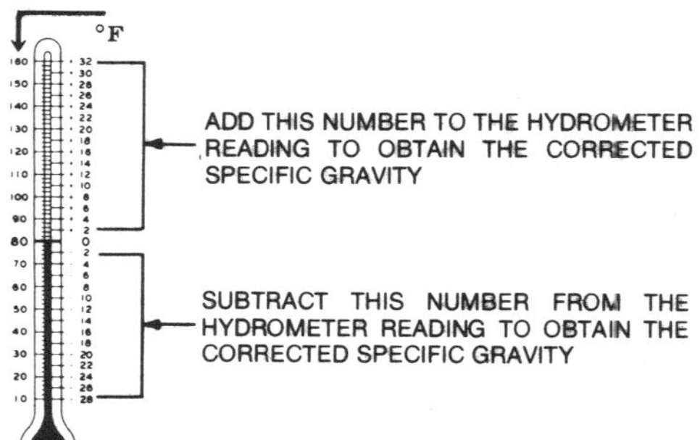

Inspect the battery case

Test and Procedure	Results and Indications	Proceed to
1.2—Check the battery cable connections: Insert a screwdriver between the battery post and the cable clamp. Turn the headlights on high beam, and observe them as the screwdriver is gently twisted to ensure good metal to metal contact.	If the lights brighten, remove and clean the clamp and post; coat the post with petroleum jelly, install and tighten the clamp:	**1.4**
	If no improvement is noted:	**1.3**

Test and Procedure	Results and Indications	Proceed to
1.3—Test the state of charge of the battery using an individual cell tester or hydrometer.	If indicated, charge the battery. **NOTE:** *If no obvious reason exists for the low state of charge (i.e., battery age, prolonged storage), proceed to:*	**1.4**

°F

ADD THIS NUMBER TO THE HYDROMETER READING TO OBTAIN THE CORRECTED SPECIFIC GRAVITY

SUBTRACT THIS NUMBER FROM THE HYDROMETER READING TO OBTAIN THE CORRECTED SPECIFIC GRAVITY

Specific Gravity (@ 80° F.)

Minimum	Battery Charge
1.260	100% Charged
1.230	75% Charged
1.200	50% Charged
1.170	25% Charged
1.140	Very Little Power Left
1.110	Completely Discharged

The effects of temperature on battery specific gravity (left) and amount of battery charge in relation to specific gravity (right)

Test and Procedure	Results and Indications	Proceed to
1.4—Visually inspect battery cables for cracking, bad connection to ground, or bad connection to starter.	If necessary, tighten connections or replace the cables:	**2.1**

Section 2—Starting System
See Chapter 3 for service procedures

Test and Procedure	Results and Indications	Proceed to
Note: Tests in Group 2 are performed with coil high tension lead disconnected to prevent accidental starting.		
2.1—Test the starter motor and solenoid: Connect a jumper from the battery post of the solenoid (or relay) to the starter post of the solenoid (or relay).	If starter turns the engine normally:	2.2
	If the starter buzzes, or turns the engine very slowly:	2.4
	If no response, replace the solenoid (or relay).	3.1
	If the starter turns, but the engine doesn't, ensure that the flywheel ring gear is intact. If the gear is undamaged, replace the starter drive.	3.1
2.2—Determine whether ignition override switches are functioning properly (clutch start switch, neutral safety switch), by connecting a jumper across the switch(es), and turning the ignition switch to "start".	If starter operates, adjust or replace switch:	3.1
	If the starter doesn't operate:	2.3
2.3—Check the ignition switch "start" position: Connect a 12V test lamp or voltmeter between the starter post of the solenoid (or relay) and ground. Turn the ignition switch to the "start" position, and jiggle the key.	If the lamp doesn't light or the meter needle doesn't move when the switch is turned, check the ignition switch for loose connections, cracked insulation, or broken wires. Repair or replace as necessary:	3.1
	If the lamp flickers or needle moves when the key is jiggled, replace the ignition switch.	3.3

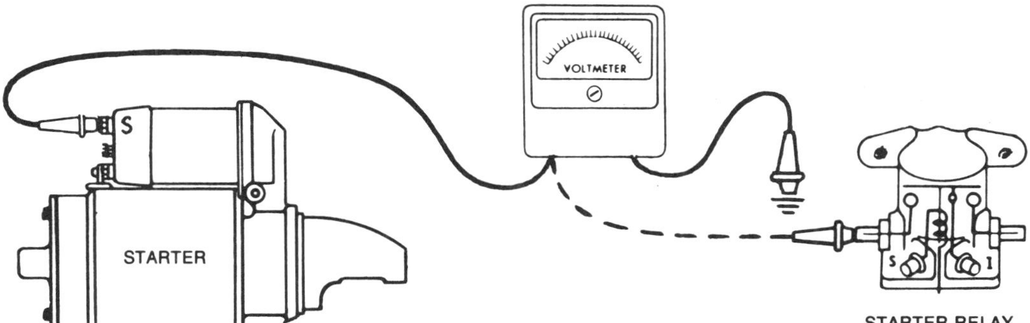

VOLTMETER

STARTER

STARTER RELAY
(IF EQUIPPED)

Checking the ignition switch "start" position

Test and Procedure	Results and Indications	Proceed to
2.4—Remove and bench test the starter, according to specifications in the engine electrical section.	If the starter does not meet specifications, repair or replace as needed:	3.1
	If the starter is operating properly:	2.5
2.5—Determine whether the engine can turn freely: Remove the spark plugs, and check for water in the cylinders. Check for water on the dipstick, or oil in the radiator. Attempt to turn the engine using an 18″ flex drive and socket on the crankshaft pulley nut or bolt.	If the engine will turn freely only with the spark plugs out, and hydrostatic lock (water in the cylinders) is ruled out, check valve timing:	9.2
	If engine will not turn freely, and it is known that the clutch and transmission are free, the engine must be disassembled for further evaluation:	Chapter 3

Section 3—Primary Electrical System

Test and Procedure	Results and Indications	Proceed to
3.1—Check the ignition switch "on" position: Connect a jumper wire between the distributor side of the coil and ground, and a 12V test lamp between the switch side of the coil and ground. Remove the high tension lead from the coil. Turn the ignition switch on and jiggle the key.	If the lamp lights:	**3.2**
	If the lamp flickers when the key is jiggled, replace the ignition switch:	**3.3**
	If the lamp doesn't light, check for loose or open connections. If none are found, remove the ignition switch and check for continuity. If the switch is faulty, replace it:	**3.3**

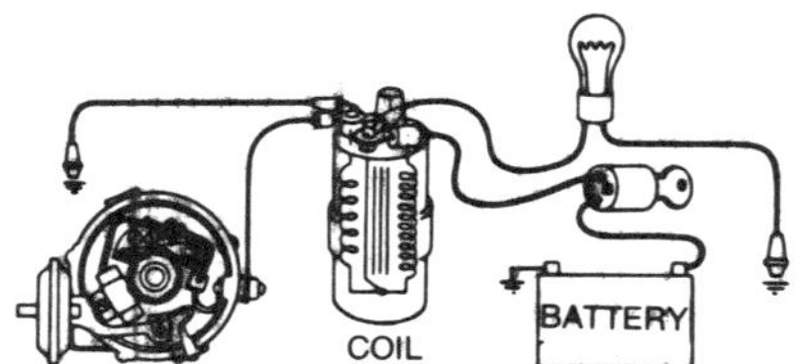

Checking the ignition switch "on" position

Test and Procedure	Results and Indications	Proceed to
3.2—Check the ballast resistor or resistance wire for an open circuit, using an ohmmeter. See Chapter 3 for specific tests.	Replace the resistor or resistance wire if the resistance is zero. **NOTE:** *Some ignition systems have no ballast resistor.*	**3.3**

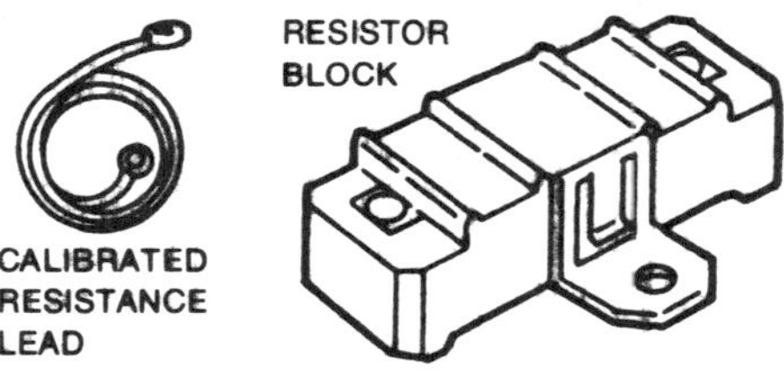

Two types of resistors

Test and Procedure	Results and Indications	Proceed to
3.3—On point-type ignition systems, visually inspect the breaker points for burning, pitting or excessive wear. Gray coloring of the point contact surfaces is normal. Rotate the crankshaft until the contact heel rests on a high point of the distributor cam and adjust the point gap to specifications. On electronic ignition models, remove the distributor cap and visually inspect the armature. Ensure that the armature pin is in place, and that the armature is on tight and rotates when the engine is cranked. Make sure there are no cracks, chips or rounded edges on the armature.	If the breaker points are intact, clean the contact surfaces with fine emery cloth, and adjust the point gap to specifications. If the points are worn, replace them. On electronic systems, replace any parts which appear defective. If condition persists:	**3.4**

Test and Procedure	Results and Indications	Proceed to
3.4—On point-type ignition systems, connect a dwell-meter between the distributor primary lead and ground. Crank the engine and observe the point dwell angle. On electronic ignition systems, conduct a stator (magnetic pickup assembly) test. See Chapter 3.	On point-type systems, adjust the dwell angle if necessary. **NOTE:** *Increasing the point gap decreases the dwell angle and vice-versa.*	3.6
	If the dwell meter shows little or no reading;	3.5
	On electronic ignition systems, if the stator is bad, replace the stator. If the stator is good, proceed to the other tests in Chapter 3.	

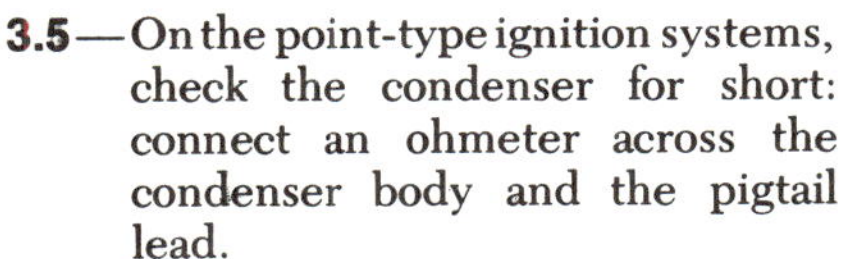

Dwell is a function of point gap

Test and Procedure	Results and Indications	Proceed to
3.5—On the point-type ignition systems, check the condenser for short: connect an ohmeter across the condenser body and the pigtail lead.	If any reading other than infinite is noted, replace the condenser	3.6

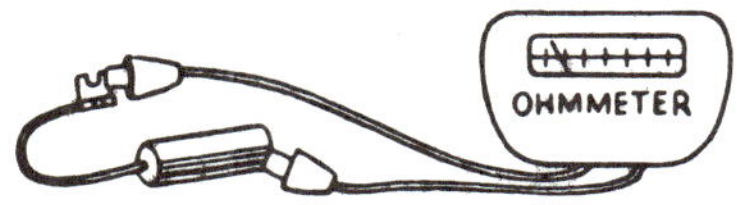

Checking the condenser for short

Test and Procedure	Results and Indications	Proceed to
3.6—Test the coil primary resistance: On point-type ignition systems, connect an ohmmeter across the coil primary terminals, and read the resistance on the low scale. Note whether an external ballast resistor or resistance wire is used. On electronic ignition systems, test the coil primary resistance as in Chapter 3.	Point-type ignition coils utilizing ballast resistors or resistance wires should have approximately 1.0 ohms resistance. Coils with internal resistors should have approximately 4.0 ohms resistance. If values far from the above are noted, replace the coil.	4.1

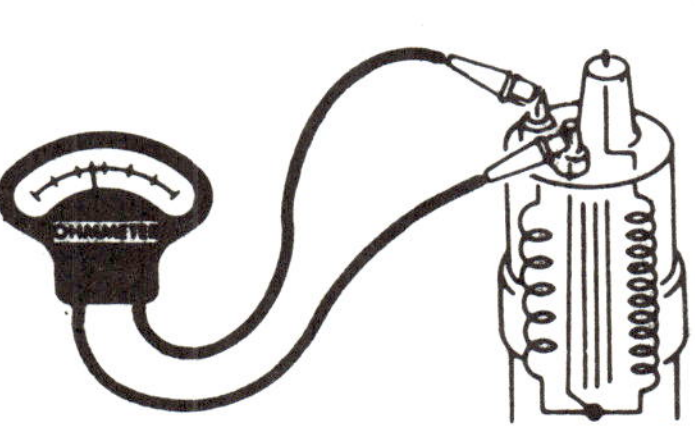

Check the coil primary resistance

Section 4—Secondary Electrical System
See Chapters 2–3 for service procedures

Test and Procedure	Results and Indications	Proceed to
4.1—Check for spark: Hold each spark plug wire approximately ¼″ from ground with gloves or a heavy, dry rag. Crank the engine, and observe the spark.	If no spark is evident:	**4.2**
	If spark is good in some cylinders:	**4.3**
	If spark is good in all cylinders:	**4.6**

Check for spark at the plugs

Test and Procedure	Results and Indications	Proceed to
4.2—Check for spark at the coil high tension lead: Remove the coil high tension lead from the distributor and position it approximately ¼″ from ground. Crank the engine and observe spark. **CAUTION:** *This test should not be performed on engines equipped with electronic ignition.*	If the spark is good and consistent:	**4.3**
	If the spark is good but intermittent, test the primary electrical system starting at 3.3:	**3.3**
	If the spark is weak or non-existent, replace the coil high tension lead, clean and tighten all connections and retest. If no improvement is noted:	**4.4**
4.3—Visually inspect the distributor cap and rotor for burned or corroded contacts, cracks, carbon tracks, or moisture. Also check the fit of the rotor on the distributor shaft (where applicable).	If moisture is present, dry thoroughly, and retest per 4.1:	**4.1**
	If burned or excessively corroded contacts, cracks, or carbon tracks are noted, replace the defective part(s) and retest per 4.1:	**4.1**
	If the rotor and cap appear intact, or are only slightly corroded, clean the contacts thoroughly (including the cap towers and spark plug wire ends) and retest per 4.1:	
	If the spark is good in all cases:	**4.6**
	If the spark is poor in all cases:	**4.5**

Inspect the distributor cap and rotor

Test and Procedure	Results and Indications	Proceed to

4.4—Check the coil secondary resistance: On point-type systems connect an ohmmeter across the distributor side of the coil and the coil tower. Read the resistance on the high scale of the ohmmeter. On electronic ignition systems, see Chapter 3 for specific tests.

The resistance of a satisfactory coil should be between 4,000 and 10,000 ohms. If resistance is considerably higher (i.e., 40,000 ohms) replace the coil and retest per 4.1. **NOTE: *This does not apply to high performance coils.***

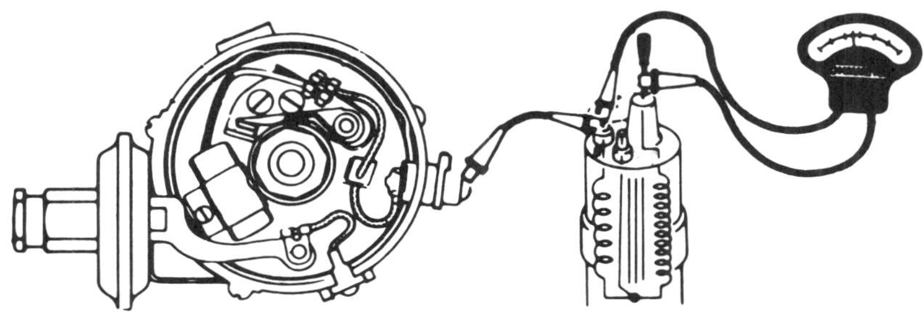

Testing the coil secondary resistance

4.5—Visually inspect the spark plug wires for cracking or brittleness. Ensure that no two wires are positioned so as to cause induction firing (adjacent and parallel). Remove each wire, one by one, and check resistance with an ohmmeter.

Replace any cracked or brittle wires. If any of the wires are defective, replace the entire set. Replace any wires with excessive resistance (over $8000\,\Omega$ per foot for suppression wire), and separate any wires that might cause induction firing.

4.6

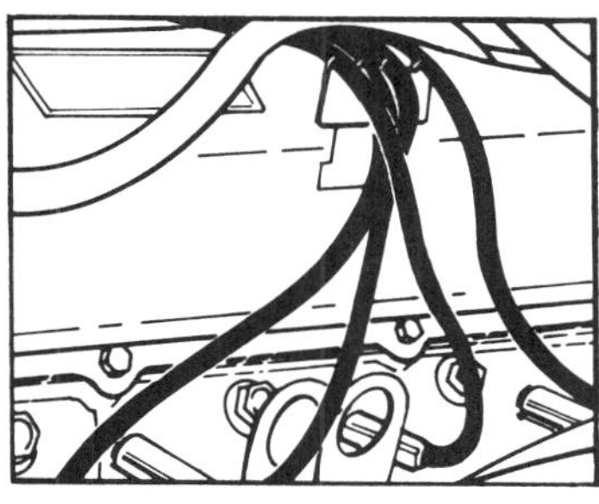

Misfiring can be the result of spark plug leads to adjacent, consecutively firing cylinders running parallel and too close together

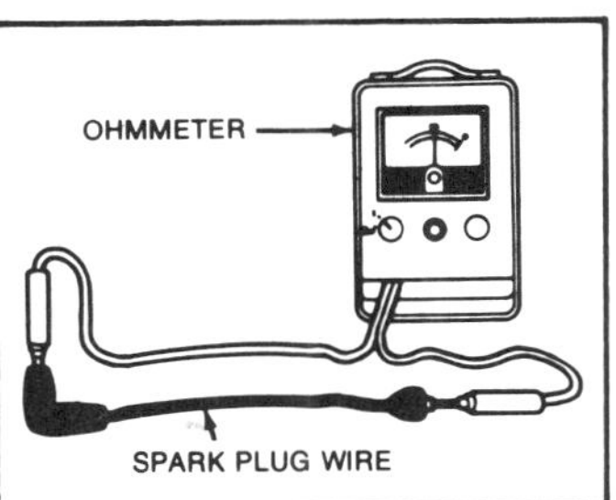

On point-type ignition systems, check the spark plug wires as shown. On electronic ignitions, do not remove the wire from the distributor cap terminal; instead, test through the cap

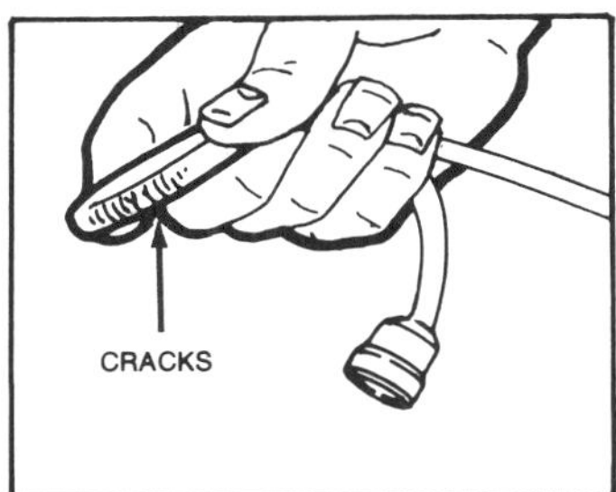

Spark plug wires can be checked visually by bending them in a loop over your finger. This will reveal any cracks, burned or broken insulation. Any wire with cracked insulation should be replaced

4.6—Remove the spark plugs, noting the cylinders from which they were removed, and evaluate according to the color photos in the middle of this book.

See following.

See following.

Test and Procedure	Results and Indications	Proceed to
4.7—Examine the location of all the plugs.	The following diagrams illustrate some of the conditions that the location of plugs will reveal.	**4.8**

Two adjacent plugs are fouled in a 6-cylinder engine, 4-cylinder engine or either bank of a V-8. This is probably due to a blown head gasket between the two cylinders

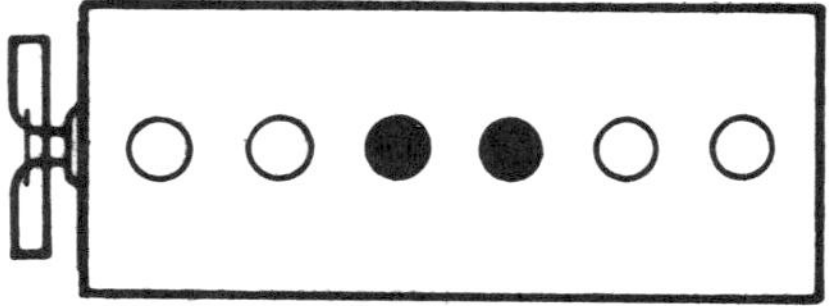

The two center plugs in a 6-cylinder engine are fouled. Raw fuel may be "boiled" out of the carburetor into the intake manifold after the engine is shut-off. Stop-start driving can also foul the center plugs, due to overly rich mixture. Proper float level, a new float needle and seat or use of an insulating spacer may help this problem

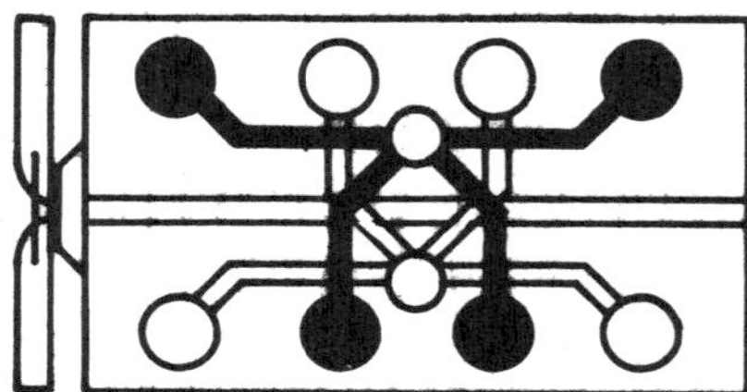

An unbalanced carburetor is indicated. Following the fuel flow on this particular design shows that the cylinders fed by the right-hand barrel are fouled from overly rich mixture, while the cylinders fed by the left-hand barrel are normal

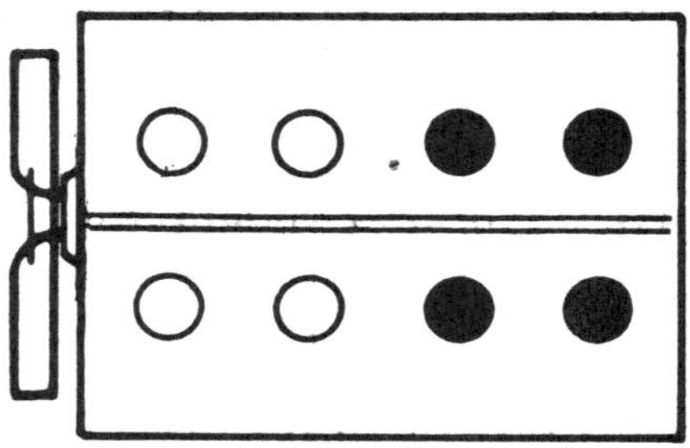

If the four rear plugs are overheated, a cooling system problem is suggested. A thorough cleaning of the cooling system may restore coolant circulation and cure the problem

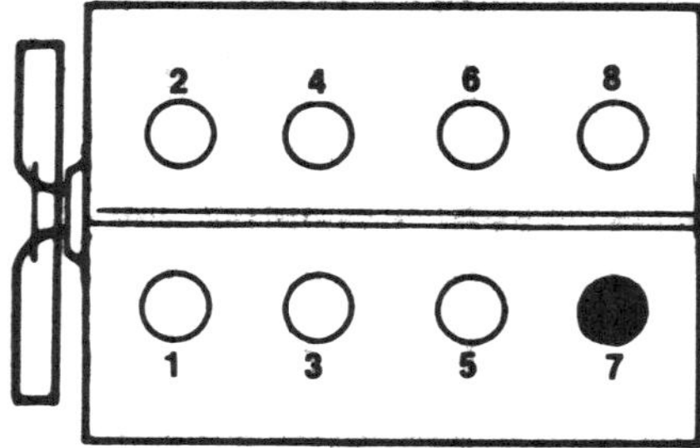

Finding one plug overheated may indicate an intake manifold leak near the affected cylinder. If the overheated plug is the second of two adjacent, consecutively firing plugs, it could be the result of ignition cross-firing. Separating the leads to these two plugs will eliminate cross-fire

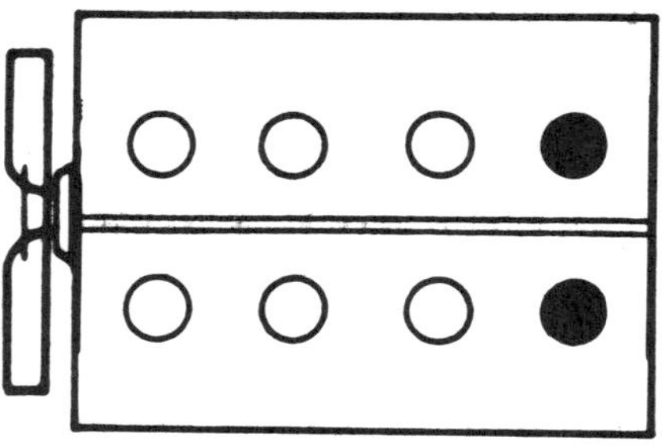

Occasionally, the two rear plugs in large, lightly used V-8's will become oil fouled. High oil consumption and smoky exhaust may also be noticed. It is probably due to plugged oil drain holes in the rear of the cylinder head, causing oil to be sucked in around the valve stems. This usually occurs in the rear cylinders first, because the engine slants that way

Test and Procedure	Results and Indications	Proceed to
4.8—Determine the static ignition timing. Using the crankshaft pulley timing marks as a guide, locate top dead center on the compression stroke of the number one cylinder.	The rotor should be pointing toward the No. 1 tower in the distributor cap, and, on electronic ignitions, the armature spoke for that cylinder should be lined up with the stator.	**4.8**
4.9—Check coil polarity: Connect a voltmeter negative lead to the coil high tension lead, and the positive lead to ground (**NOTE:** *Reverse the hook-up for positive ground systems*). Crank the engine momentarily. **Checking coil polarity**	If the voltmeter reads up-scale, the polarity is correct: If the voltmeter reads down-scale, reverse the coil polarity (switch the primary leads):	**5.1** **5.1**

Section 5—Fuel System

See Chapter 4 for service procedures

Test and Procedure	Results and Indications	Proceed to
5.1—Determine that the air filter is functioning efficiently: Hold paper elements up to a strong light, and attempt to see light through the filter.	Clean permanent air filters in solvent (or manufacturer's recommendation), and allow to dry. Replace paper elements through which light cannot be seen:	**5.2**
5.2—Determine whether a flooding condition exists: Flooding is identified by a strong gasoline odor, and excessive gasoline present in the throttle bore(s) of the carburetor. **If the engine floods repeatedly, check the choke butterfly flap**	If flooding is not evident: If flooding is evident, permit the gasoline to dry for a few moments and restart. If flooding doesn't recur: If flooding is persistent:	**5.3** **5.7** **5.5**
5.3—Check that fuel is reaching the carburetor: Detach the fuel line at the carburetor inlet. Hold the end of the line in a cup (not styrofoam), and crank the engine. **Check the fuel pump by disconnecting the output line (fuel pump-to-carburetor) at the carburetor and operating the starter briefly**	If fuel flows smoothly: If fuel doesn't flow (**NOTE:** *Make sure that there is fuel in the tank*), or flows erratically:	**5.7** **5.4**

Test and Procedure	Results and Indications	Proceed to
5.4—Test the fuel pump: Disconnect all fuel lines from the fuel pump. Hold a finger over the input fitting, crank the engine (with electric pump, turn the ignition or pump on); and feel for suction.	If suction is evident, blow out the fuel line to the tank with low pressure compressed air until bubbling is heard from the fuel filler neck. Also blow out the carburetor fuel line (both ends disconnected):	**5.7**
	If no suction is evident, replace or repair the fuel pump: NOTE: *Repeated oil fouling of the spark plugs, or a no-start condition, could be the result of a ruptured vacuum booster pump diaphragm, through which oil or gasoline is being drawn into the intake manifold (where applicable).*	**5.7**
5.5—Occasionally, small specks of dirt will clog the small jets and orifices in the carburetor. With the engine cold, hold a flat piece of wood or similar material over the carburetor, where possible, and crank the engine.	If the engine starts, but runs roughly the engine is probably not run enough. If the engine won't start:	**5.9**
5.6—Check the needle and seat: Tap the carburetor in the area of the needle and seat.	If flooding stops, a gasoline additive (e.g., Gumout) will often cure the problem:	**5.7**
	If flooding continues, check the fuel pump for excessive pressure at the carburetor (according to specifications). If the pressure is normal, the needle and seat must be removed and checked, and/or the float level adjusted:	**5.7**
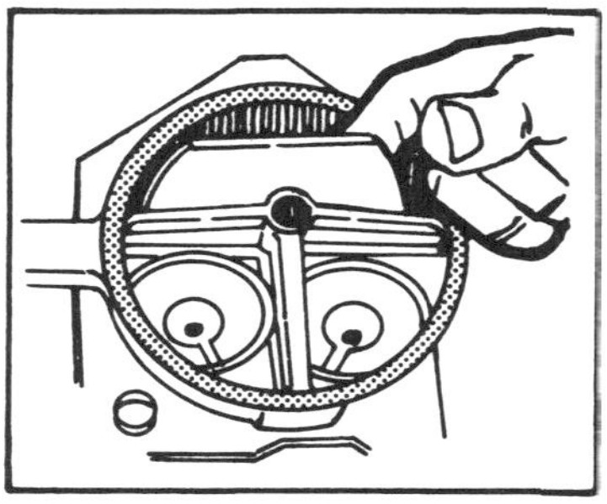 **5.7**—Test the accelerator pump by looking into the throttle bores while operating the throttle.	If the accelerator pump appears to be operating normally:	**5.8**
	If the accelerator pump is not operating, the pump must be reconditioned. Where possible, service the pump with the carburetor(s) installed on the engine. If necessary, remove the carburetor. Prior to removal:	**5.8**
5.8—Determine whether the carburetor main fuel system is functioning: Spray a commercial starting fluid into the carburetor while attempting to start the engine.	If the engine starts, runs for a few seconds, and dies:	**5.9**
	If the engine doesn't start:	**6.1**

Check for gas at the carburetor by looking down the carburetor throat while someone moves the accelerator

Test and Procedure	Results and Indications	Proceed to
5.9—Uncommon fuel system malfunctions: See below:	If the problem is solved:	6.1
	If the problem remains, remove and recondition the carburetor.	

Condition	Indication	Test	Prevailing Weather Conditions	Remedy
Vapor lock	Engine will not restart shortly after running.	Cool the components of the fuel system until the engine starts. Vapor lock can be cured faster by draping a wet cloth over a mechanical fuel pump.	Hot to very hot	Ensure that the exhaust manifold heat control valve is operating. Check with the vehicle manufacturer for the recommended solution to vapor lock on the model in question.
Carburetor icing	Engine will not idle, stalls at low speeds.	Visually inspect the throttle plate area of the throttle bores for frost.	High humidity, 32–40° F.	Ensure that the exhaust manifold heat control valve is operating, and that the intake manifold heat riser is not blocked.
Water in the fuel	Engine sputters and stalls; may not start.	Pump a small amount of fuel into a glass jar. Allow to stand, and inspect for droplets or a layer of water.	High humidity, extreme temperature changes.	For droplets, use one or two cans of commercial gas line anti-freeze. For a layer of water, the tank must be drained, and the fuel lines blown out with compressed air.

Section 6—Engine Compression
See Chapter 3 for service procedures

6.1—Test engine compression: Remove all spark plugs. Block the throttle wide open. Insert a compression gauge into a spark plug port, crank the engine to obtain the maximum reading, and record.	If compression is within limits on all cylinders:	7.1
	If gauge reading is extremely low on all cylinders:	6.2
	If gauge reading is low on one or two cylinders: (If gauge readings are identical and low on two or more adjacent cylinders, the head gasket must be replaced.)	6.2

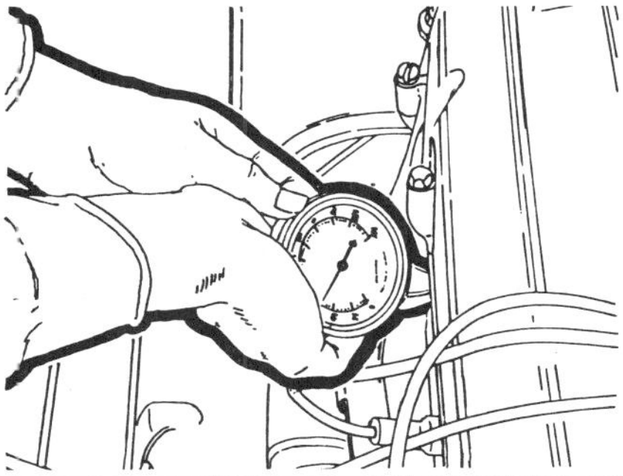

Checking compression

6.2—Test engine compression (wet): Squirt approximately 30 cc. of engine oil into each cylinder, and retest per 6.1.	If the readings improve, worn or cracked rings or broken pistons are indicated:	See Chapter 3
	If the readings do not improve, burned or excessively carboned valves or a jumped timing chain are indicated:	7.1
	NOTE: *A jumped timing chain is often indicated by difficult cranking.*	

Section 7—Engine Vacuum
See Chapter 3 for service procedures

Test and Procedure	Results and Indications	Proceed to
7.1—Attach a vacuum gauge to the intake manifold beyond the throttle plate. Start the engine, and observe the action of the needle over the range of engine speeds.	See below.	**See below**

INDICATION: normal engine in good condition

Proceed to: 8.1

Normal engine

Gauge reading: steady, from 17–22 in./Hg.

INDICATION: sticking valves or ignition miss

Proceed to: 9.1, 8.3

Sticking valves

Gauge reading: intermittent fluctuation at idle

INDICATION: late ignition or valve timing, low compression, stuck throttle valve, leaking carburetor or manifold gasket

Proceed to: 6.1

Incorrect valve timing

Gauge reading: low (10–15 in./Hg) but steady

INDICATION: improper carburetor adjustment or minor intake leak.

Proceed to: 7.2

Carburetor requires adjustment

Gauge reading: drifting needle

INDICATION: ignition miss, blown cylinder head gasket, leaking valve or weak valve spring

Proceed to: 8.3, 6.1

Blown head gasket

Gauge reading: needle fluctuates as engine speed increases

INDICATION: burnt valve or faulty valve clearance. Needle will fall when defective valve operates

Proceed to: 9.1

Burnt or leaking valves

Gauge reading: steady needle, but drops regularly

INDICATION: choked muffler, excessive back pressure in system

Proceed to: 10.1

Clogged exhaust system

Gauge reading: gradual drop in reading at idle

INDICATION: worn valve guides

Proceed to: 9.1

Worn valve guides

Gauge reading: needle vibrates excessively at idle, but steadies as engine speed increases

White pointer = steady gauge hand | Black pointer = fluctuating gauge hand

Test and Procedure	Results and Indications	Proceed to
7.2—Attach a vacuum gauge per 7.1, and test for an intake manifold leak. Squirt a small amount of oil around the intake manifold gaskets, carburetor gaskets, plugs and fittings. Observe the action of the vacuum gauge.	If the reading improves, replace the indicated gasket, or seal the indicated fitting or plug: If the reading remains low:	8.1 7.3
7.3—Test all vacuum hoses and accessories for leaks as described in 7.2. Also check the carburetor body (dashpots, automatic choke mechanism, throttle shafts) for leaks in the same manner.	If the reading improves, service or replace the offending part(s): If the reading remains low:	8.1 6.1

Section 8—Secondary Electrical System
See Chapter 2 for service procedures

Test and Procedure	Results and Indications	Proceed to
8.1—Remove the distributor cap and check to make sure that the rotor turns when the engine is cranked. Visually inspect the distributor components.	Clean, tighten or replace any components which appear defective.	8.2
8.2—Connect a timing light (per manufacturer's recommendation) and check the dynamic ignition timing. Disconnect and plug the vacuum hose(s) to the distributor if specified, start the engine, and observe the timing marks at the specified engine speed.	If the timing is not correct, adjust to specifications by rotating the distributor in the engine: (Advance timing by rotating distributor opposite normal direction of rotor rotation, retard timing by rotating distributor in same direction as rotor rotation.)	8.3
8.3—Check the operation of the distributor advance mechanism(s): To test the mechanical advance, disconnect the vacuum lines from the distributor advance unit and observe the timing marks with a timing light as the engine speed is increased from idle. If the mark moves smoothly, without hesitation, it may be assumed that the mechanical advance is functioning properly. To test vacuum advance and/or retard systems, alternately crimp and release the vacuum line, and observe the timing mark for movement. If movement is noted, the system is operating.	If the systems are functioning: If the systems are not functioning, remove the distributor, and test on a distributor tester:	8.4 8.4
8.4—Locate an ignition miss: With the engine running, remove each spark plug wire, one at a time, until one is found that doesn't cause the engine to roughen and slow down.	When the missing cylinder is identified:	4.1

Section 9—Valve Train
See Chapter 3 for service procedures

Test and Procedure	Results and Indications	Proceed to
9.1—Evaluate the valve train: Remove the valve cover, and ensure that the valves are adjusted to specifications. A mechanic's stethoscope may be used to aid in the diagnosis of the valve train. By pushing the probe on or near push rods or rockers, valve noise often can be isolated. A timing light also may be used to diagnose valve problems. Connect the light according to manufacturer's recommendations, and start the engine. Vary the firing moment of the light by increasing the engine speed (and therefore the ignition advance), and moving the trigger from cylinder to cylinder. Observe the movement of each valve.	Sticking valves or erratic valve train motion can be observed with the timing light. The cylinder head must be disassembled for repairs.	**See Chapter 3**
9.2—Check the valve timing: Locate top dead center of the No. 1 piston, and install a degree wheel or tape on the crankshaft pulley or damper with zero corresponding to an index mark on the engine. Rotate the crankshaft in its direction of rotation, and observe the opening of the No. 1 cylinder intake valve. The opening should correspond with the correct mark on the degree wheel according to specifications.	If the timing is not correct, the timing cover must be removed for further investigation.	**See Chapter 3**

Section 10—Exhaust System

Test and Procedure	Results and Indications	Proceed to
10.1—Determine whether the exhaust manifold heat control valve is operating: Operate the valve by hand to determine whether it is free to move. If the valve is free, run the engine to operating temperature and observe the action of the valve, to ensure that it is opening.	If the valve sticks, spray it with a suitable solvent, open and close the valve to free it, and retest. If the valve functions properly: If the valve does not free, or does not operate, replace the valve:	**10.2** **10.2**
10.2—Ensure that there are no exhaust restrictions: Visually inspect the exhaust system for kinks, dents, or crushing. Also note that gases are flowing freely from the tailpipe at all engine speeds, indicating no restriction in the muffler or resonator.	Replace any damaged portion of the system:	**11.1**

Section 11—Cooling System
See Chapter 3 for service procedures

Test and Procedure	Results and Indications	Proceed to
11.1—Visually inspect the fan belt for glazing, cracks, and fraying, and replace if necessary. Tighten the belt so that the longest span has approximately ½″ play at its midpoint under thumb pressure (see Chapter 1).	Replace or tighten the fan belt as necessary:	**11.2**

Checking belt tension

Test and Procedure	Results and Indications	Proceed to
11.2—Check the fluid level of the cooling system.	If full or slightly low, fill as necessary:	**11.5**
	If extremely low:	**11.3**
11.3—Visually inspect the external portions of the cooling system (radiator, radiator hoses, thermostat elbow, water pump seals, heater hoses, etc.) for leaks. If none are found, pressurize the cooling system to 14–15 psi.	If cooling system holds the pressure:	**11.5**
	If cooling system loses pressure rapidly, reinspect external parts of the system for leaks under pressure. If none are found, check dipstick for coolant in crankcase. If no coolant is present, but pressure loss continues:	**11.4**
	If coolant is evident in crankcase, remove cylinder head(s), and check gasket(s). If gaskets are intact, block and cylinder head(s) should be checked for cracks or holes. If the gasket(s) is blown, replace, and purge the crankcase of coolant:	**12.6**
	NOTE: *Occasionally, due to atmospheric and driving conditions, condensation of water can occur in the crankcase. This causes the oil to appear milky white. To remedy, run the engine until hot, and change the oil and oil filter.*	
11.4—Check for combustion leaks into the cooling system: Pressurize the cooling system as above. Start the engine, and observe the pressure gauge. If the needle fluctuates, remove each spark plug wire, one at a time, noting which cylinder(s) reduce or eliminate the fluctuation.	Cylinders which reduce or eliminate the fluctuation, when the spark plug wire is removed, are leaking into the cooling system. Replace the head gasket on the affected cylinder bank(s).	

Pressurizing the cooling system

Test and Procedure	Results and Indications	Proceed to
11.5—Check the radiator pressure cap: Attach a radiator pressure tester to the radiator cap (wet the seal prior to installation). Quickly pump up the pressure, noting the point at which the cap releases.	If the cap releases within ± 1 psi of the specified rating, it is operating properly:	**11.6**
	If the cap releases at more than ± 1 psi of the specified rating, it should be replaced:	**11.6**

Checking radiator pressure cap

Test and Procedure	Results and Indications	Proceed to
11.6—Test the thermostat: Start the engine cold, remove the radiator cap, and insert a thermometer into the radiator. Allow the engine to idle. After a short while, there will be a sudden, rapid increase in coolant temperature. The temperature at which this sharp rise stops is the thermostat opening temperature.	If the thermostat opens at or about the specified temperature:	**11.7**
	If the temperature doesn't increase: (If the temperature increases slowly and gradually, replace the thermostat.)	**11.7**
11.7—Check the water pump: Remove the thermostat elbow and the thermostat, disconnect the coil high tension lead (to prevent starting), and crank the engine momentarily.	If coolant flows, replace the thermostat and retest per 11.6:	**11.6**
	If coolant doesn't flow, reverse flush the cooling system to alleviate any blockage that might exist. If system is not blocked, and coolant will not flow, replace the water pump.	

Section 12—Lubrication
See Chapter 3 for service procedures

Test and Procedure	Results and Indications	Proceed to
12.1—Check the oil pressure gauge or warning light: If the gauge shows low pressure, or the light is on for no obvious reason, remove the oil pressure sender. Install an accurate oil pressure gauge and run the engine momentarily.	If oil pressure builds normally, run engine for a few moments to determine that it is functioning normally, and replace the sender.	—
	If the pressure remains low:	**12.2**
	If the pressure surges:	**12.3**
	If the oil pressure is zero:	**12.3**
12.2—Visually inspect the oil: If the oil is watery or very thin, milky, or foamy, replace the oil and oil filter.	If the oil is normal:	**12.3**
	If after replacing oil the pressure remains low:	**12.3**
	If after replacing oil the pressure becomes normal:	—

Test and Procedure	Results and Indications	Proceed to
12.3—Inspect the oil pressure relief valve and spring, to ensure that it is not sticking or stuck. Remove and thoroughly clean the valve, spring, and the valve body.	If the oil pressure improves: If no improvement is noted:	— **12.4**
12.4—Check to ensure that the oil pump is not cavitating (sucking air instead of oil): See that the crankcase is neither over nor underfull, and that the pickup in the sump is in the proper position and free from sludge.	Fill or drain the crankcase to the proper capacity, and clean the pickup screen in solvent if necessary. If no improvement is noted:	**12.5**
12.5—Inspect the oil pump drive and the oil pump:	If the pump drive or the oil pump appear to be defective, service as necessary and retest per 12.1: If the pump drive and pump appear to be operating normally, the engine should be disassembled to determine where blockage exists:	**12.1** **See Chapter 3**
12.6—Purge the engine of ethylene glycol coolant: Completely drain the crankcase and the oil filter. Obtain a commercial butyl cellosolve base solvent, designated for this purpose, and follow the instructions precisely. Following this, install a new oil filter and refill the crankcase with the proper weight oil. The next oil and filter change should follow shortly thereafter (1000 miles).		

TROUBLESHOOTING EMISSION CONTROL SYSTEMS

See Chapter 4 for procedures applicable to individual emission control systems used on specific combinations of engine/transmission/model.

TROUBLESHOOTING THE CARBURETOR

See Chapter 4 for service procedures

Carburetor problems cannot be effectively isolated unless all other engine systems (particularly ignition and emission) are functioning properly and the engine is properly tuned.

Condition	Possible Cause
Engine cranks, but does not start	1. Improper starting procedure 2. No fuel in tank 3. Clogged fuel line or filter 4. Defective fuel pump 5. Choke valve not closing properly 6. Engine flooded 7. Choke valve not unloading 8. Throttle linkage not making full travel 9. Stuck needle or float 10. Leaking float needle or seat 11. Improper float adjustment
Engine stalls	1. Improperly adjusted idle speed or mixture **Engine hot** 2. Improperly adjusted dashpot 3. Defective or improperly adjusted solenoid 4. Incorrect fuel level in fuel bowl 5. Fuel pump pressure too high 6. Leaking float needle seat 7. Secondary throttle valve stuck open 8. Air or fuel leaks 9. Idle air bleeds plugged or missing 10. Idle passages plugged **Engine Cold** 11. Incorrectly adjusted choke 12. Improperly adjusted fast idle speed 13. Air leaks 14. Plugged idle or idle air passages 15. Stuck choke valve or binding linkage 16. Stuck secondary throttle valves 17. Engine flooding—high fuel level 18. Leaking or misaligned float
Engine hesitates on acceleration	1. Clogged fuel filter 2. Leaking fuel pump diaphragm 3. Low fuel pump pressure 4. Secondary throttle valves stuck, bent or misadjusted 5. Sticking or binding air valve 6. Defective accelerator pump 7. Vacuum leaks 8. Clogged air filter 9. Incorrect choke adjustment (engine cold)
Engine feels sluggish or flat on acceleration	1. Improperly adjusted idle speed or mixture 2. Clogged fuel filter 3. Defective accelerator pump 4. Dirty, plugged or incorrect main metering jets 5. Bent or sticking main metering rods 6. Sticking throttle valves 7. Stuck heat riser 8. Binding or stuck air valve 9. Dirty, plugged or incorrect secondary jets 10. Bent or sticking secondary metering rods. 11. Throttle body or manifold heat passages plugged 12. Improperly adjusted choke or choke vacuum break.
Carburetor floods	1. Defective fuel pump. Pressure too high. 2. Stuck choke valve 3. Dirty, worn or damaged float or needle valve/seat 4. Incorrect float/fuel level 5. Leaking float bowl

Condition	Possible Cause
Engine idles roughly and stalls	1. Incorrect idle speed 2. Clogged fuel filter 3. Dirt in fuel system or carburetor 4. Loose carburetor screws or attaching bolts 5. Broken carburetor gaskets 6. Air leaks 7. Dirty carburetor 8. Worn idle mixture needles 9. Throttle valves stuck open 10. Incorrectly adjusted float or fuel level 11. Clogged air filter
Engine runs unevenly or surges	1. Defective fuel pump 2. Dirty or clogged fuel filter 3. Plugged, loose or incorrect main metering jets or rods 4. Air leaks 5. Bent or sticking main metering rods 6. Stuck power piston 7. Incorrect float adjustment 8. Incorrect idle speed or mixture 9. Dirty or plugged idle system passages 10. Hard, brittle or broken gaskets 11. Loose attaching or mounting screws 12. Stuck or misaligned secondary throttle valves
Poor fuel economy	1. Poor driving habits 2. Stuck choke valve 3. Binding choke linkage 4. Stuck heat riser 5. Incorrect idle mixture 6. Defective accelerator pump 7. Air leaks 8. Plugged, loose or incorrect main metering jets 9. Improperly adjusted float or fuel level 10. Bent, misaligned or fuel-clogged float 11. Leaking float needle seat 12. Fuel leak 13. Accelerator pump discharge ball not seating properly 14. Incorrect main jets
Engine lacks high speed performance or power	1. Incorrect throttle linkage adjustment 2. Stuck or binding power piston 3. Defective accelerator pump 4. Air leaks 5. Incorrect float setting or fuel level 6. Dirty, plugged, worn or incorrect main metering jets or rods 7. Binding or sticking air valve 8. Brittle or cracked gaskets 9. Bent, incorrect or improperly adjusted secondary metering rods 10. Clogged fuel filter 11. Clogged air filter 12. Defective fuel pump

TROUBLESHOOTING FUEL INJECTION PROBLEMS

Each fuel injection system has its own unique components and test procedures, for which it is impossible to generalize. Refer to Chapter 4 of this Repair & Tune-Up Guide for specific test and repair procedures, if the vehicle is equipped with fuel injection.

TROUBLESHOOTING ELECTRICAL PROBLEMS

See Chapter 5 for service procedures

For any electrical system to operate, it must make a complete circuit. This simply means that the power flow from the battery must make a complete circle. When an electrical component is operating, power flows from the battery to the component, passes through the component causing it to perform its function (lighting a light bulb), and then returns to the battery through the ground of the circuit. This ground is usually (but not always) the metal part of the car or truck on which the electrical component is mounted.

Perhaps the easiest way to visualize this is to think of connecting a light bulb with two wires attached to it to the battery. If one of the two wires attached to the light bulb were attached to the negative post of the battery and the other were attached to the positive post of the battery, you would have a complete circuit. Current from the battery would flow to the light bulb, causing it to light, and return to the negative post of the battery.

The normal automotive circuit differs from this simple example in two ways. First, instead of having a return wire from the bulb to the battery, the light bulb returns the current to the battery through the chassis of the vehicle. Since the negative battery cable is attached to the chassis and the chassis is made of electrically conductive metal, the chassis of the vehicle can serve as a ground wire to complete the circuit. Secondly, most automotive circuits contain switches to turn components on and off as required.

Every complete circuit from a power source must include a component which is using the power from the power source. If you were to disconnect the light bulb from the wires and touch the two wires together (don't do this) the power supply wire to the component would be grounded before the normal ground connection for the circuit.

Because grounding a wire from a power source makes a complete circuit—less the required component to use the power—this phenomenon is called a short circuit. Common causes are: broken insulation (exposing the metal wire to a metal part of the car or truck), or a shorted switch.

Some electrical components which require a large amount of current to operate also have a relay in their circuit. Since these circuits carry a large amount of current, the thickness of the wire in the circuit (gauge size) is also greater. If this large wire were connected from the component to the control switch on the instrument panel, and then back to the component, a voltage drop would occur in the circuit. To prevent this potential drop in voltage, an electromagnetic switch (relay) is used. The large wires in the circuit are connected from the battery to one side of the relay, and from the opposite side of the relay to the component. The relay is normally open, preventing current from passing through the circuit. An additional, smaller, wire is connected from the relay to the control switch for the circuit. When the control switch is turned on, it grounds the smaller wire from the relay and completes the circuit. This closes the relay and allows current to flow from the battery to the component. The horn, headlight, and starter circuits are three which use relays.

It is possible for larger surges of current to pass through the electrical system of your car or truck. If this surge of current were to reach an electrical component, it could burn it out. To prevent this, fuses, circuit breakers or fusible links are connected into the current supply wires of most of the major electrical systems. When an electrical current of excessive power passes through the component's fuse, the fuse blows out and breaks the circuit, saving the component from destruction.

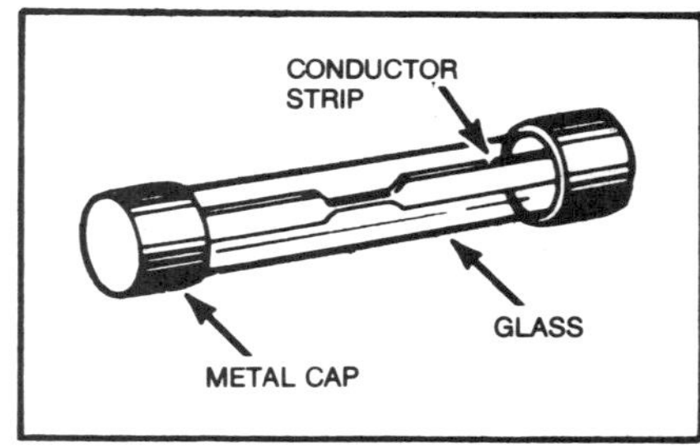

Typical automotive fuse

A circuit breaker is basically a self-repairing fuse. The circuit breaker opens the circuit the same way a fuse does. However, when either the short is removed from the circuit or the surge subsides, the circuit breaker resets itself and does not have to be replaced as a fuse does.

A fuse link is a wire that acts as a fuse. It is normally connected between the starter relay and the main wiring harness. This connection is usually under the hood. The fuse link (if installed) protects all the

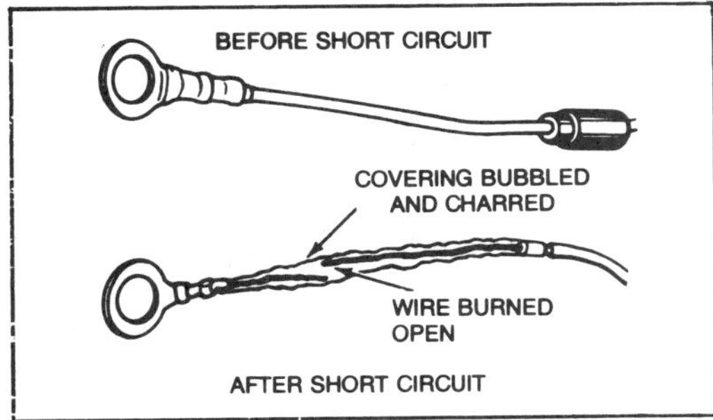

Most fusible links show a charred, melted insulation when they burn out

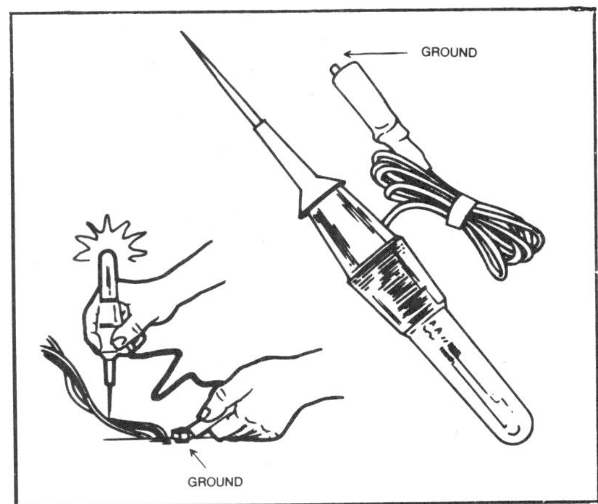

The test light will show the presence of current when touched to a hot wire and grounded at the other end

chassis electrical components, and is the probable cause of trouble when none of the electrical components function, unless the battery is disconnected or dead.

Electrical problems generally fall into one of three areas:

1. The component that is not functioning is not receiving current.

2. The component itself is not functioning.

3. The component is not properly grounded.

The electrical system can be checked with a test light and a jumper wire. A test light is a device that looks like a pointed screwdriver with a wire attached to it and has a light bulb in its handle. A jumper wire is a piece of insulated wire with an alligator clip attached to each end.

If a component is not working, you must follow a systematic plan to determine which of the three causes is the villain.

1. Turn on the switch that controls the inoperable component.

2. Disconnect the power supply wire from the component.

3. Attach the ground wire on the test light to a good metal ground.

4. Touch the probe end of the test light to the end of the power supply wire that was disconnected from the component. If the component is receiving current, the test light will go on.

NOTE: *Some components work only when the ignition switch is turned on.*

If the test light does not go on, then the problem is in the circuit between the battery and the component. This includes all the switches, fuses, and relays in the system. Follow the wire that runs back to the battery. The problem is an open circuit between the

battery and the component. If the fuse is blown and, when replaced, immediately blows again, there is a short circuit in the system which must be located and repaired. If there is a switch in the system, bypass it with a jumper wire. This is done by connecting one end of the jumper wire to the power supply wire into the switch and the other end of the jumper wire to the wire coming out of the switch. If the test light lights with the jumper wire installed, the switch or whatever was bypassed is defective.

NOTE: *Never substitute the jumper wire for the component, since it is required to use the power from the power source.*

5. If the bulb in the test light goes on, then the current is getting to the component that is not working. This eliminates the first of the three possible causes. Connect the power supply wire and connect a jumper wire from the component to a good metal ground. Do this with the switch which controls the component turned on, and also the ignition switch turned on if it is required for the component to work. If the component works with the jumper wire installed, then it has a bad ground. This is usually caused by the metal area on which the component mounts to the chassis being coated with some type of foreign matter.

6. If neither test located the source of the trouble, then the component itself is defective. Remember that for any electrical system to work, all connections must be clean and tight.

Troubleshooting Basic Turn Signal and Flasher Problems
See Chapter 5 for service procedures

Most problems in the turn signals or flasher system can be reduced to defective flashers or bulbs, which are easily replaced. Occasionally, the turn signal switch will prove defective.

F = Front   R = Rear   ● = Lights off   ○ = Lights on

Condition		Possible Cause
Turn signals light, but do not flash	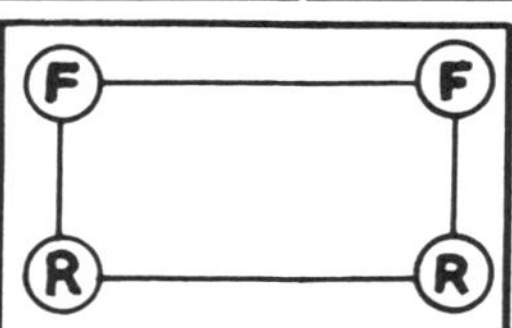	Defective flasher
No turn signals light on either side	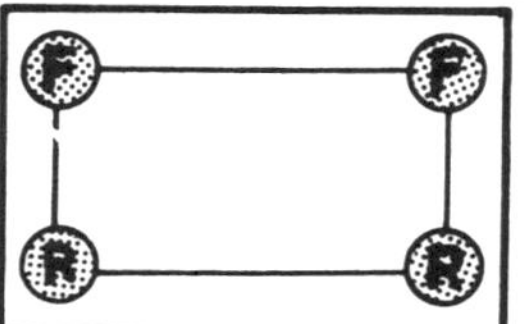	Blown fuse. Replace if defective. Defective flasher. Check by substitution. Open circuit, short circuit or poor ground.
Both turn signals on one side don't work	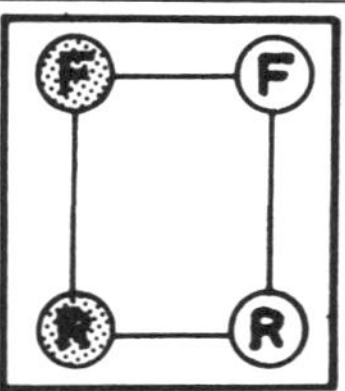	Bad bulbs. Bad ground in both (or either) housings.
One turn signal light on one side doesn't work	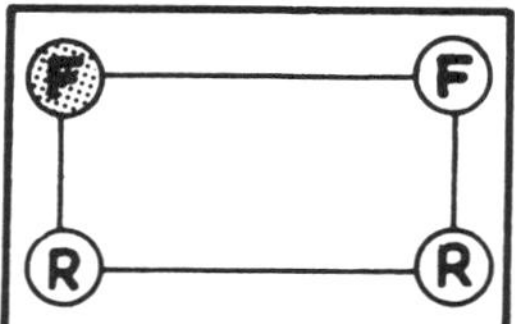	Defective bulb. Corrosion in socket. Clean contacts. Poor ground at socket.
Turn signal flashes too fast or too slowly	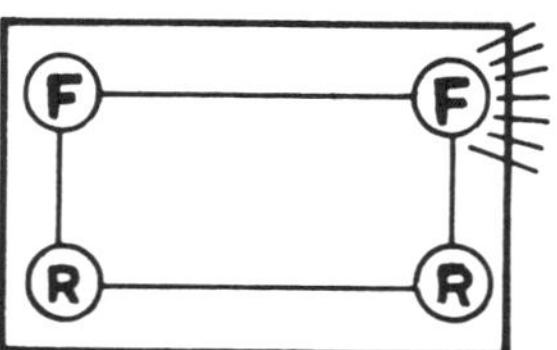	Check any bulb on the side flashing too fast. A heavy-duty bulb is probably installed in place of a regular bulb. Check the bulb flashing too slowly. A standard bulb was probably installed in place of a heavy-duty bulb. Loose connections or corrosion at the bulb socket.
Indicator lights don't work in either direction		Check if the turn signals are working. Check the dash indicator lights. Check the flasher by substitution.
One indicator light doesn't light	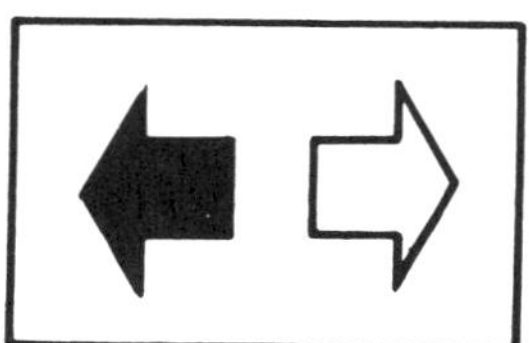	On systems with one dash indicator: See if the lights work on the same side. Often the filaments have been reversed in systems combining stoplights with tail-lights and turn signals. Check the flasher by substitution. On systems with two indicators: Check the bulbs on the same side. Check the indicator light bulb. Check the flasher by substitution.

Troubleshooting Lighting Problems
See Chapter 5 for service procedures

Condition	Possible Cause
One or more lights don't work, but others do	1. Defective bulb(s) 2. Blown fuse(s) 3. Dirty fuse clips or light sockets 4. Poor ground circuit
Lights burn out quickly	1. Incorrect voltage regulator setting or defective regulator 2. Poor battery/alternator connections
Lights go dim	1. Low/discharged battery 2. Alternator not charging 3. Corroded sockets or connections 4. Low voltage output
Lights flicker	1. Loose connection 2. Poor ground. (Run ground wire from light housing to frame) 3. Circuit breaker operating (short circuit)
Lights "flare"—Some flare is normal on acceleration—If excessive, see "Lights Burn Out Quickly"	High voltage setting
Lights glare—approaching drivers are blinded	1. Lights adjusted too high 2. Rear springs or shocks sagging 3. Rear tires soft

Troubleshooting Dash Gauge Problems
Most problems can be traced to a defective sending unit or faulty wiring. Occasionally, the gauge itself is at fault. See Chapter 5 for service procedures.

Condition	Possible Cause
COOLANT TEMPERATURE GAUGE	
Gauge reads erratically or not at all	1. Loose or dirty connections 2. Defective sending unit. 3. Defective gauge. To test a bi-metal gauge, remove the wire from the sending unit. Ground the wire for an instant. If the gauge registers, replace the sending unit. To test a magnetic gauge, disconnect the wire at the sending unit. With ignition ON gauge should register COLD. Ground the wire; gauge should register HOT.
AMMETER GAUGE—TURN HEADLIGHTS ON (DO NOT START ENGINE). NOTE REACTION	
Ammeter shows charge Ammeter shows discharge Ammeter does not move	1. Connections reversed on gauge 2. Ammeter is OK 3. Loose connections or faulty wiring 4. Defective gauge

OIL PRESSURE GAUGE

Condition	Possible Cause
Gauge does not register or is inaccurate	1. On mechanical gauge, Bourdon tube may be bent or kinked. 2. Low oil pressure. Remove sending unit. Idle the engine briefly. If no oil flows from sending unit hole, problem is in engine. 3. Defective gauge. Remove the wire from the sending unit and ground it for an instant with the ignition ON. A good gauge will go to the top of the scale. 4. Defective wiring. Check the wiring to the gauge. If it's OK and the gauge doesn't register when grounded, replace the gauge. 5. Defective sending unit.

ALL GAUGES

Condition	Possible Cause
All gauges do not operate All gauges read low or erratically All gauges pegged	1. Blown fuse 2. Defective instrument regulator 3. Defective or dirty instrument voltage regulator 4. Loss of ground between instrument voltage regulator and frame 5. Defective instrument regulator

WARNING LIGHTS

Condition	Possible Cause
Light(s) do not come on when ignition is ON, but engine is not started	1. Defective bulb 2. Defective wire 3. Defective sending unit. Disconnect the wire from the sending unit and ground it. Replace the sending unit if the light comes on with the ignition ON.
Light comes on with engine running	4. Problem in individual system 5. Defective sending unit

Troubleshooting Clutch Problems

It is false economy to replace individual clutch components. The pressure plate, clutch plate and throwout bearing should be replaced as a set, and the flywheel face inspected, whenever the clutch is overhauled. See Chapter 6 for service procedures.

Condition	Possible Cause
Clutch chatter	1. Grease on driven plate (disc) facing 2. Binding clutch linkage or cable 3. Loose, damaged facings on driven plate (disc) 4. Engine mounts loose 5. Incorrect height adjustment of pressure plate release levers 6. Clutch housing or housing to transmission adapter misalignment 7. Loose driven plate hub
Clutch grabbing	1. Oil, grease on driven plate (disc) facing 2. Broken pressure plate 3. Warped or binding driven plate. Driven plate binding on clutch shaft
Clutch slips	1. Lack of lubrication in clutch linkage or cable (linkage or cable binds, causes incomplete engagement) 2. Incorrect pedal, or linkage adjustment 3. Broken pressure plate springs 4. Weak pressure plate springs 5. Grease on driven plate facings (disc)

Troubleshooting Clutch Problems (cont.)

Condition	Possible Cause
Incomplete clutch release	1. Incorrect pedal or linkage adjustment or linkage or cable binding 2. Incorrect height adjustment on pressure plate release levers 3. Loose, broken facings on driven plate (disc) 4. Bent, dished, warped driven plate caused by overheating
Grinding, whirring grating noise when pedal is depressed	1. Worn or defective throwout bearing 2. Starter drive teeth contacting flywheel ring gear teeth. Look for milled or polished teeth on ring gear.
Squeal, howl, trumpeting noise when pedal is being released (occurs during first inch to inch and one-half of pedal travel)	Pilot bushing worn or lack of lubricant. If bushing appears OK, polish bushing with emery cloth, soak lube wick in oil, lube bushing with oil, apply film of chassis grease to clutch shaft pilot hub, reassemble. NOTE: Bushing wear may be due to misalignment of clutch housing or housing to transmission adapter
Vibration or clutch pedal pulsation with clutch disengaged (pedal fully depressed)	1. Worn or defective engine transmission mounts 2. Flywheel run out. (Flywheel run out at face not to exceed 0.005″) 3. Damaged or defective clutch components

Troubleshooting Manual Transmission Problems
See Chapter 6 for service procedures

Condition	Possible Cause
Transmission jumps out of gear	1. Misalignment of transmission case or clutch housing. 2. Worn pilot bearing in crankshaft. 3. Bent transmission shaft. 4. Worn high speed sliding gear. 5. Worn teeth or end-play in clutch shaft. 6. Insufficient spring tension on shifter rail plunger. 7. Bent or loose shifter fork. 8. Gears not engaging completely. 9. Loose or worn bearings on clutch shaft or mainshaft. 10. Worn gear teeth. 11. Worn or damaged detent balls.
Transmission sticks in gear	1. Clutch not releasing fully. 2. Burred or battered teeth on clutch shaft, or sliding sleeve. 3. Burred or battered transmission mainshaft. 4. Frozen synchronizing clutch. 5. Stuck shifter rail plunger. 6. Gearshift lever twisting and binding shifter rail. 7. Battered teeth on high speed sliding gear or on sleeve. 8. Improper lubrication, or lack of lubrication. 9. Corroded transmission parts. 10. Defective mainshaft pilot bearing. 11. Locked gear bearings will give same effect as stuck in gear.
Transmission gears will not synchronize	1. Binding pilot bearing on mainshaft, will synchronize in high gear only. 2. Clutch not releasing fully. 3. Detent spring weak or broken. 4. Weak or broken springs under balls in sliding gear sleeve. 5. Binding bearing on clutch shaft, or binding countershaft. 6. Binding pilot bearing in crankshaft. 7. Badly worn gear teeth. 8. Improper lubrication. 9. Constant mesh gear not turning freely on transmission mainshaft. Will synchronize in that gear only.

Condition	Possible Cause
Gears spinning when shifting into gear from neutral	1. Clutch not releasing fully. 2. In some cases an extremely light lubricant in transmission will cause gears to continue to spin for a short time after clutch is released. 3. Binding pilot bearing in crankshaft.
Transmission noisy in all gears	1. Insufficient lubricant, or improper lubricant. 2. Worn countergear bearings. 3. Worn or damaged main drive gear or countergear. 4. Damaged main drive gear or mainshaft bearings. 5. Worn or damaged countergear anti-lash plate.
Transmission noisy in neutral only	1. Damaged main drive gear bearing. 2. Damaged or loose mainshaft pilot bearing. 3. Worn or damaged countergear anti-lash plate. 4. Worn countergear bearings.
Transmission noisy in one gear only	1. Damaged or worn constant mesh gears. 2. Worn or damaged countergear bearings. 3. Damaged or worn synchronizer.
Transmission noisy in reverse only	1. Worn or damaged reverse idler gear or idler bushing. 2. Worn or damaged mainshaft reverse gear. 3. Worn or damaged reverse countergear. 4. Damaged shift mechanism.

TROUBLESHOOTING AUTOMATIC TRANSMISSION PROBLEMS

Keeping alert to changes in the operating characteristics of the transmission (changing shift points, noises, etc.) can prevent small problems from becoming large ones. If the problem cannot be traced to loose bolts, fluid level, misadjusted linkage, clogged filters or similar problems, you should probably seek professional service.

Transmission Fluid Indications

The appearance and odor of the transmission fluid can give valuable clues to the overall condition of the transmission. Always note the appearance of the fluid when you check the fluid level or change the fluid. Rub a small amount of fluid between your fingers to feel for grit and smell the fluid on the dipstick.

If the fluid appears:	It indicates:
Clear and red colored	Normal operation
Discolored (extremely dark red or brownish) or smells burned	Band or clutch pack failure, usually caused by an overheated transmission. Hauling very heavy loads with insufficient power or failure to change the fluid often result in overheating. Do not confuse this appearance with newer fluids that have a darker red color and a strong odor (though not a burned odor).
Foamy or aerated (light in color and full of bubbles)	1. The level is too high (gear train is churning oil) 2. An internal air leak (air is mixing with the fluid). Have the transmission checked professionally.
Solid residue in the fluid	Defective bands, clutch pack or bearings. Bits of band material or metal abrasives are clinging to the dipstick. Have the transmission checked professionally.
Varnish coating on the dipstick	The transmission fluid is overheating

TROUBLESHOOTING DRIVE AXLE PROBLEMS

First, determine when the noise is most noticeable.

Drive Noise: Produced under vehicle acceleration.

Coast Noise: Produced while coasting with a closed throttle.

Float Noise: Occurs while maintaining constant speed (just enough to keep speed constant) on a level road.

External Noise Elimination

It is advisable to make a thorough road test to determine whether the noise originates in the rear axle or whether it originates from the tires, engine, transmission, wheel bearings or road surface. Noise originating from other places cannot be corrected by servicing the rear axle.

ROAD NOISE

Brick or rough surfaced concrete roads produce noises that seem to come from the rear axle. Road noise is usually identical in Drive or Coast and driving on a different type of road will tell whether the road is the problem.

TIRE NOISE

Tire noise can be mistaken as rear axle noise, even though the tires on the front are at fault. Snow tread and mud tread tires or tires worn unevenly will frequently cause vibrations which seem to originate elsewhere; *temporarily, and for test purposes only,* inflate the tires to 40–50 lbs. This will significantly alter the noise produced by the tires, but will not alter noise from the rear axle. Noises from the rear axle will normally cease at speeds below 30 mph on coast, while tire noise will continue at lower tone as speed is decreased. The rear axle noise will usually change from drive conditions to coast conditions, while tire noise will not. Do not forget to lower the tire pressure to normal after the test is complete.

ENGINE/TRANSMISSION NOISE

Determine at what speed the noise is most pronounced, then stop in a quiet place. With the transmission in Neutral, run the engine through speeds corresponding to road speeds where the noise was noticed. Noises produced with the vehicle standing still are coming from the engine or transmission.

FRONT WHEEL BEARINGS

Front wheel bearing noises, sometimes confused with rear axle noises, will not change when comparing drive and coast conditions. While holding the speed steady, lightly apply the footbrake. This will often cause wheel bearing noise to lessen, as some of the weight is taken off the bearing. Front wheel bearings are easily checked by jacking up the wheels and spinning the wheels. Shaking the wheels will also determine if the wheel bearings are excessively loose.

REAR AXLE NOISES

Eliminating other possible sources can narrow the cause to the rear axle, which normally produces noise from worn gears or bearings. Gear noises tend to peak in a narrow speed range, while bearing noises will usually vary in pitch with engine speeds.

Noise Diagnosis

The Noise Is:	Most Probably Produced By:
1. Identical under Drive or Coast	Road surface, tires or front wheel bearings
2. Different depending on road surface	Road surface or tires
3. Lower as speed is lowered	Tires
4. Similar when standing or moving	Engine or transmission
5. A vibration	Unbalanced tires, rear wheel bearing, unbalanced driveshaft or worn U-joint
6. A knock or click about every two tire revolutions	Rear wheel bearing
7. Most pronounced on turns	Damaged differential gears
8. A steady low-pitched whirring or scraping, starting at low speeds	Damaged or worn pinion bearing
9. A chattering vibration on turns	Wrong differential lubricant or worn clutch plates (limited slip rear axle)
10. Noticed only in Drive, Coast or Float conditions	Worn ring gear and/or pinion gear

Troubleshooting Steering & Suspension Problems

Condition	Possible Cause
Hard steering (wheel is hard to turn)	1. Improper tire pressure 2. Loose or glazed pump drive belt 3. Low or incorrect fluid 4. Loose, bent or poorly lubricated front end parts 5. Improper front end alignment (excessive caster) 6. Bind in steering column or linkage 7. Kinked hydraulic hose 8. Air in hydraulic system 9. Low pump output or leaks in system 10. Obstruction in lines 11. Pump valves sticking or out of adjustment 12. Incorrect wheel alignment
Loose steering (too much play in steering wheel)	1. Loose wheel bearings 2. Faulty shocks 3. Worn linkage or suspension components 4. Loose steering gear mounting or linkage points 5. Steering mechanism worn or improperly adjusted 6. Valve spool improperly adjusted 7. Worn ball joints, tie-rod ends, etc.
Veers or wanders (pulls to one side with hands off steering wheel)	1. Improper tire pressure 2. Improper front end alignment 3. Dragging or improperly adjusted brakes 4. Bent frame 5. Improper rear end alignment 6. Faulty shocks or springs 7. Loose or bent front end components 8. Play in Pitman arm 9. Steering gear mountings loose 10. Loose wheel bearings 11. Binding Pitman arm 12. Spool valve sticking or improperly adjusted 13. Worn ball joints
Wheel oscillation or vibration transmitted through steering wheel	1. Low or uneven tire pressure 2. Loose wheel bearings 3. Improper front end alignment 4. Bent spindle 5. Worn, bent or broken front end components 6. Tires out of round or out of balance 7. Excessive lateral runout in disc brake rotor 8. Loose or bent shock absorber or strut
Noises (see also "Troubleshooting Drive Axle Problems")	1. Loose belts 2. Low fluid, air in system 3. Foreign matter in system 4. Improper lubrication 5. Interference or chafing in linkage 6. Steering gear mountings loose 7. Incorrect adjustment or wear in gear box 8. Faulty valves or wear in pump 9. Kinked hydraulic lines 10. Worn wheel bearings
Poor return of steering	1. Over-inflated tires 2. Improperly aligned front end (excessive caster) 3. Binding in steering column 4. No lubrication in front end 5. Steering gear adjusted too tight
Uneven tire wear (see "How To Read Tire Wear")	1. Incorrect tire pressure 2. Improperly aligned front end 3. Tires out-of-balance 4. Bent or worn suspension parts

HOW TO READ TIRE WEAR

The way your tires wear is a good indicator of other parts of the suspension. Abnormal wear patterns are often caused by the need for simple tire maintenance, or for front end alignment.

Excessive wear at the center of the tread indicates that the air pressure in the tire is consistently too high. The tire is riding on the center of the tread and wearing it prematurely. Occasionally, this wear pattern can result from outrageously wide tires on narrow rims. The cure for this is to replace either the tires or the wheels.

This type of wear usually results from consistent under-inflation. When a tire is under-inflated, there is too much contact with the road by the outer treads, which wear prematurely. When this type of wear occurs, and the tire pressure is known to be consistently correct, a bent or worn steering component or the need for wheel alignment could be indicated.

Feathering is a condition when the edge of each tread rib develops a slightly rounded edge on one side and a sharp edge on the other. By running your hand over the tire, you can usually feel the sharper edges before you'll be able to see them. The most common causes of feathering are incorrect toe-in setting or deteriorated bushings in the front suspension.

When an inner or outer rib wears faster than the rest of the tire, the need for wheel alignment is indicated. There is excessive camber in the front suspension, causing the wheel to lean too much putting excessive load on one side of the tire. Misalignment could also be due to sagging springs, worn ball joints, or worn control arm bushings. Be sure the vehicle is loaded the way it's normally driven when you have the wheels aligned.

Cups or scalloped dips appearing around the edge of the tread almost always indicate worn (sometimes bent) suspension parts. Adjustment of wheel alignment alone will seldom cure the problem. Any worn component that connects the wheel to the suspension can cause this type of wear. Occasionally, wheels that are out of balance will wear like this, but wheel imbalance usually shows up as bald spots between the outside edges and center of the tread.

Second-rib wear is usually found only in radial tires, and appears where the steel belts end in relation to the tread. It can be kept to a minimum by paying careful attention to tire pressure and frequently rotating the tires. This is often considered normal wear but excessive amounts indicate that the tires are too wide for the wheels.

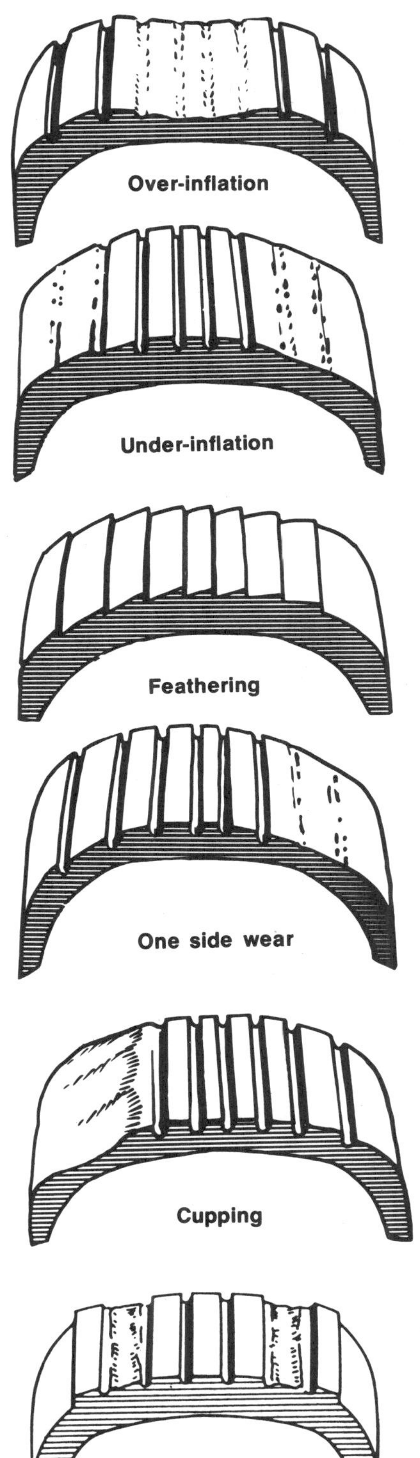

Troubleshooting Disc Brake Problems

Condition	Possible Cause
Noise—groan—brake noise emanating when slowly releasing brakes (creep-groan)	Not detrimental to function of disc brakes—no corrective action required. (This noise may be eliminated by slightly increasing or decreasing brake pedal efforts.)
Rattle—brake noise or rattle emanating at low speeds on rough roads, (front wheels only).	1. Shoe anti-rattle spring missing or not properly positioned. 2. Excessive clearance between shoe and caliper. 3. Soft or broken caliper seals. 4. Deformed or misaligned disc. 5. Loose caliper.
Scraping	1. Mounting bolts too long. 2. Loose wheel bearings. 3. Bent, loose, or misaligned splash shield.
Front brakes heat up during driving and fail to release	1. Operator riding brake pedal. 2. Stop light switch improperly adjusted. 3. Sticking pedal linkage. 4. Frozen or seized piston. 5. Residual pressure valve in master cylinder. 6. Power brake malfunction. 7. Proportioning valve malfunction.
Leaky brake caliper	1. Damaged or worn caliper piston seal. 2. Scores or corrosion on surface of cylinder bore.
Grabbing or uneven brake action—Brakes pull to one side	1. Causes listed under "Brakes Pull". 2. Power brake malfunction. 3. Low fluid level in master cylinder. 4. Air in hydraulic system. 5. Brake fluid, oil or grease on linings. 6. Unmatched linings. 7. Distorted brake pads. 8. Frozen or seized pistons. 9. Incorrect tire pressure. 10. Front end out of alignment. 11. Broken rear spring. 12. Brake caliper pistons sticking. 13. Restricted hose or line. 14. Caliper not in proper alignment to braking disc. 15. Stuck or malfunctioning metering valve. 16. Soft or broken caliper seals. 17. Loose caliper.
Brake pedal can be depressed without braking effect	1. Air in hydraulic system or improper bleeding procedure. 2. Leak past primary cup in master cylinder. 3. Leak in system. 4. Rear brakes out of adjustment. 5. Bleeder screw open.
Excessive pedal travel	1. Air, leak, or insufficient fluid in system or caliper. 2. Warped or excessively tapered shoe and lining assembly. 3. Excessive disc runout. 4. Rear brake adjustment required. 5. Loose wheel bearing adjustment. 6. Damaged caliper piston seal. 7. Improper brake fluid (boil). 8. Power brake malfunction. 9. Weak or soft hoses.

Troubleshooting Disc Brake Problems (cont.)

Condition	Possible Cause
Brake roughness or chatter (pedal pumping)	1. Excessive thickness variation of braking disc. 2. Excessive lateral runout of braking disc. 3. Rear brake drums out-of-round. 4. Excessive front bearing clearance.
Excessive pedal effort	1. Brake fluid, oil or grease on linings. 2. Incorrect lining. 3. Frozen or seized pistons. 4. Power brake malfunction. 5. Kinked or collapsed hose or line. 6. Stuck metering valve. 7. Scored caliper or master cylinder bore. 8. Seized caliper pistons.
Brake pedal fades (pedal travel increases with foot on brake)	1. Rough master cylinder or caliper bore. 2. Loose or broken hydraulic lines/connections. 3. Air in hydraulic system. 4. Fluid level low. 5. Weak or soft hoses. 6. Inferior quality brake shoes or fluid. 7. Worn master cylinder piston cups or seals.

Troubleshooting Drum Brakes

Condition	Possible Cause
Pedal goes to floor	1. Fluid low in reservoir. 2. Air in hydraulic system. 3. Improperly adjusted brake. 4. Leaking wheel cylinders. 5. Loose or broken brake lines. 6. Leaking or worn master cylinder. 7. Excessively worn brake lining.
Spongy brake pedal	1. Air in hydraulic system. 2. Improper brake fluid (low boiling point). 3. Excessively worn or cracked brake drums. 4. Broken pedal pivot bushing.
Brakes pulling	1. Contaminated lining. 2. Front end out of alignment. 3. Incorrect brake adjustment. 4. Unmatched brake lining. 5. Brake drums out of round. 6. Brake shoes distorted. 7. Restricted brake hose or line. 8. Broken rear spring. 9. Worn brake linings. 10. Uneven lining wear. 11. Glazed brake lining. 12. Excessive brake lining dust. 13. Heat spotted brake drums. 14. Weak brake return springs. 15. Faulty automatic adjusters. 16. Low or incorrect tire pressure.

Condition	Possible Cause
Squealing brakes	1. Glazed brake lining. 2. Saturated brake lining. 3. Weak or broken brake shoe retaining spring. 4. Broken or weak brake shoe return spring. 5. Incorrect brake lining. 6. Distorted brake shoes. 7. Bent support plate. 8. Dust in brakes or scored brake drums. 9. Linings worn below limit. 10. Uneven brake lining wear. 11. Heat spotted brake drums.
Chirping brakes	1. Out of round drum or eccentric axle flange pilot.
Dragging brakes	1. Incorrect wheel or parking brake adjustment. 2. Parking brakes engaged or improperly adjusted. 3. Weak or broken brake shoe return spring. 4. Brake pedal binding. 5. Master cylinder cup sticking. 6. Obstructed master cylinder relief port. 7. Saturated brake lining. 8. Bent or out of round brake drum. 9. Contaminated or improper brake fluid. 10. Sticking wheel cylinder pistons. 11. Driver riding brake pedal. 12. Defective proportioning valve. 13. Insufficient brake shoe lubricant.
Hard pedal	1. Brake booster inoperative. 2. Incorrect brake lining. 3. Restricted brake line or hose. 4. Frozen brake pedal linkage. 5. Stuck wheel cylinder. 6. Binding pedal linkage. 7. Faulty proportioning valve.
Wheel locks	1. Contaminated brake lining. 2. Loose or torn brake lining. 3. Wheel cylinder cups sticking. 4. Incorrect wheel bearing adjustment. 5. Faulty proportioning valve.
Brakes fade (high speed)	1. Incorrect lining. 2. Overheated brake drums. 3. Incorrect brake fluid (low boiling temperature). 4. Saturated brake lining. 5. Leak in hydraulic system. 6. Faulty automatic adjusters.
Pedal pulsates	1. Bent or out of round brake drum.
Brake chatter and shoe knock	1. Out of round brake drum. 2. Loose support plate. 3. Bent support plate. 4. Distorted brake shoes. 5. Machine grooves in contact face of brake drum (Shoe Knock). 6. Contaminated brake lining. 7. Missing or loose components. 8. Incorrect lining material. 9. Out-of-round brake drums. 10. Heat spotted or scored brake drums. 11. Out-of-balance wheels.

Troubleshooting Drum Brakes (cont.)

Condition	Possible Cause
Brakes do not self adjust	1. Adjuster screw frozen in thread. 2. Adjuster screw corroded at thrust washer. 3. Adjuster lever does not engage star wheel. 4. Adjuster installed on wrong wheel.
Brake light glows	1. Leak in the hydraulic system. 2. Air in the system. 3. Improperly adjusted master cylinder pushrod. 4. Uneven lining wear. 5. Failure to center combination valve or proportioning valve.

Appendix

General Conversion Table

Multiply by	To convert	To	
2.54	Inches	Centimeters	.3937
30.48	Feet	Centimeters	.0328
.914	Yards	Meters	1.094
1.609	Miles	Kilometers	.621
6.45	Square inches	Square cm.	.155
.836	Square yards	Square meters	1.196
16.39	Cubic inches	Cubic cm.	.061
28.3	Cubic feet	Liters	.0353
.4536	Pounds	Kilograms	2.2045
3.785	Gallons	Liters	.264
.068	Lbs./sq. in. (psi)	Atmospheres	14.7
.138	Foot pounds	Kg. m.	7.23
1.014	H.P. (DIN)	H.P. (SAE)	.9861
—	To obtain	From	Multiply by

Note: 1 cm. equals 10 mm.; 1 mm. equals .0394".

Conversion—Common Fractions to Decimals and Millimeters

Common Fractions	Decimal Fractions	Millimeters (approx.)	Common Fractions	Decimal Fractions	Millimeters (approx.)	Common Fractions	Decimal Fractions	Millimeters (approx.)
1/128	.008	0.20	11/32	.344	8.73	43/64	.672	17.07
1/64	.016	0.40	23/64	.359	9.13	11/16	.688	17.46
1/32	.031	0.79	3/8	.375	9.53	45/64	.703	17.86
3/64	.047	1.19	25/64	.391	9.92	23/32	.719	18.26
1/16	.063	1.59	13/32	.406	10.32	47/64	.734	18.65
5/64	.078	1.98	27/64	.422	10.72	3/4	.750	19.05
3/32	.094	2.38	7/16	.438	11.11	49/64	.766	19.45
7/64	.109	2.78	29/64	.453	11.51	25/32	.781	19.84
1/8	.125	3.18	15/32	.469	11.91	51/64	.797	20.24
9/64	.141	3.57	31/64	.484	12.30	13/16	.813	20.64
5/32	.156	3.97	1/2	.500	12.70	53/64	.828	21.03
11/64	.172	4.37	33/64	.516	13.10	27/32	.844	21.43
3/16	.188	4.76	17/32	.531	13.49	55/64	.859	21.83
13/64	.203	5.16	35/64	.547	13.89	7/8	.875	22.23
7/32	.219	5.56	9/16	.563	14.29	57/64	.891	22.62
15/64	.234	5.95	37/64	.578	14.68	29/32	.906	23.02
1/4	.250	6.35	19/32	.594	15.08	59/64	.922	23.42
17/64	.266	6.75	39/64	.609	15.48	15/16	.938	23.81
9/32	.281	7.14	5/8	.625	15.88	61/64	.953	24.21
19/64	.297	7.54	41/64	.641	16.27	31/32	.969	24.61
5/16	.313	7.94	21/32	.656	16.67	63/64	.984	25.00
21/64	.328	8.33						

Conversion—Millimeters to Decimal Inches

mm	inches	mm	inches	mm	inches	mm	inches	mm	inches
1	.039 370	31	1.220 470	61	2.401 570	91	3.582 670	210	8.267 700
2	.078 740	32	1.259 840	62	2.440 940	92	3.622 040	220	8.661 400
3	.118 110	33	1.299 210	63	2.480 310	93	3.661 410	230	9.055 100
4	.157 480	34	1.338 580	64	2.519 680	94	3.700 780	240	9.448 800
5	.196 850	35	1.377 949	65	2.559 050	95	3.740 150	250	9.842 500
6	.236 220	36	1.417 319	66	2.598 420	96	3.779 520	260	10.236 200
7	.275 590	37	1.456 689	67	2.637 790	97	3.818 890	270	10.629 900
8	.314 960	38	1.496 050	68	2.677 160	98	3.858 260	280	11.032 600
9	.354 330	39	1.535 430	69	2.716 530	99	3.897 630	290	11.417 300
10	.393 700	40	1.574 800	70	2.755 900	100	3.937 000	300	11.811 000
11	.433 070	41	1.614 170	71	2.795 270	105	4.133 848	310	12.204 700
12	.472 440	42	1.653 540	72	2.834 640	110	4.330 700	320	12.598 400
13	.511 810	43	1.692 910	73	2.874 010	115	4.527 550	330	12.992 100
14	.551 180	44	1.732 280	74	2.913 380	120	4.724 400	340	13.385 800
15	.590 550	45	1.771 650	75	2.952 750	125	4.921 250	350	13.779 500
16	.629 920	46	1.811 020	76	2.992 120	130	5.118 100	360	14.173 200
17	.669 290	47	1.850 390	77	3.031 490	135	5.314 950	370	14.566 900
18	.708 660	48	1.889 760	78	3.070 860	140	5.511 800	380	14.960 600
19	.748 030	49	1.929 130	79	3.110 230	145	5.708 650	390	15.354 300
20	.787 400	50	1.968 500	80	3.149 600	150	5.905 500	400	15.748 000
21	.826 770	51	2.007 870	81	3.188 970	155	6.102 350	500	19.685 000
22	.866 140	52	2.047 240	82	3.228 340	160	6.299 200	600	23.622 000
23	.905 510	53	2.086 610	83	3.267 710	165	6.496 050	700	27.559 000
24	.944 880	54	2.125 980	84	3.307 080	170	6.692 900	800	31.496 000
25	.984 250	55	2.165 350	85	3.346 450	175	6.889 750	900	35.433 000
26	1.023 620	56	2.204 720	86	3.385 820	180	7.086 600	1000	39.370 000
27	1.062 990	57	2.244 090	87	3.425 190	185	7.283 450	2000	78.740 000
28	1.102 360	58	2.283 460	88	3.464 560	190	7.480 300	3000	118.110 000
29	1.141 730	59	2.322 830	89	3.503 903	195	7.677 150	4000	157.480 000
30	1.181 100	60	2.362 200	90	3.543 300	200	7.874 000	5000	196.850 000

To change decimal millimeters to decimal inches, position the decimal point where desired on either side of the millimeter measurement shown and reset the inches decimal by the same number of digits in the same direction. For example, to convert 0.001 mm to decimal inches, reset the decimal behind the 1 mm (shown on the chart) to 0.001; change the decimal inch equivalent (0.039″ shown) to 0.000039″.

Tap Drill Sizes

National Fine or S.A.E.

Screw & Tap Size	Threads Per Inch	Use Drill Number
No. 5	44	37
No. 6	40	33
No. 8	36	29
No. 10	32	21
No. 12	28	15
1/4	28	3
5/16	24	1
3/8	24	Q
7/16	20	W
1/2	20	29/64
9/16	18	33/64
5/8	18	37/64
3/4	16	11/16
7/8	14	13/16
1 1/8	12	1 3/64
1 1/4	12	1 11/64
1 1/2	12	1 27/64

Tap Drill Sizes

National Coarse or U.S.S.

Screw & Tap Size	Threads Per Inch	Use Drill Number
No. 5	40	39
No. 6	32	36
No. 8	32	29
No. 10	24	25
No. 12	24	17
1/4	20	8
5/16	18	F
3/8	16	5/16
7/16	14	U
1/2	13	27/64
9/16	12	31/64
5/8	11	17/32
3/4	10	21/32
7/8	9	49/64
1	8	7/8
1 1/8	7	63/64
1 1/4	7	1 7/64
1 1/2	6	1 11/32

Decimal Equivalent Size of the Number Drills

Drill No.	Decimal Equivalent	Drill No.	Decimal Equivalent	Drill No.	Decimal Equivalent
80	.0135	53	.0595	26	.1470
79	.0145	52	.0635	25	.1495
78	.0160	51	.0670	24	.1520
77	.0180	50	.0700	23	.1540
76	.0200	49	.0730	22	.1570
75	.0210	48	.0760	21	.1590
74	.0225	47	.0785	20	.1610
73	.0240	46	.0810	19	.1660
72	.0250	45	.0820	18	.1695
71	.0260	44	.0860	17	.1730
70	.0280	43	.0890	16	.1770
69	.0292	42	.0935	15	.1800
68	.0310	41	.0960	14	.1820
67	.0320	40	.0980	13	.1850
66	.0330	39	.0995	12	.1890
65	.0350	38	.1015	11	.1910
64	.0360	37	.1040	10	.1935
63	.0370	36	.1065	9	.1960
62	.0380	35	.1100	8	.1990
61	.0390	34	.1110	7	.2010
60	.0400	33	.1130	6	.2040
59	.0410	32	.1160	5	.2055
58	.0420	31	.1200	4	.2090
57	.0430	30	.1285	3	.2130
56	.0465	29	.1360	2	.2210
55	.0520	28	.1405	1	.2280
54	.0550	27	.1440		

Decimal Equivalent Size of the Letter Drills

Letter Drill	Decimal Equivalent	Letter Drill	Decimal Equivalent	Letter Drill	Decimal Equivalent
A	.234	J	.277	S	.348
B	.238	K	.281	T	.358
C	.242	L	.290	U	.368
D	.246	M	.295	V	.377
E	.250	N	.302	W	.386
F	.257	O	.316	X	.397
G	.261	P	.323	Y	.404
H	.266	Q	.332	Z	.413
I	.272	R	.339		

Anti-Freeze Chart

Temperatures Shown in Degrees Fahrenheit +32 is Freezing

Cooling System Capacity Quarts	Quarts of ETHYLENE GLYCOL Needed for Protection to Temperatures Shown Below													
	1	2	3	4	5	6	7	8	9	10	11	12	13	14
10	+24°	+16°	+ 4°	−12°	−34°	−62°								
11	+25	+18	+ 8	− 6	−23	−47								
12	+26	+19	+10	0	−15	−34	−57°							
13	+27	+21	+13	+ 3	− 9	−25	−45							
14			+15	+ 6	− 5	−18	−34							
15			+16	+ 8	0	−12	−26							
16			+17	+10	+ 2	− 8	−19	−34	−52°					
17			+18	+12	+ 5	− 4	−14	−27	−42					
18			+19	+14	+ 7	0	−10	−21	−34	−50°				
19			+20	+15	+ 9	+ 2	− 7	−16	−28	−42				
20				+16	+10	+ 4	− 3	−12	−22	−34	−48°			
21				+17	+12	+ 6	0	− 9	−17	−28	−41			
22				+18	+13	+ 8	+ 2	− 6	−14	−23	−34	−47°		
23				+19	+14	+ 9	+ 4	− 3	−10	−19	−29	−40		
24				+19	+15	+10	+ 5	0	− 8	−15	−23	−34	−46°	
25				+20	+16	+12	+ 7	+ 1	− 5	−12	−20	−29	−40	−50°
26					+17	+13	+ 8	+ 3	− 3	− 9	−16	−25	−34	−44
27					+18	+14	+ 9	+ 5	− 1	− 7	−13	−21	−29	−39
28					+18	+15	+10	+ 6	+ 1	− 5	−11	−18	−25	−34
29					+19	+16	+12	+ 7	+ 2	− 3	− 8	−15	−22	−29
30					+20	+17	+13	+ 8	+ 4	− 1	− 6	−12	−18	−25

For capacities over 30 quarts divide true capacity by 3. Find quarts Anti-Freeze for the ⅓ and multiply by 3 for quarts to add.

For capacities under 10 quarts multiply true capacity by 3. Find quarts Anti-Freeze for the tripled volume and divide by 3 for quarts to add.

To Increase the Freezing Protection of Anti-Freeze Solutions Already Installed

Cooling System Capacity Quarts	Number of Quarts of ETHYLENE GLYCOL Anti-Freeze Required to Increase Protection													
	From +20° F. to					From +10° F. to					From 0° F. to			
	0°	−10°	−20°	−30°	−40°	0°	−10°	−20°	−30°	−40°	−10°	−20°	−30°	−40°
10	1¾	2¼	3	3½	3¾	¾	1½	2¼	2¾	3¼	¾	1½	2	2½
12	2	2¾	3½	4	4½	1	1¾	2½	3¼	3¾	1	1¾	2½	3¼
14	2¼	3¼	4	4¾	5½	1¼	2	3	3¾	4½	1	2	3	3½
16	2½	3½	4½	5¼	6	1¼	2½	3½	4¼	5¼	1¼	2¼	3¼	4
18	3	4	5	6	7	1½	2¾	4	5	5¾	1½	2½	3¾	4¾
20	3¼	4½	5¾	6¾	7½	1¾	3	4¼	5½	6½	1½	2¾	4¼	5¼
22	3½	5	6¼	7¼	8¼	1¾	3¼	4¾	6	7¼	1¾	3¼	4½	5½
24	4	5½	7	8	9	2	3½	5	6½	7½	1¾	3½	5	6
26	4¼	6	7½	8¾	10	2	4	5½	7	8¼	2	3¾	5½	6¾
28	4½	6¼	8	9½	10½	2¼	4¼	6	7½	9	2	4	5¾	7¼
30	5	6¾	8½	10	11½	2½	4½	6½	8	9½	2¼	4¼	6¼	7¾

Test radiator solution with proper hydrometer. Determine from the table the number of quarts of solution to be drawn off from a full cooling system and replace with undiluted anti-freeze, to give the desired increased protection. For example, to increase protection of a 22-quart cooling system containing Ethylene Glycol (permanent type) anti-freeze, from +20° F. to −20° F. will require the replacement of 6¼ quarts of solution with undiluted anti-freeze.

Index